THE HUBBLE SPACE TELESCOPE

AND

THE HIGH REDSHIFT UNIVERSE

THE HUBBLE SPACE TELESCOPE

AND The 37th Herstmonceux Conference

THE HIGH REDSHIFT UNIVERSE

Cambridge, United Kingdom
July 1–5, 1996

Editors

Nial R. Tanvir
Institute of Astronomy
University of Cambridge

Alfonso Aragón-Salamanca
Institute of Astronomy
University of Cambridge

Jasper V. Wall
Royal Greenwich Observatory
Cambridge

World Scientific
Singapore • New Jersey • London • Hong Kong

Published by

World Scientific Publishing Co. Pte. Ltd.

P O Box 128, Farrer Road, Singapore 912805

USA office: Suite 1B, 1060 Main Street, River Edge, NJ 07661

UK office: 57 Shelton Street, Covent Garden, London WC2H 9HE

THE HUBBLE SPACE TELESCOPE AND THE HIGH REDSHIFT UNIVERSE

ISBN 981-02-2896-1

Printed in Singapore.

CONTENTS

Section 2: Galaxies in Distant Clusters

Section 3: Gravitational Lensing

Section 4: Quasar Absorbers

Section 5: Quasars & Active Galactic Nuclei

Section 6: The Future of HST & the Next Generation Space Telescopes

FOREWORD

The arrival of Alec Boksenberg's 60th birthday together with floods of superb data from a rejuvenated Hubble Space Telescope is a coincidence which may have a causal connection. If so, this connection escaped the organisers of the 37th Herstmonceux Conference. However the opportunity provided by the coincidence, causal or otherwise, was not lost. We happily gathered together under the twin motivations, and devoted the week of 1 to 5 July 1996 at Cambridge to 'HST and the High-Redshift Universe'. A total of 236 registered participants attended the meeting.

There are standard phrases used in prefaces for cosmology conferences for the past 20 years, which include 'impressive new observational facilities', 'wealth of data', and 'great progress in our understanding of the history of the Universe'. Hx37 lived up to and surpassed all this. In particular the Hubble Deep Field produced startling results bearing on galaxy formation and evolution. Equally impressive (in a conference devoted primarily to the HST) were the new complementary and supporting results from modern ground-based telescopes such as Keck.

If mists on the epoch of star and galaxy formation were lifted by magnificent results reported here, it must be said —as Richard Ellis did in his opening remarks— that the epoch of UK observer-formation has been well-defined. It has largely happened around Alec Boksenberg, his Image Photon Counting System, his disciples and his students. There are tributes to him in these proceedings. There were many more in the course of this lively meeting.

The organising committee was Alfonso Aragón-Salamanca, Brian Boyle, Robin Catchpole, Richard Ellis, Gill Harrison, Steve Maddox, Richard McMahon, Max Pettini, Nial Tanvir, Roberto Terlevich and Jasper Wall. Local organisation was carried out superbly well by Gill Harrison with help from Antony Peake and Beccy Wright. We also had extra help from Vernon Bailey, Audra Baleisis, Jarle Brinchmann, Kimberly Ennico, Stefano Ettori and Matthew Kenworthy.

The meeting was supported by the RGO (PPARC) and the IoA, with additional support from the European Community Human Capital and Mobility Euroconferences scheme and World Scientific Publishing.

PARTICIPANTS

Roberto Abraham	Insitute of Astronomy, Cambridge	UK
Jeremy Allington-Smith	Durham University	UK
Alessandra Aloisi	Bologna University	Italy
Alfonso Aragón-Salamanca	Institute of Astonomy, Cambridge	UK
Itziar Aretxaga	Royal Greenwich Observatory	UK
David Axon	Space Telescope Science Institute	USA
Arif Babul	New York University	USA
John Bahcall	Institute for Advance Study	USA
Neta Bahcall	Princeton University	USA
Vernon Bailey	Institute of Astronomy, Cambridge	UK
Joanne Baker	Mullard Radio Astronomy Observatory	UK
Audra Baleisis	Institute for Astronomy, Cambridge	UK
Sandro Bardelli	Osservatorio Astronomico, Trieste	Italy
Amy Barger	Institute of Astronomy, Cambridge	UK
Brian Bates	Queen's University	Ireland
Paola Belloni	Institut für Astronomie & Asprophysik	Germany
Piero Benvenuti	ST-ECF	Germany
Jacqueline Bergeron	European Southern Observatory	Germany
Rebecca Bernstein	Caltech	USA
Carlo Bertoni	Istituto di Radioastronomia del CNR	Italy
Saskia Besier	University of New South Wales	Australia
Philip Best	MRAO, Cambridge	UK
Jocelyn Bezecourt	Observatoire Midi-Pyrénées	France
Adam Black	Cambridge University Press	UK
Chris Blades	Space Telescope Science Instutute	USA
Andrew Blain	Cavendish Laboratory	UK
Jonathan Bland-Hawthorn	Oxford University (AAO)	UK
Katherine Blundell	University of Oxford	UK
Alec Boksenberg	Institute of Astronomy, Cambridge	UK
Ulf Borgeest	Geo-Redaktion	Germany
David Bowen	Royal Observatory Edinburgh	UK
Brian Boyle	Royal Greenwich Observatory	UK
Christian Bracco	Observatoire de Haute Provence	France

Tereasa Brainerd	Boston University	USA
Terry Bridges	Royal Greenwich Observatory	UK
Jarle Brinchmann	Institute of Astronomy, Cambridge	UK
Margaret Burbidge	University of California, San Diego	USA
Ana Campos	University of Durham	UK
Alberto Cappi	Osservatorio Astronomico di Bologna	Italy
Richard Carson	University of St Andrews	UK
Bob Carswell	Institute of Astronomy, Cambridge	UK
Robin Catchpole	Royal Greenwich Observatory	UK
Paolo Ciliegi	Istituto di Radio Astronomia - CNR	Italy
Neil Clark	Sheffield University	UK
Dave Clements	European Southern Observatory	Germany
Roger Clowes	University of Central Lancashire	UK
Matthew Colless	Mount Stromlo & Siding Spring Obs.	Australia
Andrew Connolly	Johns Hopkins University	USA
Ian Corbett	PPARC	UK
Garret Cotter	Mullard Radio Astronomy Observatory	UK
Len Cowie	University of Hawaii	USA
David Crampton	Dominion Astrophysical Observatory	Canada
Carolin Crawford	Institute of Astronomy, Cambridge	UK
Sandro D'Odorico	European Southern Observatory	Germany
Roger Davies	University of Durham	UK
Eugene de Geus	Kluwer Academic Publishers	Netherlands
Ana de la Varga	Hamburger Sternwarte	Germany
Kiyomi Denda	University of Tokyo	Japan
Jane Dennett-Thorpe	Mullard Radio Astronomy Observatory	UK
Arjun Dey	NOAO	USA
Mark Dickinson	Space Telescope Science Institute	USA
Mike Disney	University College Cardiff	UK
Nigel Douglas	Kapteyn Institute	Netherlands
Alan Dressler	Carnegie Observatories	USA
James Dunlop	Institue for Astronomy, Edinburgh	UK
Tim Ebbels	Institute of Astronomy, Cambridge	UK
Mike Edmunds	University of Wales	UK
George Efstathiou	University of Oxford	UK
David Elbaz	Service d'Astrophysique	France

Richard Ellis	Institute of Astronomy, Cambridge	UK
Rebecca Elson	Institute of Astronomy, Cambridge	UK
Kimberly Ennico	Institute of Astronomy, Cambridge	UK
Stefano Ettori	Institute of Astronomy, Cambridge	UK
Andrew Fabian	Institute of Astronomy, Cambridge	UK
Michael Fall	Space Telescope Science Institute	USA
Henry Ferguson	Space Telescope Science Institute	USA
Bob Fosbury	European Southern Observatory	Germany
Carlos Frenk	University of Durham	UK
Andrew Fruchter	Space Telescope Science Institute	USA
Jonathan Gardner	University of Durham	UK
Bernhard Geiger	Max-Planck-Institut fuer Astrophysik	Germany
Mauro Giavalisco	Observatories of the Carnegie Institution	USA
Karl Glazebrook	Anglo-Australian Observatory	Australia
Matthew Graham	University of Central Lancashire	UK
Richard Griffiths	Johns Hopkins University	USA
Carlotta Gruppioni	Istituto di Radio Astronomia - CNR	Italy
Einar Gudmundsson	University of Iceland	Iceland
Stephen Gwyn	University of Victoria, Canada	Canada
Wolfgang Hackenberg	MPI fuer extraterrestrische Physik	Germany
Martin Haehnelt	Max Planck Institute fur Astrophysik	Germany
Fred Hamann	University of California, San Diego	USA
Gill Harrison	Royal Greenwich Observatory	UK
Sara Heap	NASA/Goddard Space Flight Center	USA
Timothy Heckman	The Johns Hopkins University	USA
Sune Hermit	Astronomical Observatory	Denmark
Jens Hjorth	Institute of Astronomy, Cambridge	UK
David Hogg	Caltech	USA
Anthony Holloway	University of Manchester	UK
Ulrich Hopp	Universitaetssternwarte Muenchen	Germany
Esther Hu	Institute for Astronomy, Honolulu	USA
Mike Hudson	University of Victoria	Canada
Shaun Hughes	Royal Greenwich Observatory	UK
Garth Illingworth	UCO / Lick Observatory	USA
Chris Impey	Steward Observatory	USA
Masanori Iye	National Astronomical Observatory	Japan

Pascale Jablonka	Observatoire de Paris-Meudon	France
John Jackson	University of Northumbria	UK
Carole Jackson	Institute of Astronomy, Cambridge	UK
Peter Jakobsen	European Space Agency	Netherlands
Buell Jannuzi	NOAO	USA
Mike Jones	Mullard Radio Astronomy Observatory	UK
Vesa Junkkarinen	UCSD / CASS 0111	USA
Nobunari Kashikawa	National Astronomical Observatory	Japan
Guinevere Kauffmann	MPI fur Astrophysik	Germany
Ken Kellermann	National Radio Astronomy Observatory	USA
Douglas Kelly	University of Wyoming	USA
Daniel Kelson	UCO / Lick Observatory	USA
Julia Kennefick	Ohio State University	USA
Matthew Kenworthy	Institute of Astronomy, Cambridge	UK
David King	Royal Greenwich Observatory	UK
Suzanne Koehler	Hamburger Sternwarte	Germany
David Koo	Lick Observatory	USA
Cedric Lacey	Theoretical Astrophysics Center	Denmark
Ofer Lahav	Institute of Astronomy. Cambridge	UK
Olivier Le Fèvre	Observatoire de Paris-Meudon	France
Anthony Leonard	Oxford University	UK
Geraint Lewis	Institute of Astronomy, Cambridge	UK
Christopher Lidman	European Southern Observatory	Germany
Malcolm Longair	Mullard Radio Astronomy Observatory	UK
Sebastian Lopez	Hamburger Sternwarte	Germany
James Lowenthal	UCO / Lick Observatory	USA
Limin Lu	California Institute of Technology	USA
Lori Lubin	Carnegie observatories	USA
Paul Lynam	Liverpool John Moores University	UK
Donald Lynden-Bell	Institute of Astronomy, Cambridge	UK
Piero Madau	Space Telescope Science Institute	USA
Steve Maddox	Royal Greenwich Observatory	UK
Matthew Malkan	UCLA Astronomy	USA
Eliot Malumuth	CSC/Goddard Space Flight Center	USA
Filippo Mannucci	CAISMI - CNR	Italy
Claudia Maraston	Universita di Bologna	Italy

Enrique Martinez-Gonzalez	Instituto de Fisica de Cantabria	Spain
Ronald Marzke	Dominion Astrophysical Observatory	Canada
Sophie Maurogordato	CNRS	France
Conal Mc Keith	Queen's University of Belfast	Ireland
Henry McCracken	University of Durham	UK
Richard McMahon	Institute of Astronomy, Cambridge	UK
Yannick Mellier	Institut d'Astrophysique de Paris	France
Georges Meylan	European Southern Observatory	Germany
Joan-Marc Miralles	Observatoire Midi-Pyrénées	France
Leonidas Moustakas	University of California	USA
Jan Müecket	Astrophysikalisches Institut Potsdam	Germany
Carole Mundell	Nuffield Radio Astronomy Lab	UK
Avi Naim	The Johns Hopkins University	USA
Joan Najita	Harvard-Smithsonian CfA	USA
Priya Natarajan	Institute of Astronomy, Cambridge	UK
Peter Newman	University of Central Lancashire	UK
Robert O'Connell	University of Virginia	USA
Stephen Odewahn	Arizona State University	USA
Augustus Oemler	Carnegie Observatories	USA
Patrick Osmer	Ohio State University	USA
Bernard Pagel	NORDITA	Denmark
Neil Parker	Royal Greenwich Observatory	UK
John Peacock	Royal Observatory Edinburgh	UK
Antony Peake	The Royal Greenwich Observatory	UK
Roser Pelló	Observatoire Midi-Pyrenees	France
Margaret Penston	Royal Greenwich Observatory	UK
Clovis Peres	Institute of Astronomy, Cambridge	UK
Joseph Pesce	Space Telescope Science Institute	USA
Max Pettini	Royal Greenwich Observatory	UK
Diana Pisani	Liverpool John Moores University	UK
Bianca Poggianti	Kapteyn Instituut	Netherlands
Lucia Pozzetti	Universita di Bologna	Italy
Isabella Prandoni	Istituto di Radio Astronomia del CNR	Italy
Chris Pritchet	University of victoria	Canada
Michael Rauch	California Institute of Technology	USA
Martin Rees	Institute of Astronomy, Cambridge	UK

Claudia Rola	Royal Greenwich Observatory	UK
Michael Rowan-Robinson	Blackett Laboratory	UK
Jose Luis Sanz	Instituto de Fisica de Cantabria	Spain
Vicki Sarajedini	Steward Observatory	USA
Wal Sargent	Palomar Observatory	USA
Bill Saslaw	Institute of Astronomy, Cambridge	UK
Katsuhiko Sato	The University of Tokyo	Japan
Roberto Scaramella	Osservatorio Astronomico di Roma	Italy
David Schade	University of Toronto	Canada
David Schlegel	University of Durham	UK
Peter Schneider	Max Planck Institut fur Astrophysik	Germany
Tom Shanks	University of Durham	UK
Joseph Shields	Steward Observatory	USA
Remco Slijkhuis	Sterrewacht Leiden	Netherlands
Adam Stanford	IGPP / Livermore Labs	USA
Erik Stengler	Instituto de Fisica de Cantabria	Spain
Daniel Stern	University of Berkley	USA
Nic Stewart	Serco Group plc	UK
Lisa Storrie-Lombardi	University of California, San Diego	USA
Yoshiaki Taniguchi	Tohoku University	Japan
Nial Tanvir	Institute of Astronomy, Cambridge	UK
Roger Tayler	University of Sussex	UK
Andy Taylor	Institute for Astronomy, Edinburgh	UK
Eduardo Telles	Universidade de Sao Paulo	Brasil
Alejandro Terlevich	University of Durham	UK
Patricia Tissera	Universidad Autónoma de Madrid	Spain
Kenji Tomita	Yukawa Institute for Theoretical Physics	Japan
Laurence Tresse	Institute of Astronomy, Cambridge	UK
Keith Tritton	Royal Greenwich Observatory	UK
David Tytler	University of California, San Diego	USA
David Valls-Gabaud	Observatoire to Strasbourg	France
Wil van Breugel	University of California	USA
Harry van der Laan	Utrecht University	Netherlands
Michiel van Haarlem	University of Durham	UK
Eelco Van Kampen	Royal Observatory, Edinburgh	UK
Giovanni Vladilo	Osservatorio Astronomico di Trieste	Italy

1 D Lynden-Bell
2 J Hjorth
3 S Maddox
5 C Lacey
6 N Tanvir
7 B Geiger
8 S Besier
9 A Terlevich
10 G Lewis
11 D King
12 D Pisani
13 W van Breugel
15 R Griffiths
16 S White
17 M Hudson
18 C Lidman
19 C Frenk
20 B Ziegler
21 E de Geus
22 A Holloway
23 N Clark
24 P Lynam
26 D Valls-Gabaud
27 A Blain
28 P Jakobsen
29 B Woodgate
30 A Connolly
31 P Madau
32 D Hogg
33 J Peacock
34 C Impey
35 M Rauch
36 J-M Delthorn
37 S Odewahn
38 R Marzke
39 D Stern
40 J Brinchmann
41 P Newman
42 J Willis
43 M Malkan
44 C Peres
45 S D'Odorico
46 R Scaramella
47 G Meylan
48 D Kelly
49 M Fall
50 T Ebbels
51 D Schlegel
52 H McCracken
53 U Hopp
54 J Pesce
55 E Stengler
56 A Cappi
57 K Tomita
58 J Lowenthal
59 I Aretxaga
60 P Natarajan
61 C Crawford
62 M Treyer
63 M Graham
64 B Jannuzi
65 A Stanford
66 S Heap
67 E Malumuth
68 A Leonard
69 I Waddington
70 W Webb
71 L Lu
72 A Fruchter
73 A Barger
74 O Lahav
75 H Ferguson
76 A Babul
77 J Miralles
78 L Cowie
79 P Osmer
80 M Disney
81 A Franceschini
82 T Shanks
83 S Gwyn
84 V Sarajedini
85 D Crampton
86 D Koo
87 N Vogt
88 R Pelló
89 M Haehnelt
90 G Vladilo
91 F Mannucci
92 A Baleisis
93 S Hermit
94 P Jablonka
95 Y Taniguchi
96 A de la Varga
97 S Köhler
98 J Gardner
99 S Maurogordato
100 K Wu
101 R Carson
102 J Kennefick
103 R Bernstein
104 H Yee
105 A Zezulová
106 E Gudmundsson
107 J Bezecourt
108 D Kelson
109 R Abraham
110 T Brainerd
111 T Small
112 A Nusser
113 E Telles
114 C Rola
115 V Nazarova
116 K Denda
117 K Sato
118 R Slijkhuis
119 L Storrie-Lombardi
120 L Junkkarinen
121 J Naim
122 M Davies
123 M Kenworthy
124 V Bailey
125 K Kellermann
126 K Blundell
127 D Clements
128 C Mundell
129 M Iye
130 J Jackson
131 Y Zitelli
132 A Campos
133 M Rees
134 R Clowes
135 M Rowan-Robinson
137 Y Mellier
138 P Schneider
139 R O'Connell
140 L Williams
141 N Kashikawa
142 A Aloisi
143 C Gruppioni
144 O Le Fèvre
145 B Saslaw
146 M Colless
147 I Parry
148 K Ennico
149 J Dunlop
150 C Blades
151 G Kauffmann
152 R Williams
153 G Efstathiou
154 N Bahcall
155 M Giavalisco
156 J Wall
157 W Sargent
158 A Boksenberg
159 R Ellis
160 H van der Laan
161 A Dressler
162 G Illingworth
163 B Heckman
164 M Pettini
165 B Wright
166 G Harrison
167 A Peake
168 H Fosbury
169 B Pagel
170 B Boyle
171 R Elson
172 D Schade
173 K Glazebrook
174 C Pritchet
175 C Jackson
176 J Baldwin
177 B Carswell
178 D Van Kampen
179 A Taylor
181 M Moustakas
182 M van Haarlem
183 A Aragón-Salamanca
184 D Bowen
185 R Catchpole
186 S Bardelli
187 E Zucca
188 J Shields
189 F Hamann
190 A Oemler
191 J Allington-Smith
192 E Zirbel
193 M Dickinson
194 S Hughes
195 P Benvenuti
196 J Bergeron
197 E Martinez-Gonzalez
198 J L Sanz
199 L Pozzetti
200 C Maraston
201 B Poggianti

INTRODUCTION

RICHARD S ELLIS

*Institute of Astronomy, Madingley Road,
Cambridge, CB3 OHA, England*

Welcome to the 37th Herstmonceux Conference on 'HST and the High Redshift Universe' to celebrate the 60th birthday of Alec Boksenberg!

As the name suggests, this annual conference series began in the 1960s when the Royal Greenwich Observatory was at Herstmonceux, Sussex but it has been arranged jointly with the Institute of Astronomy since the RGO moved to Cambridge in 1989. This year's meeting is supported by the RGO, IoA and the European Economic Community.

I should first admit that, although nominally a member of the Local Organising Committee, my involvement has largely been to grunt approval at decisions made by more hard-working individuals at various meetings. Nonetheless the organisers have a hard time selecting from a heavily oversubscribed meeting which reflects something which I think we all are aware of, namely that the topic of the High Redshift Universe is going through a tremendously exciting phase. We were sufficiently oversubscribed with impressive results that we could easily have filled a two week meeting with talks. Bearing this in mind, we have gone to great troubles to ensure as many posters as possible can be presented. This is where most of the action is (and it is very satisfying to see so many of them featuring in the final proceedings).

Clearly, graduate students and postdocs working in this area today have unprecedented opportunities. Indeed, I sense a mood in many of our minds that we are close to resolving some of the grand cosmological questions. Lest we get too optimistic, perhaps we should heed the words of Zel'dovich who said at an IAU Symposium in 1977 that "extrapolating to the next symposium in the early eighties, one can be pretty sure that the question of the formation of galaxies and clusters will be solved".

Although rapid progress is being made theoretically and computationally in understanding the high redshift Universe, we must surely agree the main developments are the acquisition of exciting new data. HST is opening up the morphology of the intermediate and high redshift Universe; we can see resolved galaxies to redshifts beyond 3. However, we have also learnt how to use our ground-based telescopes to complement HST (and the impact of Keck became abundantly clear at this meeting – particularly to those of us who don't have access to it!).

In selecting key scientific questions for the meeting, I found it helpful to consider the conclusions of the meeting entitled 'The Epoch of Galaxy Formation' held in Durham in 1988. That meeting was an important landmark in observational cosmology just prior to the launch of HST. In fact, several of this week's invited speakers were at that meeting (some now greyer and with less hair but still in fine form).

With gentle guidance from Carlos Frenk, the participants of that meeting were asked to vote on when galaxies formed, both in terms of the era (z^*) when half the stars hard formed and that when half the assembled mass within 5 kpc came together. Interestingly, the majority voted for z^*=2.5-5 with a sizeable minority preferring z^*=0-2.5 and a few going for z^* >5. .

New developments in 1996 include evidence that some galaxies are truly old (both cluster ellipticals and distant radio galaxies). It seems *some* star formation occurred at very high redshifts. But what fraction of the present population are we talking about? Does this contribute significantly to the ionisation of the high z Universe? On the other hand, one is struck by the rarity of faint field galaxies ($\simeq$5%) whose Lyman limits have entered the optical. This result was suggested in 1990 from the work of Guhathakurta et al and has now been confirmed spectroscopically using the Keck. It would seem to me from various arguments that the bulk of the star formation occurred below z=2-3. But perhaps we are being deluded by biased evidence?

So my personal key questions for the meeting would include:

1. How robust are the conclusions that some massive galaxies are very old? Are these representative galaxies? Is the standard model in danger as some suggest?

2. Have we found all the z >3 galaxies? Are we missing any of the population by virtue of selection effects? What are the prospects for finding higher redshift sources?

3. Can we reconcile the era of apparent rapid star-formation in field galaxies (z <2) with other diagnostics such as the gas content and composition of the QSO absorption lines and the history of the Milky Way? What leads to this decline in activity (mergers/fading) and are 'extra populations' required to explain the faint blue excess?

4. How and when did structure form ? Can the growth of structure explain the morphology-density relation through environmentally-driven processes ('nurture' as opposed to 'nature')?

5. What are the prospects with NICMOS and STIS, and with forthcoming instrumentation on ground-based telescopes? What should we plan as a successor to HST?

I would like to finish this introduction with just a few brief words on the other equally important aspect of the meeting, namely the 60th birthday of Alec Boksenberg. Alec was, of course, Director of the RGO and the Royal Observatories for over 15 years and thus played a key role in almost half of the Herstmonceux conferences. He retired from that position last year and is now a Research Professor with us at the Institute. There will be many tributes to Alec and I would just add one brief personal note.

I've known Alec since the late 1960's. He was my undergraduate lecturer at UCL and taught me Statistical Mechanics (he did such a splendid job at teaching theoretical physics that I decided immediately to become an observer like him). As the UK's leading observer, we should congratulate Alec on his recently-awarded CBE in Her Majesty's birthday honours list. For non-Brits, I should explain that CBE stands for Commander of the British Empire. Royal sources confirm that this prestigious award entitles Alec to unlimited telescope time on all colonial telescopes defined as those where the Union Jack still flies. As this includes Hawaii, he gets access to both Kecks, CFHT and shortly Subaru!

However let me finish with a more serious comment on the contribution Alec made to UK astronomy. His Image Photon Counting System was largely responsible for the renaissance of UK astronomy at the AAT and enabled many of us to start our careers. I had a quick look at the CV's of the current generation of observational professors in the UK and found almost all the prominent papers which propelled observers like Roger Davies, George Efstathiou, Andy Lawrence and myself to senior positions were based on Alec's IPCS. In short, without Alec, an entire generation of UK observers would not have been formed. He also, of course, played a crucial part in developing the Faint Object Camera on HST which many of you have used.

I am delighted that this important meeting is being held in Cambridge in recognition of Alec's achievements and continuing interest in the high z Universe. I hope you all find the meeting stimulating and that your time in Cambridge will be productive.

AFTER DINNER MINCE

J.V. WALL

Royal Greenwich Observatory, Madingley Road,
Cambridge, CB3 0EZ, UK

1 Introduction

Ladies and Gentlemen,

2 Observations

Alec Boksenberg is a Child of the Universe. How do I know this?

- The Hubble constant is 71 ± 15.

  So is Alec.

- Moreover the Hubble constant has been claimed to be 51.

  So did Alec, and for some years.

  However, it must be admitted that the Hubble constant was once 550; Alec has never been 550 to my knowledge, although re-incarnation and/or parallel universes are not my forte.

- A Hubble constant of 71 gives a problem in that some portions/bits of the Universe have turned out to be substantially older than the main body.

  It's been the same with Alec.

  The only difference here is that plumbers fixed Alec while theorists will fix the Universe. But theorists and plumbers are the same in my book; both are overpaid, the plumbers perhaps a bit more so.

3 Conclusions

Ladies and Gentlemen, as the stories I can add about Alec carry even less political correctness than those already related at the conference, let me forego them and summarize without viewgraphs:

- Alec, eternal optimist,

- Alec, inveterate plumbicon twiddler, and experienced non-reducer of Wal Sargent's data,

- Alec, the man of endless resource (just ask his seriously-ageing former finance officers),

- Alec of the ever-present smile, described to me by one media guru as so perfect for radio,

- Alec of the ever-absent memory for names (and let me remind you Alec in case you wish to write this down for future reference that *my* name is Dave Axon),

- Alec, good friend....

Let me wish you on behalf of your assembled friends,

1. many years of the continued smile,

2. and may this mean many years of happiness, in science and amongst us.

Ladies and gentlemen - please rise and drink with me to Alec Boksenberg.

SECTION 1

FIELD GALAXY EVOLUTION &

THE HUBBLE DEEP FIELD

REFLECTIONS ON THE HUBBLE DEEP FIELD CAMPAIGN

ROBERT WILLIAMS

Space Telescope Science Institute,
Baltimore, MD 21218, USA

Let me begin by paying tribute to Alec Boksenberg on this occasion. I first met Alec in 1971 when I was on sabbatical leave from the University of Arizona at University College London. The Steward Observatory 2.3m telescope had been in operation a few years by that time, and our workhorse detector was one of the RCA Carnegie image tubes used in a photographic mode for both imaging and spectroscopy. I was advised of a chap named Boksenberg who was developing a new digital system at UCL that might interest us at Steward. So, I sought out Alec in his lab in the basement of the UCL Physics/Astronomy building and got him to explain to me exactly what he was doing, and a friendship ensued.

Now, 25 years later, we can say that the instrument Alec was building, the IPCS, turned out to be a landmark instrument in astronomy. As a spectroscopist, I believe that Joe Wampler's Image Tube Scanner and Alec's Image Photon Counting System, with their spectacular success in acquiring spectra of faint objects previously unreachable, were the two instruments that led astronomical spectroscopy from photography into the new age of digital data. There are numerous other accomplishments that Alec can look back to in his career, but this one was indeed very important to our science. I am happy to wish Alec well in what I understand is a mandated administrative, but certainly not an intellectual!, retirement. We all look forward to his continued involvement in astronomy in the future.

1   Formulation of the HDF Project

The Hubble Deep Field was a compelling observation that was certain to be done with HST sooner or later, and those of us at STScI were fortunate in being in a position to insure that such a project was carried out sooner rather than later. It was compelling because of the success of the imaging projects with HST that preceded it, such as the Medium Deep Survey of Griffiths and colleagues which established even with spherical aberration the ability of HST to resolve structure in distant galaxies. The images of the z=0.4 cluster CL0939+4713 obtained by Dressler et al. both before and after the first servicing mission, and the 32-orbit study of the cluster(s) at z=1.2 associated with the 3C 324 field by Dickinson and colleagues established the capability of

HST to address galaxy evolution at epochs when significant evolution should be taking place. Especially noteworthy were the findings of numerous spiral galaxies in CL0939, no spiral galaxies in the higher redshift 3C 324 clusters, and the fact that the radial light distributions of the red, apparent elliptical systems followed the $r^{1/4}$-law at these early epochs.

Based upon these results I felt that a program should be undertaken immediately to exploit the power of HST to study distant galaxies while the telescope was functioning so superbly. For such a program, three features seemed desirable to those of us at the Institute who discussed how such observations should be carried out. The first was to target a few known galaxies at high redshift. The second was to make use of the HST continuous viewing zone (CVZ) in order to gain a factor of two in the total imaging time available. The third was to waive the normal proprietary rights to the HST data so that they could be made immediately available to the community as a stimulus for follow-up study of the observations.

An Advisory Committee was convened to discuss these ideas, and in March 1995 the group debated the best way in which to use HST time to study galaxies at early epochs. Initially divided between devoting the HST time to targeted sources or to undistinguished areas of the sky, the Advisory Committee eventually gave a recommendation to "take deep images of 1-2 undistinguished/untargeted fields in 2-3 colors." This recommendation owed as much to the inability of those advocates of targeted objects to agree on exactly WHICH galaxies to image as it did to the forcefulness of those who were arguing the opposite point of view!

Following the Advisory Committee deliberations, a working group of Institute staff was constituted to devise an observing program for what we began calling the Hubble Deep Field. This group, subsequently augmented by scientists from the HST European Coordinating Facility in Garching and who comprise the co-authors of the HDF paper in the Astronomical. Journal (August 1996), deserve most of the credit for the crafting of the HDF program of observations and data reduction. Since much information about the HDF has already appeared elsewhere, I will not repeat here the key facets of the field or the observations. A number of the papers in this present volume deal with the interpretation of the data.

2 Lessons Learned

Based upon the early results from the HDF various conclusions can already be drawn from the campaign, and I will here single out the following:

1. The combined HDF images in the nominal V and I bands reach to almost 30th magnitude (5 sigma over background). The sky brightness for HST at these wavelengths is fully 5 magnitudes brighter, i.e., not that much better than that from the ground. The depth of the HDF is therefore determined not so much by the detection of faint galaxy photons as it is by the accuracy of determining and subtracting the far higher background level. The superior ability of HST to observe to fainter levels than large ground-based telescopes is due to the better temporal stability of the HST sky.

2. There is value in devoting a fraction of time on a multi-purpose telescope like HST to a data gathering, survey mode, even though this decreases the observing time available to individual PI's. The tendency of the peer review process for unique and expensive facilities like the Hubble, whose success depends upon broad community support, is to try to give as large a segment of the community as possible a piece of the action, especially when data reduction funds accompany the observing time. TACs whittle away observing time from programs in order to fit more programs in. This tendency is undoubtedly successful in minimizing community discontent over the allocation of telescope time, but it almost certainly does not maximize scientific discovery. I believe that strong arguments exist for reserving of the order of 1/3 of the observing time on any telescope to surveys, especially if the data is made public immediately. This requires that the data be archived and easily accessible.

3. The value of proprietary data rights for PI's using public facilities is a complex issue that deserves renewed discussion. Exclusive access to a set of data which one has given independent thought to defining and obtaining is certainly appropriate for a period of time. But, it is less clear to what extent this benefits the PI as opposed to providing a greater benefit to astronomy if it were made public.

   The HDF has demonstrated that data made immediately available to the public does not necessarily diminish the role of the PIs in its use by others. And, depending upon the nature of the data, it will serve to stimulate further research on the topic from many sources. The conditions under which PIs retain sole access to data from public facilities, and for how long, needs to be reconsidered by the major funding agencies, and the concept that this is a merited benefit/reward for the PI must be re-thought in light of the beneficial effects dissemination of the data has on further development of the field.

3 A Southern Deep Field

The results of the HDF have been sufficiently interesting for a wide range of topics that the question should be asked: should another HDF campaign be undertaken? What worked once can work again. On the other hand, there can be too much of a good thing, and another deep field in another location might not have the value of the first field, especially in light of other large programs that one could think of undertaking on other topics with HST.

One of the most important successes of the HDF has been its demonstration of the power of HST imaging followed by spectroscopy from large ground-based telescopes such as the Keck 10m telescope. Within a few years a number of large telescopes will begin operation in the southern hemisphere, including the four VLT 8.2m, the Gemini southern 8m, and the twin 6.5m Magellan telescopes. The availability of large amounts of time on these telescopes for follow-up spectroscopy of a southern deep field, and the additional stimulus this would give not just to the topic of galaxy evolution but also to co-ordinated multi-wavelength space-and ground-based follow up, argues strongly for another deep field. Furthermore, study of at least one other deep field is needed to demonstrate that the first field is not atypical in some important characteristic.

For these reasons, I am pleased to announce here the decision to undertake another deep field campaign in the southern sky, to be called the HDF-S. It will be very similar to the recently completed HDF(-N), and will be carried out in late 1997 as a Cycle 7 Director's Discretionary program. Since this will occur after the second servicing mission of HST, parallel observations of WFPC2 and NICMOS (and STIS) will be possible, thus providing for two adjacent fields with one in the visible and the other in the IR.

In order to benefit from U-band images, which are not affected by the increased scattered light in the daylight part of the orbit, a field of low extinction and H I column density will be selected in the southern CVZ. An area with virtually the same properties as the HDF-N will be sought with one possible addition, and that is the presence of a high redshift quasar in the field. A QSO would offer a long sightline against which absorption at various redshifts from intervening galaxies might be detected, and these galaxies might then appear imaged within the HDF-S. The detection of various galaxies at different impact parameters from the quasar line of sight would be very useful in understanding the gaseous outer component of the galaxies, compensating for the fact that the presence of a quasar with associated galaxies in the field might skew the statistics of galaxy number counts in the field.

Selection of an optimum HDF-S is now underway, and it will require identification of appropriate candidate quasars to be found from a grism search within the southern CVZ. The radio characteristics of the candidate fields will be determined by R. Norris and colleagues using the Australian Telescope, and a final field selection will be made in the near future from the available data.

4 Prologue

Two groups deserve much credit for the impact of the HDF on studies of galaxy evolution. The first is the working group of Institute and ECF scientists who carried out the observing and data reduction campaign that produced the HDF images, and whose names appear as the co-authors on the resultant HDF manuscript. The second group is the Advisory Committee that debated with me the concept of the HDF and whose input provided the real impetus for what resulted. With my thanks, they are L. Cowie, G. Djorgovski, A. Dressler, R. Ellis, S. Faber, J. Huchra, G. Illingworth, K. Kellermann, S. Lilly, G. Miley, F. Owen, H. Spinrad, and E. Turner.

TESTING GALAXY EVOLUTION MODELS WITH DEEP HST IMAGES

H. C. FERGUSON

Space Telescope Science Institute
3700 San Martin Drive, Baltimore, MD 21218, USA

Among the key tests of galaxy evolution models is the attempt to match the statistical distribution of brightnesses, redshifts, sizes, and colors of faint galaxies. The status of these tests is briefly reviewed. The recent Hubble Deep Field observations are then compared to simulations from two different models: one with an open cosmology ($q_0 = 0.01, \Lambda_0 = 0$) and fairly standard assumptions galaxy spectral evolution, and one with an Einstein-de Sitter cosmology and delayed dwarf-galaxy formation. Both models are largely successful at meeting constraints from ground-based data, but fail to reproduce the observed distributions in the Hubble Deep Field.

1 Review

Deep images of the sky provide both a source of data and a source of inspiration to those of us trying to decipher clues to the origin of galaxies. It is a simple fact that wherever we look in the night sky, if we look hard enough, we can find miriads of galaxies, more than one million per square degree, with vast array of colors shapes and sizes. The history of the universe is set out before us upon this rich tapestry, if we can only make sense of it.

The classical tests of cosmology, set out in detail by Sandage,[1] involve comparing the distributions of faint galaxies in number, redshift, and size to the predictions of models which include both cosmological curvature and galaxy evolution. Such tests were among the defining goals of the 5-m Hale telescope and the Hubble Space Telescope. Rapid observational progress in the last two decades has improved the data to the point where the statistical uncertainties are small. However, these observations have also revealed that the problem is considerably more complex than originally appreciated.

The complexity of the problem can be illustrated by considering the following measurements of the statistical properties of faint galaxies.

- *Galaxy counts from the ultraviolet to the infrared.* At blue wavelengths, the $N(m)$ relation is still rising steeply at the limits of groundbased observations, exceeding by a factor of ten the surface density expected for a $q_0 = 0.5$ non-evolving universe. In the K band the counts are flatter and more consistent with mild evolution and $q_0 = 0.5$. [2,3,4]

- *Galaxy redshift distributions.* Broadly speaking, the observed distributions are unimodal with a median redshift consistent with no-evolution models. Galaxies at $z < 0.1$ or $z > 3$ constitute less than 20% of the counts to $B < 24, I < 22.5$. [5,6,7,8]

- *The amplitude of the angular correlation function.* The faintest galaxies appear to be very weakly clustered to ground-based limits. [9,10,11,12,13]

- *The space-density of luminous galaxies at $z \sim 1$.* The K-band luminosity function of absorption-line selected galaxies at these moderate redshifts appears consistent with the bright end of the local luminosity function.[14] The luminosity function of galaxies with colors redder than Sbc appears to evolve slowly or not at all at to $z \sim 0.7$. [15]

- *The angular size distribution of faint galaxies.*[16] At $V = 28$ the typical galaxy half-light radius is less than 0.2 arcsec, significantly smaller than expected if the population is dominated by non-evolving L^* galaxies, even for low values of q_0.

Clearly there are other measurable quantities as well, but these five will serve to illustrate the problems. A successful model must reproduce these observations, as well as providing a match to the global properties of present-day galaxies. Recent investigations have focused on several different classes of models.

Pure Luminosity Evolution (PLE) models, with a standard luminosity function. These models start with the local measurements of the luminosity and color distributions and model only the photometric evolution of galaxies assuming a constant co-moving density between $z = 0$ and some redshift of formation z_f. [17,18,19,20] To match galaxy counts, such models favor low values of q_0 and/or a non-zero cosmological constant. Because much of the action happens at high redshift, these models are consistent with the slow evolution in the space density of massive galaxies, and with the weak amplitude of the angular correlation function. However, they tend to predict a high-redshift tail that is not observed in deep redshift surveys. It is not clear that the tail can be hidden (e.g. by invoking dust or moving z_f higher) without degrading the fit to the counts, since much of the success of this class of model can be attributed to the large volumes at high z inherent in low-q_0 cosmologies. It should be noted that PLE models, while they do not explicitly incorporate merging, are compatible with strong number-density evolution, so long as the merging process does not significantly affect star formation. [21]

Merger models. These models include rapid merging of galaxies at $z < 1$, sometimes coupled with triggered star formation. [22,21,23,24,16] Such models are able to match the counts and redshift distributions reasonably well for $q_0 = 0.5$ (as well as for low-density universes). However they appear to conflict with the low amplitude of the angular correlation function,[25] and with the mild evolution in the space-density of luminous galaxies. Furthermore, it is not clear that such models can reproduce the stellar populations of nearby elliptical galaxies[26] or allow disk galaxies to remain sufficiently stable. [27]

Models with a modified luminosity function. In these models,[28,29,30] the luminosity function is more heavily weighted to low luminosities than indicated by measurements of the local field-galaxy luminosity function,[31,32,33,34] which are surely incomplete *at some level* for either compact or low surface-brightness galaxies. The most recent deep redshift surveys have not detected the low-redshift peak anticipated in such dwarf-dominated models. [6,7,8]

Evolving-dwarf models. In these models,[35,36] a large dwarf-galaxy population undergoes star formation at relatively late epochs ($z \lesssim 1$) and subsequently fades. Such a population arises naturally within the CDM context if photoionization suppresses star formation in low-mass halos at high redshifts. These models are able to match the ground-based counts, colors, and redshift distributions. The dwarf galaxy population must be very weakly clustered in order to satisfy the measured angular correlation function, but that is in keeping with the expectations from biased galaxy formation.

Semi-analytic hierarchical models. These models start from an adopted fluctuation spectrum, adopt various prescriptions for cooling, star formation, feedback, and merging, and attempt to match the statistical properties of both nearby and distant galaxies. [37,38,39,40,41,42,43] A generic difficulty of such models is the inability to reproduce both the the local luminosity function and the Tully-Fisher relation, suggesting either that the underlying cosmological model is wrong, or that some additional piece of physics governing star-formation and feedback is missing. It should be noted that in these models merging has a huge influence on star-formation, either by quenching it or by triggering it; this likely to be the place where the physics is least certain. As a group, these models are reasonably successful in matching the counts and redshift distributions. They are less successful in reproducing the quiescent evolution of luminous galaxies to $z \sim 1$, and they probably face the same problem as the more heuristic merger models in accounting for the angular correlation function.

2 The Hubble Deep Field

With counts, colors, and morphologies to the faintest limits yet achieved, the Hubble Deep Field (HDF) is now a key testing ground for models of galaxy evolution. Obtained with the Wide Field Planetary Camera (WFPC-2) during December 1995, the HDF images reach 5-σ limiting AB magnitudes [44] of roughly 27.7, 28.6, 29.0, and 28.4 (for an aperture area of 0.2 square arcsec) in the F300W, F450W, F606W, and F814W bandpasses (the number corresponds to the central wavelength in nm), respectively. [45]

The focus of the remainder of this contribution is the comparison of two models to the HDF data. The purpose is (1) to show preliminary constraints on models available from this new set of deep counts, sizes, and colors; and (2) to demonstrate the importance of selection effects in trying to make model comparisons to the HDF data. More details of this work can be found elsewhere.[46] For both models, the analysis is carried out by making simulated images with noise properties that match the HDF and analyzing them with the same software (FOCAS) as the real images.

The first model is a fairly standard PLE model, with $q_0 = 0.01$ and $H_0 = 50\,\mathrm{km\,s^{-1}\,Mpc^{-1}}$. The luminosity functions for the giants (E through Sdm) are Gaussian, tuned to approximate the type-dependent luminosity functions given by Bingelli et al.[47] The faint end of the luminosity function is modeled as a power-law with a slope $\alpha = -1.3$, populated with dwarf irregular galaxies with star-formation rates that are roughly constant over a Hubble time. The bulge/disk total mass ratios and scatter are tuned for each type to match the observations of Simien & de Vaucouleurs [48] for local galaxies. Galaxies are assumed to start forming stars at $z_f = 5$, with different timescales for bulges and disks (tuned to fit present-day colors) and with inclusion of extinction due to dust during the rapid phases of formation of the spheroid components.

The second model has $q_0 = 0.5$. The luminosity function has the same parameters for the giants, but the faint end is populated with dwarf galaxies that begin forming stars at $z \sim 1$. Star formation is assumed to be delayed until this epoch due to photoionization by the UV background. [49,50] The mass function of low-mass galaxies comes from the CDM power spectrum and is not a free parameter. The star-formation episode in the dwarfs lasts 10^7 yr, but there is an exponentially declining probability for dwarf galaxies to undergo these star bursts. This model is described in detail by Babul & Ferguson, [36] who refer to this dwarf population as "boojums" (for Blue Objects Observed Just Undergoing a Moderate Starburst).

Spectral evolution for both models is computed using the Padova [51] isochrones and stellar-atmosphere models. The intrinsic spectra of the galaxies

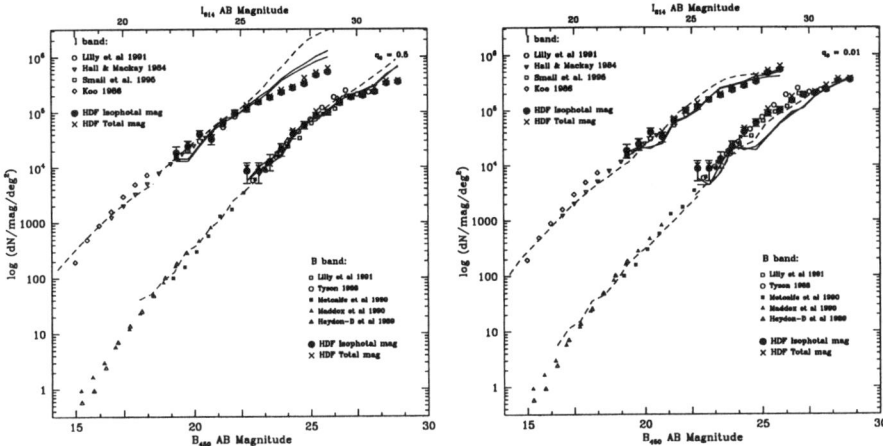

Figure 1: Galaxy counts as a function of AB magnitude in the F450W and F814W bands, together with a compilation of existing ground-based data. FOCAS isophotal and total magnitudes are shown as solid dots and X's, respectively. Model predictions based on total magnitudes (i.e. *without* accounting for observational selection or photometric biases) are shown as dashed lines. Model predictions from the FOCAS-generated catalogs of simulated galaxies are shown as solid lines. No color corrections have been applied to the ground-based data. For the typical colors of galaxies in the HDF, the color corections are less than 0.1 mag.

are attenuated by the mean expected intergalactic HI absorption.[52,53] This attenuation has a significant effect on the predicted numbers and colors of galaxies with $z > 2.5$, and has not been incorporated into previous attempts to fit galaxy counts.

The predicted redshift distributions of the two models are quite similar at the depth of the CFRS survey,[7] but differ dramatically at the depth of the HDF. At $I \sim 28$ the $q_0 = 0.01$ model has a fairly uniform redshift distribution from $0 < z < 5$, while the boojums model at this depth is completely dominated by galaxies with $z < 1$.

Counts for the two models are shown in Fig. 1. The boojums model overpredicts the faint counts in the I band, while the low q_0 model underpredicts the counts in the range $25 < B_{450} < 28$. For both models, the counts based on total magnitudes in the underlying models are significantly different from the counts measured by FOCAS from the simulated images. These differences are partly due to a ~ 0.5 magnitude offset in FOCAS isophotal magnitudes relative to total magnitudes, and partly due to incompleteness. The effect is only marginally reduced if FOCAS "total" magnitudes are used. Clearly a quantitative comparison of models to the counts must take these systematic effects into account; simulating the images is the most straightforward way to accomplish this.

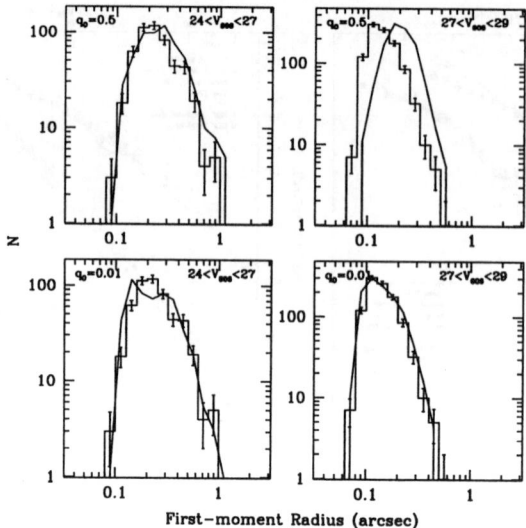

Figure 2: Distribution of radii for both models, compared to the data. The radii for both the models and the data are the first-moment radii measured by FOCAS. The HDF measurements are shown as histograms with Poisson error bars. The model predictions are the light solid curves.

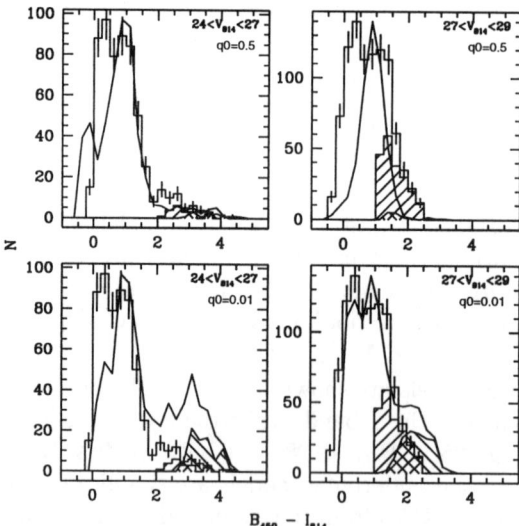

Figure 3: Distribution of FOCAS-measured $B_{450} - I_{814}$ colors for the models (solid curves) and the HDF data (histograms). For both the models and the data there are galaxies that are detected in the F814W image but not in the F450W image. For these galaxies, the color is assigned from the 1σ lower limit to the B_{450} magnitude. The hashed regions indicate portion of the diagrams populated by these lower limits. Hashes slant in opposite directions for the models and the data.

The distribution of radii are compared in Figure 2. The radii plotted are first-moment radii measured by FOCAS within the detection isophotes. The observed distribution peaks at 0.2 arcsec in the magnitude range $24 < V_{606} < 27$, and at about 0.15 arcsec in the range $27 < V_{606} < 29$. These distributions of radii are almost perfectly matched by both models at the brighter magnitudes, and by the low q_0 model in the fainter magnitude bin. However, in spite of being dwarf-dominated, the $q_0 = 0.5$ model predicts galaxies that are systematically *larger* than observed at the faintest magnitudes.

Figure 3 compares the $B_{450} - I_{814}$ colors for the models and data. While the peaks in the model color distributions roughly agree with the observations, neither model provides a very good match to the overall distributions. The $q_0 = 0.5$ model has a blue peak in the brighter magnitude bin that is not observed, while it predicts a red tail similar to that seen in the real HDF images. At fainter magnitudes ($27 < I_{814} < 29$) the model predicts a narrower distribution than observed. The $q_0 = 0.01$ model does not have enough blue galaxies at the brighter magnitudes, and greatly overpredicts the number of very red galaxies ($B_{450} - I_{814} > 2$). At fainter magnitudes the distribution is a better match, but there is still an excess of very red galaxies relative to the observations.

The differences reflect both the different redshift distributions of the models, and the different proportions of galaxies undergoing star-formation at moderate redshifts. During the starburst phase, the dwarfs in the $q_0 = 0.5$ model show up as nearly flat spectrum objects in the magnitude range $24 < I_{814} < 27$. At fainter magnitudes, the blue tail is missing because there are no dwarf galaxies in the model with star-formation rates less than 1.8 $M_\odot$ per year. This is due to the cutoff of the dwarf-galaxy mass function at 15 kms^{-1}. Lower mass potentials are not deep enough to retain 10^4 K gas during the formation epoch, and the adopted star-formation efficiency of 30%, and star-formation timescale of 10^7 years thus fixes the minimum star-formation rate. As neither the star-formation efficiency nor the duration is highly constrained, the lack of flat-spectrum objects in the faint magnitude bin may not be fundamental problem. The reddish ($B_{450} - I_{814} \approx 1$) objects seen in the faint-magnitude cut are for the most part faded, lower-redshift remnants of the starburst epoch.

For the low q_0 model, the most serious discrepancy is the prediction that there should be a large population of very red galaxies, primarily in the brighter magnitude bin, but extending also to the faintest magnitudes seen in the HDF. This red tail is due to bulges and ellipticals forming stars at $z > 3.5$. It is an extremely robust prediction of the models that such galaxies should have red $B_{450} - I_{814}$ colors, since the colors are determined largely by the intrinsic Lyman limits in galaxies, and by absorption due to intervening Lyman α clouds.

[52,53] In the bright magnitude bin, the low q_0 model, even ignoring the very red tail, is skewed to the red of the observed distribution. It is apparently underabundant in star-forming galaxies at moderate redshifts. At fainter magnitudes, the color distribution is a more reasonable match to the observations, with a higher proportion of very blue objects. This is in part explained by the virtual disappearance of passively evolving ellipticals from the sample at moderate redshifts, because the observations have gone beyond the peak of the assumed Gaussian luminosity functions. The counts are thus dominated star-forming galaxies.

The HDF observations clearly provide new and important constraints on galaxy-evolution models. At the faint magnitudes probed by these observations, the problems with the simple models discussed here are particularly acute. It is difficult to tell at present whether the problems lie in the details (e.g. the IMF, dust content, metallicity distributions) or in the fundamental assumptions (e.g. that merging is not a major regulator of star formation, or that giant galaxies all began forming at roughly the same time, with different star-forming rates). It is clear that further modeling of the growing database of galaxy properties in this small patch of sky, with careful attention to selection effects, has the potential to discriminate between widely different world models that heretofore seemed possible.

Acknowledgments

This work is based in part on work with Arif Babul. I would like to acknowledge Bob Williams and the members of the HDF team for their dedicated efforts in making the HDF possible.

References

1. A. Sandage. *ApJ*, 133:355, 1961.

2. J. A. Tyson. *AJ*, 96:1, 1988.

3. I. Smail, D. W. Hogg, L. Yan, and J. Cohen. *ApJ*, 449:105, 1995.

4. N. Metcalfe, T. Shanks, R. Fong, and N. Roche. *MNRAS*, 273:257, 1995.

5. M. M. Colless, R. S. Ellis, T. J. Broadhurst, K. Taylor, and B. A. Peterson. *MNRAS*, 261:19, 1993.

6. K. Glazebrook, R. Ellis, M. Colless, T. Broadhurst, J. Allington-Smith, and N. Tanvir. *MNRAS*, 273:157, 1995.

7. S. J. Lilly, O. Le Fevre, D. Crampton, F. Hammer, and L. Tresse. *ApJ*, 455:50, 1995.

8. L. L. Cowie, A. Songaila, E. M. Hu, and J. G. Cohen. *AJ*, 112:839, 1996.

9. G. Efstathiou, G. Bernstein, J. A. Tyson, N. Katz, and P. Guhathakurta. *ApJ*, 380:L47, 1991.

10. N. Roche, T. Shanks, N. Metcalfe, and R. Fong. *MNRAS*, 263:360, 1993.

11. L. W. Neuschaefer and R. A. Windhorst. *ApJ*, 439:14, 1995.

12. T. G. Brainerd, I. Smail, and J. Mould. *MNRAS*, 275:781, 1995.

13. N. Roche, T. Shanks, N. Metcalfe, and R. Fong. *MNRAS*, 280:397, 1996.

14. C. Steidel, M. Dickinson, and E. Persson. *ApJ*, 437:L75, 1994.

15. S. J. Lilly, L. Tresse, F. Hammer, D. Crampton, and O. Le Fevre. *ApJ*, 455:108, 1995.

16. M. Im, S. Casertano, R. E. Griffiths, and K. U. Ratnatunga. *ApJ*, 441:494, 1995.

17. Y. Yoshii and F. Takahara. *ApJ*, 326:1, 1988.

18. M. Fukugita, K. Yamashita, F. Takahara, and Y. Yoshii. *ApJ*, 361:L1, 1990.

19. B. Guiderdoni and B. Rocca-Volmerange. *A&A*, 227:362, 1990.

20. Y. Yoshii and B. Peterson. *ApJ*, 372:8, 1991.

21. T. J. Broadhurst, R. S. Ellis, and K. Glazebrook. *Nature*, 355:55, 1992.

22. B. Guiderdoni and B. Rocca-Volmerange. *A&A*, 252:435, 1991.

23. R. G. Carlberg. *ApJ*, 399:L31, 1992.

24. R. G. Carlberg and S. Charlot. *ApJ*, 397:5, 1993.

25. B. Roukema and Y. Yoshii. *ApJ*, 418:L1, 1993.

26. J. D. Dalcanton. *ApJ*, 415:L87, 1993.

27. G. Tóth and J. P. Ostriker. *ApJ*, 389:5, 1992.

28. S. P. Driver, S. Phillipps, J. I. Davies, I. Morgan, and M. J. Disney. *MNRAS*, 266:155, 1994.

29. D. C. Koo, C. Gronwall, and G. A. Bruzual. *ApJ*, 415:L21, 1993.

30. C. Gronwall and D. C. Koo. *ApJ*, 440:L1, 1995.

31. J. Loveday, B. A. Peterson, G. Efstathiou, and S. Maddox. *ApJ*, 390:338, 1992.

32. R. O. Marzke, J. P. Huchra, and M. J. Geller. *ApJ*, 428:43, 1994.

33. H. Lin, R. P. Kirshner, S. A. Schectman, S. D. Landy, A. Oemler, D. L. Tucker, and P. L. Schechter. *ApJ*, 464:60, 1996.

34. H. C. Ferguson and S. S. McGaugh. *ApJ*, 440:470, 1995.

35. S. Phillipps and S. Driver. *MNRAS*, 274:832, 1995.

36. A. Babul and H. C. Ferguson. *ApJ*, 458:100, 1996.

37. S. D. M. White and C. S. Frenk. *ApJ*, 379:52, 1991.

38. C. Lacey and J. Silk. *ApJ*, 381:14, 1991.

39. C. Lacey, B. Guiderdoni, B. Rocca-Volmerange, and J. Silk. *ApJ*, 402:15, 1993.

40. G. Kauffmann, S. D. M. White, and B. Guiderdoni. *MNRAS*, 264:201, 1993.

41. G. Kauffmann, B. Guiderdoni, and S. D. M. White. *MNRAS*, 267:981, 1994.

42. S. Cole, A. Aragon-Salamanca, C. S. Frenk, J. F. Navarro, and S. E. Zepf. *MNRAS*, 271:781, 1994.

43. J. S. Heyl, S. Cole, and C. S. Frenk. *MNRAS*, 274:755, 1995.

44. J. B. Oke. *ApJS*, 27:21, 1974.

45. R. E. Williams, B. Blacker, M. Dickinson, W. V. D. Dixon, H. C. Ferguson, A. S. Fruchter, M. Giavalisco, R. L. Gilliland, I. Heyer, R. Katsanis, Z. Levay, R. Lucas, D. McElroy, L. Petro, M. Postman, H.-M. Adorf, and R. N. Hook. *AJ*, 112:1335, 1996.

46. H. C. Ferguson and A. Babul. , in preparation, 1996.

47. B. Binggeli, A. Sandage, and G. A. Tammann. *ARA&A*, 26:509, 1988.

48. F. Simien and G. de Vaucouleurs. *ApJ*, 302:564, 1986.

49. A. Babul and M. Rees. *MNRAS*, 255:346, 1992.

50. G. Efstathiou. *MNRAS*, 256:43P., 1992.

51. G. Bertelli, A. Bressan, C. Chiosi, F. Fagotto, and E. Nasi. *A&AS*, 106:275, 1994.

52. P. Madau. *ApJ*, 441:18, 1995.

53. Y. Yoshii and B. Peterson. *ApJ*, 436:551, 1994.

STAR-FORMING GALAXIES AT REDSHIFTS $z \sim 3$

M. GIAVALISCO

The Observatories of the Carnegie Institution of Washington,
813 Santa Barbara St., Pasadena
CA, 91101, USA

C.C. STEIDEL, K.L. ADELBERGER

Palomar Observatory, Caltech,
Mail Stop 105-24, Pasadena,
CA, 91125, USA

M. PETTINI

Royal Greenwich Observatory,
Madingley Road,
Cambridge, CB3 0EZ, England

M.E. DICKINSON

Space Telescope Science Institute,
3700 San Martin Dr.,
Baltimore, MD 21218, USA

We review the properties of the population of $z \sim 3$ galaxies that we have recently discovered. These galaxies have been color-selected from ground-based multi-band imaging targeting the Lyman continuum discontinuity and Lyman-α forest as clear spectroscopic features of anything forming stars at $z \sim 3$, where the UV is shifted into the optical window. The current sample includes ~ 400 sources with redshifts now secured for more than 80 in the range $2.5 < z < 4.1$ and many more coming in the next few months. The efficiency of the method in going from photometrically selected candidates to spectroscopically confirmed galaxies is very high, with about 90% of the attempted sources confirmed at high redshifts. Our current data set, which includes far-UV morphology from *HST*-WFP2, space density, star formation rates, and preliminary estimates of mass, metallicity and age, suggests that these galaxies may be the progenitors of the present-day bright spirals and ellipticals observed when they were assembling the stars of their bulges and cores.

1 Introduction

This 37th Herstmonceaux Conference vividly testifies how the repair of *HST* and the commissioning of the Keck-1 telescope have accelerated the progress in understanding galaxy evolution to a pace of almost explosive character. In this paper, we review the work that resulted in the discovery of a massive population of star-forming, but otherwise apparently normal galaxies at redshifts $z \sim 3$ (Steidel et al. 1996 a,b; Giavalisco et al. 1996a) and discuss their properties in the context of galaxy evolution. To be sure, this survey started several years ago, at a time when *HST* still suffered problems of aberration and Keck was not yet an operating telescope. A large fraction of the ~ 3 galaxies were identified back then, using telescopes and instrumentation of the 4-meter epoch. However, it has been thanks to Keck and *HST* that the $z \sim 3$ population could be spectroscopically confirmed and properties such as morphology at the *kpc scale*, studied in detail.

It is indeed on detailed and quantitative studies of these early galaxies that the research must concentrate now. At $z \sim 3$ predictions of galaxy properties such as clustering, mass size, age of the stellar populations are extremely sensitive to the cosmology, and therefore powerfully discriminative. For these works, large-aperture and space telescopes will, obviously, be the only way to continue.

2　The Evolution of Massive Galaxies

There now appears to be general agreement among a number of diverse studies of galaxy evolution, including deep redshift surveys and morphological studies using deep *HST* WFPC2 images, that galaxies which comprise the normal Hubble sequence of morphologies from mid–type spirals to ellipticals (i.e. the brightest and most massive systems, characterized by a predominant spheroidal component) must have been in place and evolving relatively quiescently by $z \sim 1$ and very likely by $z \sim 1.5$ (e.g., Steidel et al. 1994; Lilly et al. 1995; Ellis et al. 1996; Cowie et al. 1996; Glazebrook et al. 1995; Driver et al. 1995). While these studies differ in the details of describing the evolution of the overall galaxy population, the conclusion that massive galaxies must have formed considerably before $z \sim 1$ seems well-established.

With $\sim 50\%$ of all the stars presently segregated in old spheroidal systems, such as ellipticals and bulges of spirals (Schechter & Dressler 1987), an obviously important question, then, is when and how these systems actually formed. It is not yet clear how the early formation of massive galaxies fits in with the generally accepted view that structure on the scale of galaxies formed by the hierarchical merging of sub–galactic lumps via gravitational instability. However, the appearance of these "proto-galaxies" (e.g. Partridge & Peebles 1967; Silk & Norman 1981; Baron & White 1987; Ikeuchi & Norman 1991), and some working definition of what constitutes a "formed" galaxy under the hierarchical structure formation paradigm, are not parameters that emerge from most of the models. In any case, the data seem to show that the key to understanding the formation of the massive galaxies of the present epoch is beyond the redshifts probed to date.

Early formation of massive galaxies is hardly a new idea. For example, Partridge & Peebles (1967) proposed that massive spheroids assembled quite early on, forming the bulk of their stars in monolithic bursts of star formation in an inital spherical collapse. The expectation would be that an entire galaxy's worth of stars would form in roughly a free fall time (a few $\times 10^8$ years) so that these objects should be easily recognizable by such signatures as extremely strong Lyα emission lines characteristic of photoionization by massive stars. Such "primeval galaxies" have been the subject of a huge amount of observational effort over the last ~ 20 years (see, for example, Djorgovski 1995, and references therein); the general null results of these surveys have led some to believe either that the important epoch for the formation of spheroids has not yet been probed, or that forming galaxies are obscured from traditional observational methods by a large quantity of dust. However, in the light of more recent theoretical work on galaxy formation, with galaxies generally forming much more gradually (e.g. Frenk, this volume), the expectations for the appearance of early galaxies are not entirely clear.

3   Lyman-Limit Galaxies

An alternative approach which we adopted in searching for examples of galaxies at very high redshifts was to use a broad-band color selection to target the Lyman continuum break as the clearest spectral signature for anything forming stars at $2.8 \lesssim z \lesssim 3.4$. We implemented the technique in a way that attempted to answer the question "do high redshift galaxies exist at all, with even $\sim L^*$ luminosity?"–i.e., with rest-frame B-band luminosity directly comparable to that of the present-day luminous types of the Hubble sequence. In other words, the search was geared toward detection of "average" objects, rather than spectacular ones. In fact, broad–band color selection turns out to be a highly efficient alternative to flux-limited and line–emission surveys for very high redshift galaxies, chiefly because of the dramatically reduced contamination from foreground sources, the very large volume which can be surveyed simultaneously, the lack of dependence on assumptions about the intrinsic spectral energy distributions or appearance of the galaxies, and the fact that (as it turns out) the bulk of star forming systems at high redshift do not have strong Lyα emission lines.

The selection technique we have used is described in Steidel et al. (1995). It involves a simple 3–band photometric system (U_n, G, and $\mathcal{R}$) which is able to isolate objects having UV flux (whether it be from stars or from AGN activity) at redshifts $z \gtrsim 3$, where the Lyman continuum is redshifted to the optical window (Fig.1a).

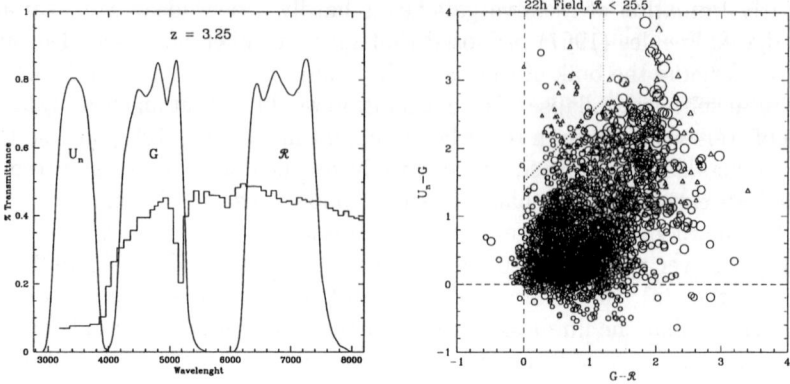

Figure 1: **a)** The idea behind the multi-band $U_nG\mathcal{R}$ imaging to color-select star-forming galaxies at $z \sim 3$. The transmittances of the filters are plotted. The histogram shows a theoretical model of rest-frame UV SED of unreddened continuous star formation placed at $z = 3.25$, with the flux in an arbitrary scale. **b)** Color-color plot obtained from deep $U_nG\mathcal{R}$ imaging in the SSA22 field. Magnitudes are in the AB scale, so that the horizontal and vertical lines are the loci of sources with flat SED ($f_\nu \propto \nu^0$). The selection window for star-forming galaxies at $2.8 \lesssim z \lesssim 3.5$ is also shown.

The method takes advantage of the ubiquitous presence of the Lyman "edge" discontinuity at 912 Å in an object's rest frame, which arises from a combination of the instrinsic discontinuity in the spectra of hot stars, the opacity of the galaxy to its own Lyman continuum radiation, and the photoelectric and line absorption due to intervening H I along the line of sight (Steidel & Hamilton 1993; Madau 1995).

High-redshift candidates are selected from color-color diagrams (Fig.1b), where the expected region occupied by relatively unextinguished high-redshift star-forming galaxies has been calculated using synthetic spectra and models of intervening photoelectric and line absorption (Madau 1995) as well as data from QSO spectra. Because of the broad–band nature of the selection, we are able to easily isolate galaxies in a fairly large range of redshifts, mostly $2.8 \lesssim z \lesssim 3.4$, from the dominant populations of foreground "faint blue" galaxies (only 1.5-2% of all field galaxies to an apparent magnitude limit of $\mathcal{R} = 25.0$ lie at these high redshifts).

Implementing these ideas with the slighlty different photometric system of WFPC2, some of us have been designing the HDF observations in the same fashion (see Ferguson, this volume. See also Madau et al. 1996), and with similar results (Steidel et al. 1996b; Lowenthal et al. this volume). For us, however, the importance of the HDF, which extends the flux limit for color selected high-z galaxies ~ 0.7 mag over our ground-based data, was not as much in providing more cases of high redshifts galaxies (which would have been a meager catch compared to the vastly larger ground-based survey) as it was to reliably probe a fainter portion of the luminosity distribution of the $z \sim 3$ population (Dickinson et al. in prep.) even for galaxies for which spectroscopy is not practical.

4 Spectroscopy and Morphology of the Lyman-Limit Galaxies

The turning point in this work has been our recent success in spectroscopically confirming our color–selected candidates using the Keck telescope (Steidel et al. 1996 a,b) down to a flux limit $\mathcal{R} \sim 25.5$ and the fact that the photometric selection of high redshift galaxies turned out to be highly efficient, with more than 90% of the candidates lying within the expected redshift range.

At the time of writing, we have secured 80 redshifts in the interval $2.5 < z < 4.1$ (out of a current candidate list of more than 400 galaxies), with many more expected in the next several months. For a subsample of the galaxies (≈ 30) we have WFPC2 (rest far–UV) morphology (not including those within the Hubble Deep Field), and for 12 galaxies we also have K-band photometry ($\sim$rest-frame V band). The spectra of the $z \sim 3$ galaxies are diverse, but are broadly similar to those of local star-forming galaxies (Figure 2 shows 2 examples compared to the Wolf-Rayet galaxy NGC 4214 observed with HST+GHRS) and exhibit an abundance of high-ionization stellar lines and interstellar low-ionization absorption lines.

The Lyα emission line is often absent, and when present it is generally rather weak ($W_\lambda \approx$ 5-20 Å), despite the intense star-formation rates. As in local star-forming galaxies, the stellar lines often exhibit P-Cygni profiles, indicative of the presence of winds from massive stars. However, the stellar lines are weaker than in local galaxies, a fact that could reflect a lower metal content. The interstellar metal lines (FeII, OI, CII, CiII, Si IV, Al III, etc.) show that the ISM has undergone chemical enrichment, although a measure of abundances is not possible using the low resolution discovery spectra due to line saturation.

HST/WFPC2 images of the $z \sim 3$ galaxies (Giavalisco et al. 1996a, Steidel et al. 1996b) show that the UV morphology at rest-frame ~ 1600 Å, which directly maps out the star formation, is typically characterized by a compact central concentration ("core") that contains $\sim 90\%$ of the light, surrounded by a "halo" with irregular structures in the form of nebulosities and sub-lumps (Figure 3).

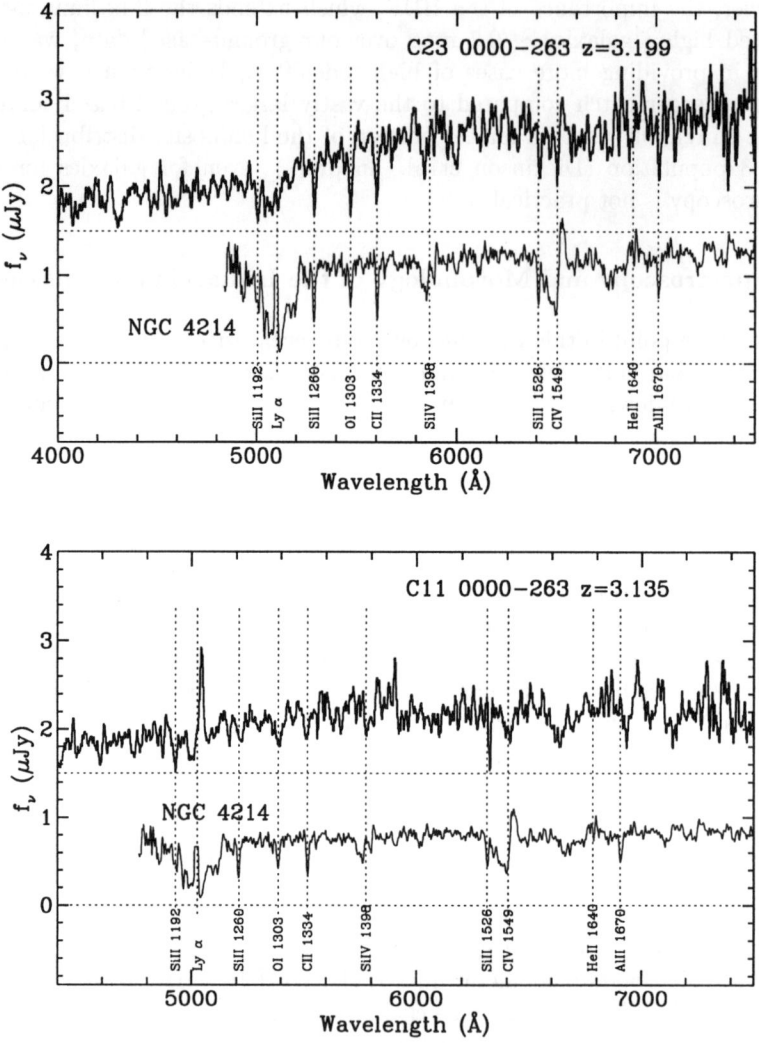

Figure 2: Two examples of spectra of $z \sim 3$ galaxies obtained with the Keck telescope and the LRIS instrument. In the top panel Lyα is observed in absorption, in the lower one in emission. The heavily saturated interstellar lines are also clearly visible. The top galaxy shows the CIV P-Cygni profile with both the emission and absorption component, while in the bottom one only the absortion is observed.

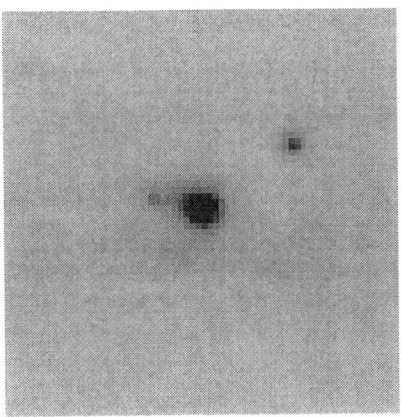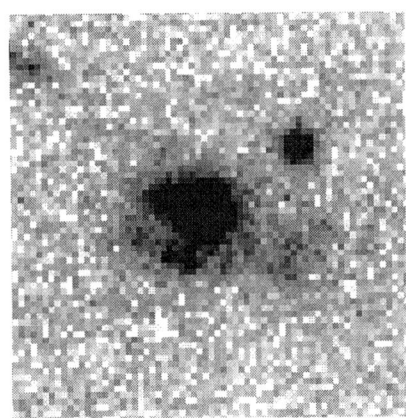

Figure 3: WFPC2/PC image of galaxy 0000-263-D6 at redshift $z = 2.961$. a) The contrast is adjusted to show the compact and regular "core" of the galaxy, which is very well modeled by a $r^{1/4}$ radial profile, with $r_e = 0.24$ arcsec (Fig.4a). b) A different contrast shows the diffuse and structured "halo", where disturbances are observed as a blob to the NE (actually merged with the core isophotes, but clearly visible in the left panel), a plume to the S and an extended nebulosity to the W. The compact source to the NW of the galaxy is at a significantly lower redshift. Both panels are 3.13 arcsec in size

The cores tend to have relatively regular morphologies, often characterized by light profiles suggestive of relaxed structures, with half-light radii of ~ 0.3 arcsec (Figure 4) which, at a mean redshift $< z > = 3.2$ corresponds to $2.1h_{50}^{-2}$ ($3.7h_{50}^{-2}$) kpc if $q_0 = 0.5$ (0.05). The observed isophotal diameters from the HST images, lower limits to the intrinsic sizes due to the extreme $(1 + z)^4$ surface brightness dimming, are typically 1.5–2 arcsec or $10.5h_{50}^{-2}$–$21h_{50}^{-2}$ ($18.5h_{50}^{-2}$–$37h_{50}^{-2}$) kpc.

There are a number of pieces of evidence that plausibly connect the $z \sim 3$ population to the bright galaxies of the present epoch:

• The often extremely strong interstellar lines are indicative of very substantial velocity fields within the galaxies, and if they reflect motions induced by gravity, then velocity dispersions of $180 - 320$ km s^{-1} (typical of $L \gtrsim L^*$ galaxies today) are implied. We are currently attempting to obtain nebular emission line widths in the near–IR (rest–frame optical) to test the dynamical interpretation of the interstellar absorption line strengths.

• The surface density of the $z \sim 3$ galaxies is relatively high, ≈ 1 galaxy per arcmin^2 to $\mathcal{R} = 25.5$, corresponding to lower limits on the volume density that already accounts for $1/2$ ($1/10$)h_{50}^3 of the volume density of present-day galaxies with $L/L^* > 1$. The K-band photometry we have obtained for a

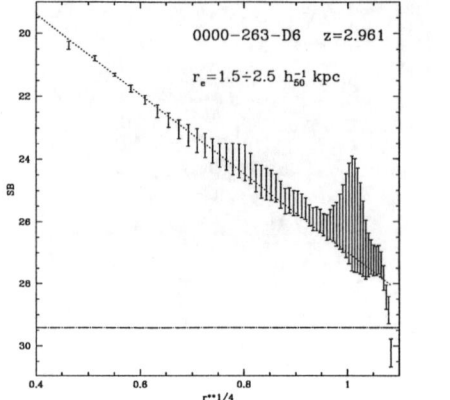 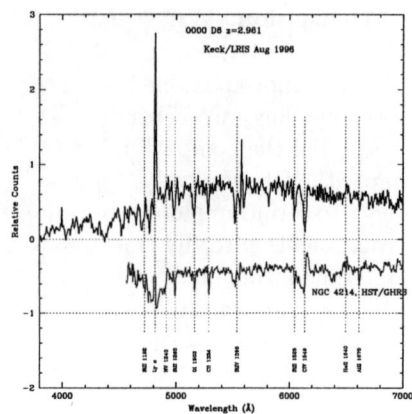

Figure 4: **a)**. Light profile of galaxy 0000-263-D6 extracted from the WFPC2/PC image of Fig.3, plotted with the fitted $r^{1/4}$ law. The presence of the irregular halo is clearly observed as deviations from the the smooth profile. The two values of r_e are for $q_0 = 0.5$ and $q_0 = 0.05$, respectively. **b)**. Keck spectrum of 0000-263-D6. Lyα is here observed in emission. Note the intervening damped Lyα system at $z \sim 2.67$ clearly detected blueward of the Lyα emission line. The spectrum is not flux calibrated.

subset of the $z \sim 3$ sample indicates rest-frame optical luminosities $L \gtrsim L^*$, comparable to bright galaxies today.

- The star formation rates per object are actually in line with the most actively star–forming galaxies at *any* redshift observed so far. From the UV continuum flux, we estimate star-formation rates in the range 4–100 (12–300)h_{50}^{-2} M$_\odot$ yr^{-1}, where $q_0 = 0.5$ (0.05), with a typical value being ~ 10 (30)h_{50}^{-2} M$_\odot$ yr^{-1}. The distribution of the UV color $G - \mathcal{R}$ and the UV–optical colors, which are in the range $2.5 \lesssim \mathcal{R} - K \lesssim 3.5$, are all consistent with continuous star formation and ages of as much as ~ 1 Gyr in the absence of reddening by dust. We can set limits on the reddening from the observed optical and optical/IR colors of E(B-V) < 0.3, although we believe that the reddening is less severe in most cases.

- Most of the stars are being formed in regions whose observed UV half-light radii and isophotal diameters are directly comparable to the effective radii r_e and sizes of present-day elliptical galaxies and bulges. This agrees well with the hints that continuous star formation is ongoing in these objects, and may have begun significantly before the epoch at which we observe them. This would imply that during the cosmic time probed by the survey they would have assembled a stellar mass comparable to that observed locally in luminous galaxies.

5 Discussion: What are the $z \sim 3$ galaxies?

All of the above possible links to present–day bright galaxies, which we believe are compelling, are nevertheless still unavoidably circumstantial. Some have interpreted the compact UV morphologies as evidence that these objects are simply the (theoretically) expected sub–galactic fragments that would one day merge to form luminous galaxies, observed during a short-lived phase of star-formation. Due to the transient character of such a UV-luminous phase, only a fraction of the underlying population would be detectable, at any given time with the color selection. Regardless of one's interpretation, however, the fact that the $z \sim 3$ galaxies are carrying the bulk of the star formation in the high redshift Universe unavoidably leads to the conclusion that they must be intimately related to the formation of the oldest stellar populations observed in normal galaxies, which are found in bulges and in elliptical galaxies.

The "small bursting fragment" scenario differs from the massive-galaxy interpretation because it requires a substantial amount of subsequent merging to assemble the final galaxies. The two interpretations, however, agree that the stars we see beeing formed at $z \sim 3$ are destined to be found, at the present epoch, in luminous, massive galaxies. The total amount of star formation is also similar in either picture. If ΔT_o is the interval of cosmic time probed by the survey and δt the duration of the star-bursting phase of the "small fragments", the total comoving stellar mass densities produced during ΔT_o in the two scenarios are $\rho^*_{cont} \approx n \times f \times \Delta T_o$ and $\rho^*_{burst} \approx n \frac{\Delta T_o}{\delta t} \times f \times \delta t_o = n \times f \times \Delta T_o$, respectively, where n is the observed comoving volume density of $z \sim 3$ galaxies and f is their average star-formation rate.

The mass function of the $z \sim 3$ galaxies, however, is clearly a fundamental quantity to understand the mechanisms that led to the formation of bright galaxies and additional empirical investigation is necessary to contrain it. But even with the limitations of the current data, is the "small bursting fragment" hypothesis actually suggested by the observations and necessary to interpret them, or is it, rather, the product of theoretical expectations?

As reviewed above, in the UV light the $z \sim 3$ galaxies have the same sizes as local bright galaxies, not of small fragments. Galaxy morphology depends on wavelength and star-formation activity (Giavalisco et al. 1996b), and the appearance of normal galaxies in the far-UV during a phase of intense star making is largely unknown. In the "small bursting fragment" scenario, one would have then to postulate that the stars are forming spread over the same large volumes that currently define massive galaxies, although the real mass and sizes of the fragments are much smaller. Clearly, NICMOS morphological studies of the $z \sim 3$ population will help clarify this issue.

Is it at least plausible that the $z \sim 3$ population mostly consists of transient bursters? Even if selection effects do not bias the survey against blue sources, not a single galaxy among the ~ 400 in our present sample has the extremely blue colors expected for an instantaneous "burst" of star formation ($G - \mathcal{R} \sim -0.5$ at $z = 3.1$), but rather the colors (after correction for blanketing in the G band from the Lyman α forest) occupy a relatively narrow range characterstic of unobscured, continuous star formation. Of course, dust could explain this, but then one needs to postulate a very specific reddening distribution function, such that all galaxies have similar observed UV colors. In the local universe bursts of star formation are observed with a range of dust reddening, including galaxies practically unreddened (Calzetti et al. 1994). If the $z \sim 3$ objects are undergoing their initial bursts, advocating a systematic and uniform reddening by dust throughout the entire population seems physically implausible.

In conclusion, the current data, although preliminary in character, are naturally explained if the $z \sim 3$ population mostly consists of massive galaxies in an early evolutionary phase. A bursting scenario of small fragments destined to merge, though certainly not ruled out by the observations, is not needed to interpret them and in fact requires some *ad hoc* assumptions to make it physically plausible.

Acknowledgments

M.G. acknowledges the support of the Hubble Fellowship Program through grant number HF-01071.01-94A, awarded by the Space Telescope Science Institute, which is operated by the Association of Universities for Research in Astronomy, Inc. under NASA contract NAS5-26555.

References

1. Baron, E., & White, S., ApJ, 322, 585, 1987.
2. Calzetti, D., Kinney, A., & Storchi-Bergman, T., ApJ, 429, 582, 1994.
3. Cowie, L., Songaila, A., Hu, E., & Cohen, J. AJ, in press, 1996.
4. Djorgovski, G. in *IAU Symposium "New Light on Galaxy Evolution"*, eds R. Bender & R. Davies (Dordrecht: Kluwer, 1996, in press)
5. Driver, S., Windhorst, R., Griffith, R., ApJ, 453, 48, 1995.
6. Ellis, R., Colless, M., Broadhurst, T., Heyl, J., & Glazebrook K., MNRAS, in press, , 1996.
7. Giavalisco, M., Steidel, C., & Macchetto, D., ApJ, 470, 189, 1996a.
8. Giavalisco, M., Livio, M., Bohlin R., Macchetto, D., & Stecher, T., AJ, 112, 369, 1996b.

9. Glazebrook, K, Ellis, R., Santiago, B., & Griffith, R. MNRAS, 275, L19, 1996.

10. Ikeuchi, S., & Norman, C., ApJ, 375, 479, 1991.

11. Lilly, S., Tresse, L., Hammer, F., Crampton, D., & Le Fevre, O., ApJ, 455, 108, 1995.

12. Madau, P., ApJ, 441, 18, 1995.

13. Partridge, R., & Peebles, P. J., ApJ, 147, 868, 1967.

14. Schechter, P., & Dressler, A., AJ, 94, 56, 1987.

15. Silk, J. & Norman, C., ApJ, 247, 59, 1981.

16. Steidel, C., & Hamilton, D., AJ, 104, 941, 1992.

17. Steidel, C., & Hamilton, D., AJ, 105, 2017, 1993.

18. Steidel, C., Dickinson, M., & Persson, E., ApJ, 437, L75, 1994.

19. Steidel, C., Pettini, M., & Hamilton, D., AJ, 110, 2519, 1995.

20. Steidel, C., Giavalisco, M., Pettini. M., Dickinson, M., & Adelberger, K., ApJ, 462, L17, 1996a.

21. Steidel, C., Giavalisco, M., Dickinson, M. & Adelberger, K., AJ, 112, 352, 1996b.

22. Van den Berg, S., Abraham, R., Ellis, R., Tanvir, N., Santiago, B., & Glazebrook, K., AJ, 112, 357, 1996.

KECK SPECTROSCOPY OF THE HUBBLE DEEP FIELD

G.D ILLINGWORTH, J. GALLEGO, R. GUZMÁN, J.D. LOWENTHAL,
A.C. PHILLIPS, N.P. VOGT, D.C. KOO & S.M. FABER
UCO/Lick Observatory, University of California, Santa Cruz, CA, 95064, USA

Spectra of 115 HDF galaxies to I ∼ 25 mag were obtained by the Santa Cruz DEEP
group with the Keck LRIS multislit spectrograph and combined with detailed
structural measurements from the HDF images to (i) derive evolution for a sample
of disk galaxies at $z \sim 0.5$ using the Tully-Fisher relation, (ii) characterize, from
size and velocity width data, the global properties to $z \sim 1$ of a sample of ∼60
"compact" galaxies, (iii) determine star formation rates in that "compact" sample,
and (iv) derive structure and redshifts for a sample of U and B-band "dropout"
objects, of which 11 have been confirmed to be at high redshift ($z = 2.2 - 3.5$).

1 DEEP

The DEEP Project[a] was developed with the goal of elucidating the physical
processes and time scales of the formation and evolution of galaxies. The Keck
telescopes and their multi-object spectrographs (LRIS and later DEIMOS) are
central to this program. Of equal importance is the high spatial resolution
imaging capability provided by HST and its WFPC2 camera (and NICMOS,
plus the Advanced Camera, the ACS, in 1999). The structural and morphological information from wide-field HST imaging has revolutionized the study of
distant galaxies. The importance of such HST-imaged regions for the study of
galaxies at intermediate to high redshift cannot be overstated. Ground-based
telescopes, even with adaptive optical systems, will not provide comparable
capability for the foreseeable future.

2 HDF Results

For a specific example of the importance of HST for clarifying the nature of
galaxies in the young universe, one needs only to think of the impact of the
Hubble Deep Field (Williams *et al.* 1996), and of the projects that the HDF
images have precipitated.
The first results from the DEEP team's Keck LRIS observations of the HDF
and its flanking fields are available in four papers, and are summarized below.
The results are based on samples of galaxies selected from the HDF to meet a
number of scientific goals. Structural parameters, colors, surface brightnesses

---

[a]Deep Extragalactic Evolutionary Probe; for further information see the DEEP home
page at http://www.ucolick.org/~deep/home.html.

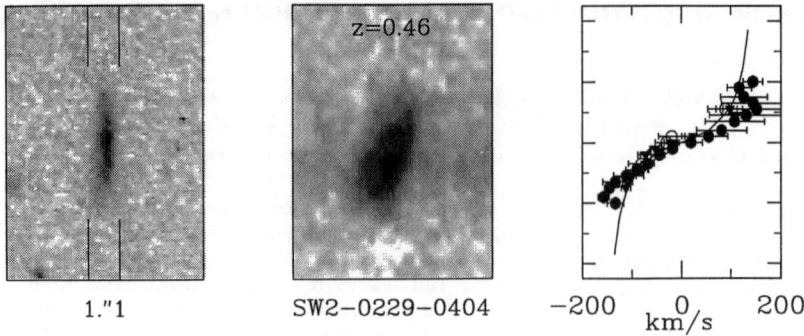

Figure 1: An example of a galaxy at $z = 0.48$ is shown with its Keck LRIS slit position, the observed emission line and the measured and model rotation curve. Note that any underestimate of the rotational velocity will lead to an *overestimate* of the brightening! The Tully-Fisher relation derived from such measurements is shown in Vogt *et al.* (1997) and Vogt & Phillips (this volume).

and morphological attributes were determined for 140 galaxies from the HST HDF images. Of this sample, redshifts were ultimately determined for 115 galaxies (now available on the web). Further measurements such as rotation curves, velocity widths and line ratios were made for subsets of these galaxies.

2.1 Tully-Fisher at Intermediate Redshift

In a recently published paper using HST data largely from the high latitude WFPC GTO field (the "Groth" strip), Vogt *et al.* (1996) demonstrated that it was possible to combine measurements from HST WFPC2 images with ground-based Keck multislit observations at moderate spectral resolution to derive rotational velocities for suitably-inclined spiral galaxies to redshifts $z \sim 1$ (see Figure 1). Based on a comparison of the velocity widths in these nine galaxies with the local Tully-Fisher (TF) relations, a limit of $\Delta M_B < 0.6$ mag could be set on the brightening at intermediate redshift. The more recent Tully-Fisher work on the disk/spiral galaxy sample in the HDF in Vogt *et al.* (1997; see also Vogt & Phillips, this volume) greatly improves the reliability of this estimate, first by adding a comparable number of galaxies, and second, adding galaxies of lower absolute luminosity. The distant galaxy TF relation is now much more clearly defined, and provides the strongest evidence to date for modest evolutionary brightening of $\Delta M_B \sim 0.4$ at $<z> \sim 0.5$.

2.2 Nature of Compact Galaxies: I.

One of the striking results of HST images of faint, distant field galaxies has been the prevalance of compact, relatively high surface brightness galaxies (with effective radii r_e typically 0.2″-0.3″). Our early work on CNELGs (compact narrow emission line galaxies – Koo *et al.* 1995; Guzmán *et al.* 1996) showed that these objects had kinematical, structural and photometric properties comparable to the more extreme examples at the current epoch of low mass, active star forming galaxies, the HII galaxies. But the question remained for the more typical objects – "what are the current-day analogues of the compact galaxies seen at redshifts $0.3 \lesssim z \lesssim 1$?"

This question was addressed in two parts. First, Phillips *et al.* (1997; see also Koo *et al.* – this volume) used the HST HDF images to derive sizes, magnitudes and morphologies. These were then combined with the Keck redshifts to derive luminosities and length scales. They are luminous, with median $M_B \sim -20$ ($H_0 = 50$ km s^{-1} Mpc^{-1}). The Keck spectra also yielded velocity widths (σ), allowing us to estimate their masses. The velocity widths ranged from $\sigma < 35$ km s^{-1} (the measurement limit) to 150 km s^{-1}. Combined, these data allowed us to compare and characterize the high redshift compact sample (see Figure 2). While a small fraction appear to be ellipticals, the vast majority are emission-line galaxies whose global properties resemble nearby HII galaxies, i.e., compact, late-type, star forming galaxies.

2.3 Nature of Compact Galaxies: II.

The second aspect, the spectroscopic properties of the emission-line "compact" sample galaxies, is discussed in Guzmán *et al.* (1997). Star formation rates were derived from measurements of the [OII] equivalent widths; these ranged from 5Å to 95Å. The corresponding star formation rates are <1-15 $M_\odot$ yr^{-1} (based on procedures and calibrations given in Guzmán *et al.*). Based on all the available data, namely velocity widths, excitations, Hβ luminosities, star formation rates, and M/L estimates, a substantial fraction of these "compact" emission-line galaxies ($\sim 60\%$) appear to be like local, low mass, extreme star forming HII galaxies. The remainder are actively star forming, but appear to be a more heterogeneous sample of late-type galaxies.

An interesting result of this effort is that some additional constraints can be now placed on the evolution of the star formation rate per unit volume with redshift (see Figure 3 and the discussion in Guzmán *et al.* 1997). The data are consistent with a large decline in the SFR from $z \sim 1$ to the present epoch.

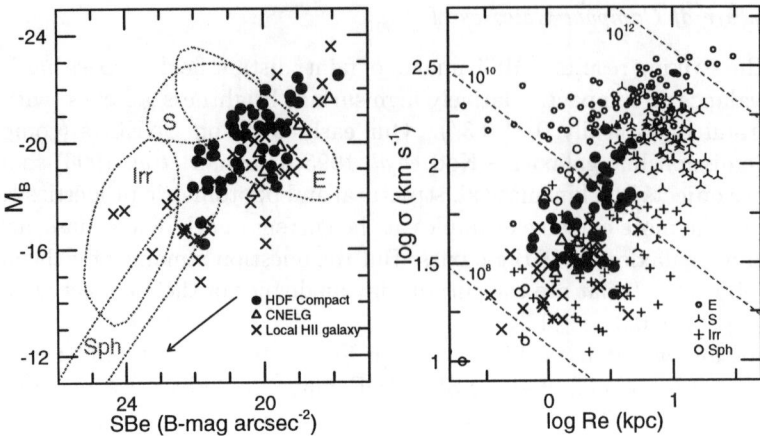

Figure 2: The luminosity-surface brightness relation for local samples of galaxies with the high redshift "compact" sample overplotted, on the left panel, from Phillips *et al.* (1997). The right panel again shows the relationship of the distant sample with local samples in the velocity width (σ)–effective radius plane. Typical masses are indicated. The "compact" objects are the filled circles. The properties of the high redshift galaxies overlap those of local, late-type galaxies.

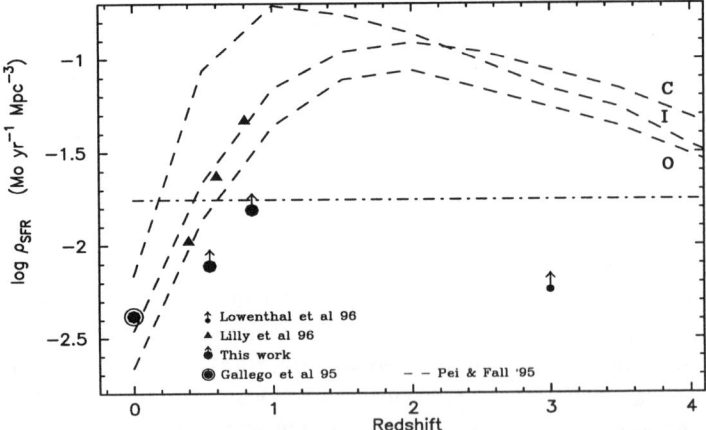

Figure 3: Estimates of the star formation rate per unit volume at several redshifts (from Guzmán *et al.* 1997). The sources are given on the figure. Models (closed, infall and outflow) from Pei and Fall (1995) are shown, as is the average at -1.75 derived from the local luminosity density (from Madau *et al.* 1996). Note that "compact" galaxies make a substantial contribution at intermediate redshifts, and that the SFR decline from $z \sim 1$ to the current epoch is large – a factor ~ 10.

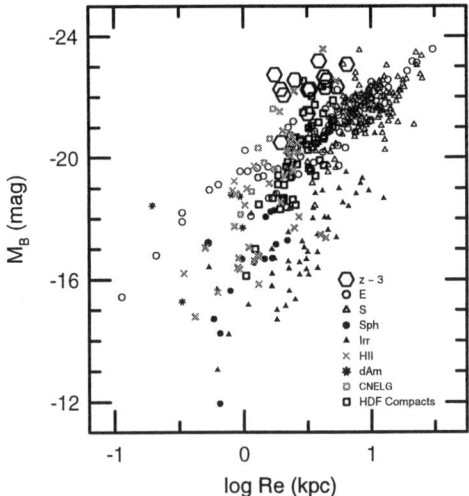

Figure 4: The location of the high redshift sample in the $M_B - r_e$ plane, compared to local galaxies (Lowenthal *et al.* 1997; see also this volume). The large symbols are the $z \sim 3$ galaxies. Even though they are very luminous, the high redshift galaxies do overlap with the local, bright, star forming HII galaxies. A key issue is whether the high redshift objects have comparable velocity widths (and hence masses).

2.4 Galaxies at very High Redshift

The use of U-band "dropout" galaxies to identify potential high redshift objects was beautifully exemplified by Steidel *et al.* (1996a) in their spectroscopic observations from Keck. This was followed by one of the very first HDF papers utilizing spectroscopic data (Steidel *et al.* 1996b); they found 5 high redshift ($z \sim 3$) galaxies. The "dropout" technique relies upon the large flux decrements that occur from line-of-sight and source absorption as both Lyα and the Lyman-limit pass through the filter bandpass. Decrements in the U-band relative to other filters typically indicate sources at redshifts $z \gtrsim 2.5$, while decrements in the B-band suggest even higher redshift sources ($z \gtrsim 3$).

This technique, applied through the use of both U and B-band "dropout" objects, was the means by which we selected a sample of 24 galaxies for observation at Keck with LRIS. Of this sample 11 were confirmed to have redshifts in the range $2.2 < z < 3.4$. The remaining galaxies, with one exception, are consistent with being high redshift objects, but could not be confirmed spectroscopically. The high redshift objects in the HDF are quite small, but luminous ($1.8 < r_{\frac{1}{2}} < 6.5h_{50}^{-1}$ kpc and $-21.5 < M_B < -23$), and show a wide diversity of forms. Interestingly, they are not as morphologically uniform as the brighter sample studied by Giavalisco *et al.* (1996).

Combining these new data with the 5 in the HDF from Steidel *et al.* (1996b), we have derived co-moving volume densities and star formation rates (Lowenthal *et al.* 1997; this volume also). It is striking that these HDF data already indicate that the volume density is comparable to the local volume density of all galaxies brighter than L^* (and 3-4× higher than the first estimate of Steidel *et al.* (1996a) based on a brighter sample of "dropout" candidates. If the remaining objects in the original "dropout" sample prove to be at high redshift, the co-moving volume density could be several times higher still. Star formation rates have been estimated from the rest-frame UV continuum fluxes. They are found to be typical of local HII galaxies and many spiral galaxies (at $6 - 27h_{50}^{-2}$ $M_\odot$ yr^{-1} for $q_0 = 0.5$, or $2 - 9h_{50}^{-2}$ $M_\odot$ yr^{-1} for $q_0 = 0.05$).

Acknowledgments

Funding was provided by the CfPA (Center for Particle Astrophysics) and by NSF grants AST-91-20005, and AST-9529098, and NASA grants AR-6337.08-94A, AR-6337.21-94A, GO-5994.01-94A, AR-5801.01-94A, and AR-6402.01-95A.

References

1. Gallego, J. *et al.*, ApJ, 455, L1, 1995.
2. Giavalisco, M. *et al.*, ApJ, in press, 1996.
3. Guzmán, R. *et al.*, ApJ, 460, L5, 1996.
4. Guzmán, R. *et al.*, ApJ, submitted, 1997.
5. Koo, D. C. *et al.*, ApJ, 440, L49, 1995.
6. Lilly, S. J. *et al.*, ApJ, 460, L1, 1996.
7. Lowenthal, J. D. *et al.*, ApJ, submitted, 1997.
8. Madau, P. *et al.*, MNRAS, in press, 1996.
9. Pei, Y. C. & Fall, S. M., ApJ, 464, 69, 1995.
10. Phillips, A. C. *et al.*, ApJ, submitted, 1997.
11. Steidel, C. C. *et al.*, ApJ, 462, L17, 1996a.
12. Steidel, C. C. *et al.*, ApJ, in press, 1996b.
13. Vogt, N. P. *et al.*, ApJ, submitted, 1997.
14. Vogt, N. P. *et al.*, ApJ, 465, L15, 1996.
15. Williams, R. *et al.*, AJ, in press, 1996.

GALAXY FORMATION AND EVOLUTION: OUTSTANDING PROBLEMS

C.S. FRENK, C.M. BAUGH, S. COLE

Physics Department, University of Durham, Durham DH1 3LE, UK

C. LACEY

Theoretical Astrophysics Center, DK-2100 Copenhagen 0, Denmark

We review recent developments in theoretical studies of galaxy formation and evolution. In combination with new data from HST, Keck and other telescopes, numerical and semi-analytic modelling is beginning to build up a coherent picture
of galaxy formation. We summarize the current status of modelling of various
galactic properties such as the structure of dark matter halos, the galaxy luminosity function, the Tully-Fisher relation, the colour-magnitude relation for ellipticals,
the gross morphological properties of galaxies and the counts of faint galaxies as
a function of magnitude, redshift and morphology. Many of these properties can
be explained, at least at some level, within a broad class of CDM cosmologies.
A number of fundamental issues, however, remain unresolved. As an example of
the power of semi-analytic modelling, we carry out a study of the evolutionary
status of the Lyman-break galaxies at $z \simeq 3.5$ recently discovered by Steidel *et al.*
The abundance and global properties of these objects are compatible with model
predictions in a variety of CDM cosmologies, including the standard version. In
all cases, these galaxies are among the very first objects in which appreciable star
formation is taking place; they thus signal the onset of galaxy formation. We
present three sample evolutionary calculations of Lyman-break galaxies which illustrate that these objects are the precursors of ordinary present day spirals and
ellipticals.

1 Introduction

The spectacular new data from HST, Keck and other telescopes presented at
this meeting are finally giving substance to the word "few" in Y.B. Zel'dovich's
famous 1977 statement: "It will only be a few years before the origin and
evolution of galaxies is understood." A full understanding of the origin and
evolution of galaxies, however, will require a great deal of detailed theoretical
modelling to uncover the physical processes manifested in the data. At present
such modelling is lagging behind observational advances partly because of the
breathtaking pace of these advances and partly because some of the physical
processes at work, particularly those involving gas dynamics and star formation, are intrinsically very complex.

Two interrelated techniques are available for theoretical modelling of galaxy
formation and evolution: numerical simulations and semi-analytic modelling.
The overall strategy is the same in both cases: to calculate how density per-

turbations emerging from the Big Bang turn into visible galaxies. This requires following a number of processes: (i) the growth of dark matter halos by accretion and mergers, (ii) the dynamics of cooling gas, (iii) the transformation of cold gas into stars, (iv) the spectrophotometric evolution of the resulting stellar populations, (v) the feedback from star formation and evolution on the properties of prestellar gas and (vi) the build-up of large galaxies by mergers. Numerical simulations so far have focussed on a small subset of these processes which are treated in as realistic a way as is allowed by current algorithms and computing power. The semi-analytic approach, on the other hand, considers the combined effect of all these processes which are simplified into parametric rules distilled from simulations or analytic considerations.

The numerical and semi-analytic approaches are clearly complementary and have different strengths and weaknesses. The simulations generally attempt to model the relevant physics from first principles, but still require various approximations and free parameters. For example, when dealing with gas dynamics one needs to chose between Lagrangian methods like "Smooth Particle Hydrodynamics (SPH)" [1,2,3,4] or Eulerian methods [5]. There are also choices to be made regarding the cooling processes to be included, the mechanism to lay down initial conditions, the resolution of the calculation, etc. For more realistic modelling of galaxies, it is also necessary to include ad hoc algorithms for turning cold gas into stars and for coupling to the gas the energy liberated by stellar winds and supernovae.

In the semi-analytic approach [6,7,11,16] the required approximations and free parameters are more readily apparent. The backbone of this technique is a Monte-Carlo implementation of the "extended Press-Schechter theory" [8,9] used to describe the formation of dark matter halos by hierarchical clustering and merging [10]. An attractive feature of this approach is that within a fairly general framework, a full model of galaxy formation is specified by a surprisingly small number of free parameters. This a common feature of the two main semi-analytic models in existence today, that of G. Kauffmann and collaborators [11,12,13,14] and that of the present authors [15,17,18,19].

We summarize here the free parameters that appear in our semi-analytic model since this will be used in the remainder of this paper. It must be emphasized that these are not fitting parameters but rather parameters that describe various astrophysical processes, mostly related to star formation, that are only partially understood. Lacking a full physical understanding of these processes, it seems sensible to adjust the parameters so as to obtain as good a match as possible to a few basic observational data, in our case, to the local galaxy luminosity functions in the B and K bands. Within a given cosmology, the model requires fixing five parameters (see Cole et al. [15] for further details):

(i) the star formation timescale, i.e. the timescale on which gas that has cooled inside a dark matter halo is turned into stars; (ii) a feedback parameter which determines the efficiency with which energy liberated from supernovae and stellar winds reheats gas cooling inside a halo; (iii) the initial mass function of the stars that form; (iv) the timescale on which a galaxy falling onto a halo merges with the central galaxy and (v) the fraction of the stellar mass in stars above the hydrogen burning limit. To describe the broad morphology of a galaxy (i.e. its bulge-to-disk ratio) a sixth parameter is required: (vi) a threshold mass fraction for a merger to turn a disk into a spheroid.

In this article, we summarize some of the lessons learned from numerical and semi-analytic models of galaxy formation, highlighting a number of unresolved issues (Section 2). We then deploy our semi-analytic tools to explore the implications of the recent discovery by Steidel et al. [22] of a population of star forming galaxies at redshift $z > 3$ (Section 3). We conclude with a brief discussion in Section 4.

2 A summary of current theoretical issues

Most theoretical work on galaxy formation is carried out within the framework of hierarchical clustering and gravitational instability (eg Peebles [23]). Within this general picture, the various relevant processes are understood at different levels. Progress in several of these areas may be summarized as follows.

• *Dark matter halos.* Processes associated with the gravitational evolution of dark matter halos are reasonably well understood. This subject has progressed significantly in the past 15 years as a result of the increased sophistication of N-body simulations allied to some degree of analytic insight. Thus, in a given cosmological model, the abundance of dark matter halos, their merging history and their internal structure can be predicted reliably [24,25,26,28]. For example, recent high resolution N-body simulations by Navarro, Frenk & White [28] have established that independently of the cosmological model, dark matter halos of all masses develop a mass density profile that follows a simple, two-parameter form, scaling like r^{-1} in the central regions and like r^{-3} near the virial radius. The two parameters of the fit, which can be expressed as the mass and characteristic density of each halo, turn out to be strongly correlated: low-mass halos are significantly denser than more massive halos because, on average, they form earlier. Thus, in effect, the spherically averaged density profiles of dark matter halos are described by a universal one-parameter function.

• *The shape of the luminosity function.* Both numerical simulations and analytic considerations indicate that the mass function of galactic halos has a much steeper slope at the low-mass end ($\alpha \simeq 2$) than the observed field galaxy luminosity function ($\alpha \simeq 1$)[37]. The semi-analytic models, however, have demonstrated that the faint end of the galaxy luminosity function is determined by the combined effect of mergers and feedback but, in general, no model so far has succeeded in producing a faint end slope much flatter than $\alpha \simeq 1.5$. This discrepancy with observations remains an unsolved problem. At the bright end, the galaxy luminosity function cuts off exponentially, much as observed, as a result of the large cooling time of gas in massive halos.

• *The Tully-Fisher relation.* The Tully-Fisher relation predicted in semi-analytic models in a variety of cosmologies has a slope and scatter quite similar to those observed[6,11,15]. However, so far it has proved impossible to match simultaneously the zero-point of this relation and the amplitude of the galaxy luminosity function. The overall luminosity normalization of the models (parameter (v) above) can be chosen to match one or the other, but not both. This is another outstanding problem and reflects an overabundance of galactic dark halos predicted in the models. The problem is particularly severe for standard CDM, but it is still present in low-density variants of this cosmology.

• *The colours of galaxies.* Standard stellar population synthesis models and a standard IMF are sufficient to produce model galaxies with the spread of colours observed in the local population. This is true in most popular cosmologies except in the mixed dark matter model in which galaxies are much too young and thus much too blue compared to observations[17].

• *The morphologies of galaxies.* N-body/gas dynamic simulations produce galaxies with spiral disks and bulges[1,3] and merger remnants that resemble ellipticals[29,30]. However, gaseous disks in simulations with realistic initial conditions rotate more slowly than observed disks because the fragments from which they form loose angular momentum to the halo as they merge[27]. Thus, the origin of the angular momentum of spiral disks is not yet fully understood. Incorporating a simple prescription for merger-induced transformations of disks into spheroids in semi-analytic models reproduces the Dressler[31] morphology-density relation[13,18]. This success provides suggestive support for the view that accretion of rotating gas and mergers are the key ingredients in understanding the broad morphological characteristics of galaxies. The same environmental effects in clusters that produce the morphology-density relation today are responsible for the Butcher-Oemler effect[13,19].

• *The colour-magnitude relation of cluster ellipticals.* Semi-analytic models tend to produce colour-magnitude relations with an approximately flat slope and small scatter[14,19]. This is a counterintuitive result in hierarchical clustering models and arises because star formation in subgalactic fragments generally precedes the assembly of the galaxy by mergers. Furthermore, elliptical galaxies tend to form from fragments whose mass is biased towards large values. The traditional argument that the small scatter in the colour-magnitude relation requires ellipticals to be old and mergers to be unimportant thus appears to be incorrect. However, the observed colour-magnitude relation has a small but non-negligible slope [32]. In the context of current models, this must arise from metallicity effects which are neglected at present. It remains a major challenge for the models to reproduce the observed slope while retaining a small scatter.

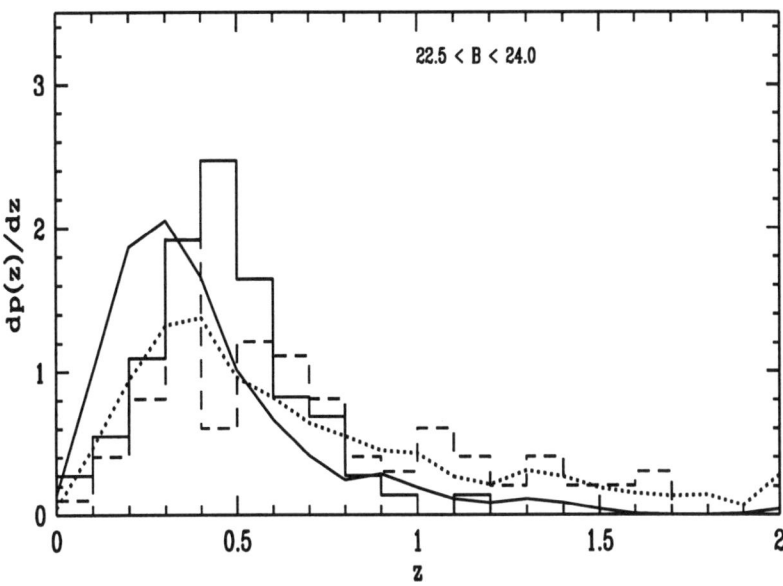

Figure 1: The redshift distribution of galaxies with magnitudes in the range $22.5 < B < 24.0$. The solid histogram shows the data of Glazebrook et al[33], while the dashed histogram shows the data from Cowie et al[34]. The lines show the predictions of the model of Cole et al[15] for a Scalo IMF (solid line) and a Miller-Scalo IMF (dotted line).

• *Counts of faint galaxies as a function of magnitude, redshift and morphology.* A notable success of the semi-analytic models is the excellent match they provide to the counts of faint galaxies as a function of magnitude, redshift and morphology. The supporting data are presented in the papers by White & Frenk [6], Kauffmann et al. [12], Cole et al. [15] and Baugh et al [18]. The agreement is particularly good in the standard CDM model but it is also acceptable in low-density CDM models. Particularly noteworthy is the match to the morphological data from the Hubble Deep Field discussed by Baugh *et al.* [18] and the prediction that faintwards of $I \simeq 25$ the galaxy counts should become increasingly dominated by irregulars. Also noteworthy is the successful prediction of the redshift distribution of $B \sim 24$ mag galaxies. The model predictions of Cole *et al.* [15], published before the observations were made, are compared with the recent data of Cowie *et al.* [34] in Figure 1. This agreement is the most striking indication so far that the models contain some element of truth. A consequence of these successes is that the models of Baugh *et al.* [19] also give a reasonable match to the redshift evolution of the luminosity function recently measured from the CFRS survey by Lilly *et al.* [35] and from other surveys by Ellis *et al* [36]. It should be noted, however, that the good match to faint data is due, in part, to the steep faint end slope in the model luminosity function for local field galaxies.

3 The Lyman break galaxies

Steidel *et al.* [22] have recently discovered a population of star forming galaxies at redshift $z \simeq 3.5$. In the context of the models discussed here, these galaxies are among the first objects in which appreciable star formation has taken place. Because of their great importance in understanding galaxy formation, we discuss them here in some detail.

Candidate high-z galaxies were identified spectroscopically, using U_n, G and R filters [21]. At $z \simeq 3$ the 912 Å break produced by the Lyman limit shifts into the U_n filter passband while, for the roughly flat spectrum characteristic of a star-forming object, the fluxes in the two other filters are comparable. Followup spectroscopy at Keck revealed that the objects so identified are indeed starforming galaxies at $3.0 \lesssim z \lesssim 3.5$. Steidel *et al.* find that these galaxies represent 1.3% of the faint counts brighter than $R = 25$, corresponding to a comoving number density comparable to that of present day L_* galaxies. The spectra of these galaxies are similar to those of nearby star-forming regions; their velocity dispersions are $180 \leq \sigma \leq 320$ km/s (if the line widths of saturated interstellar lines are assumed to reflect the velocity dispersion of the galaxy); and their inferred star formation rates are in the range $\sim 1 - 6h^{-2} M_\odot$ yr^{-1} (where h is Hubble's constant in units of 100 km s^{-1} Mpc^{-1}).

At first sight, the existence of a sizeable population of massive star-forming galaxies at such high redshifts may appear surprising, particularly in the context of the standard $\Omega = 1$, $h = 0.5$ CDM cosmology in which, as has been emphasized for a number of years, galaxy formation is a relatively recent phenomenon [20]. Our semi-analytic machinery allows us to investigate in detail whether galaxies with the required properties occur in a given cosmological model. Here we present results only for the standard CDM model which, apart from the mixed dark matter model, is the most extreme example of the family of currently popular cosmologies based on CDM. Further details of this analysis and results for other cosmologies will be presented in a forthcoming paper.

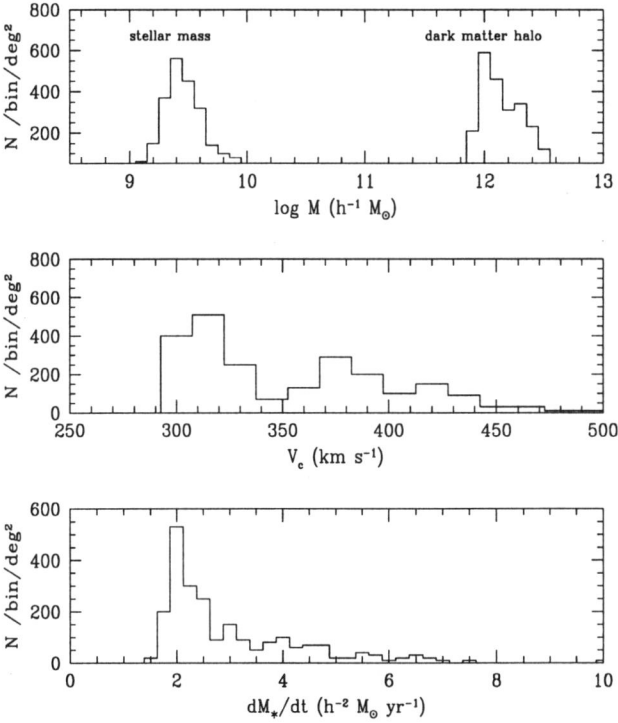

Figure 2: The properties of Lyman-break or "UV drop-out" galaxies identified in our model using identical selection criteria to those used in the observations of Steidel *et al* [22]. The top panel shows the distribution of stellar mass and halo mass for the model galaxies; the middle panel shows the distribution of halo circular velocities; and the bottom panel shows the distribution of star formation rates.

Following our general philosophy, we adopt our fully specified model, i.e. the model in which all free parameters have been previously fixed by reference to local galaxy data, in essence the model published by Cole *et al* [15]. The only change we have made is to assume a Miller-Scalo rather than a Scalo IMF. As Cole *et al.* discuss, the choice of IMF has little effect on the properties of the local galaxy population but it does influence the faint counts. At high redshift our model works much better with a Miller-Scalo IMF. We first selected galaxies using exactly the same filters and colour criteria as Steidel *et al.* These criteria did indeed pick out galaxies at $2.8 \lesssim z \lesssim 3.5$. The models produced 5560 galaxies per square degree with $23.5 < R < 25.0$ satisfying the colour criteria, of which 2800 lie in the redshift interval 3-3.5. Steidel *et al.* found 2127 galaxies in this redshift interval, a number which, they argue, might be an underestimate of the true abundance.

The properties of our model "Lyman-break" galaxies are displayed in Figure 2. Their typical stellar masses are a few times $10^9 h^{-1} M_\odot$ and these galaxies inhabit dark matter halos with typical mass $10^{12} h^{-1} M_\odot$. The velocity dispersions of the model galaxies are similar or, if anything, somewhat larger, than those inferred by Steidel *et al.* Finally, the star formation rates in the model galaxies agree remarkably well with those inferred by Steidel *et al.* (Star formation rates are not directly measured in the data but inferred from the R magnitude assuming an IMF and a stellar population synthesis model which are similar, but not identical, to those in our galaxy formation model.)

The success of our fiducial CDM model of galaxy formation in accounting for the observed abundance and overall properties of the Steidel *et al.* galaxies is both striking and surprising. However, several caveats are in order. Firstly, the predicted abundance of star-forming galaxies at high redshift depends sensitively on at least two model assumptions: the IMF and the normalisation of the linear density fluctuation spectrum, σ_8. Adopting a Scalo rather than a Miller-Scalo IMF reduces the total number of faint R-band counts by only 20% but it reduces the number of Lyman-break galaxies in the redshift interval of interest by about a factor of 10. Similarly, reducing σ_8 from our adopted value of 0.67 to 0.5 reduces the number of high redshift galaxies also by a factor of approximately 10. These uncertainties dwarf the changes produced by varying the cosmological parameters of the model.

Regardless of the uncertainties just discussed, the Lyman-break galaxies of Steidel *et al.* correspond to the first objects in our models in which significant star formation is taking place. Figure 3 shows how the overall stellar population builds up in three models, standard CDM and two low-density CDM variants with $\Omega = 0.3$, $\Lambda = 0.7$, $h = 0.6$, $\sigma_8 = 1$ and different primordial fluctuation spectra. In all cases, only a small fraction of the final stellar component of the Universe has formed by redshift $z = 3.5$. The standard CDM model and one of the low-density models have almost identical star formation histories and both have formed less than 5% of the total stellar population by $z = 3.5$. In the low-density model with more large scale power this fraction is still less than 10%. Thus, in the class of models we are considering, the redshift $z \simeq 3.5$ at which the Steidel *et al.* Lyman break galaxies are found is close to the onset of galaxy formation. Very few bright objects exist beyond this redshift.

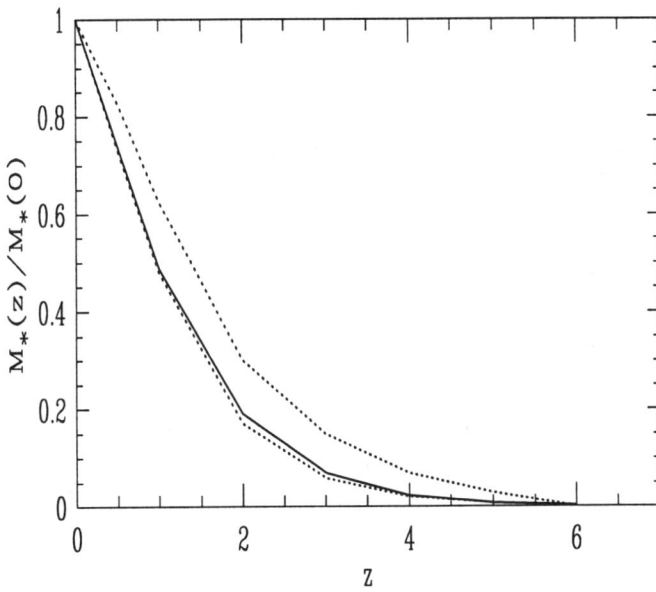

Figure 3: The mass in stars formed by redshift z as a fraction of the final mass in stars at redshift zero. The solid line shows results for the standard CDM model, while the dotted lines show results for flat cosmological models with $\Omega_0 = 0.3$ and two different intial power spectra. In all cases, less than 10% of the total mass of stars has formed by $z = 3.5$.

We can interrogate our galaxy formation model to find out what sort of objects the Lyman-break galaxies eventually turn into. Two examples are shown in the "tree diagrams" of Figure 4. Redshift decreases downward in these plots and the width of the shaded region is proportional to the mass in stars at each epoch. Stars generally form in subgalactic fragments at high redshift which grow larger as gas cools onto a disk and turns into stars. Fragments can merge together and if the merger is massive enough, the disks turn into a spheroid; a new disk may grow by subsequent accretion of gas [19]. The galaxy on the left of Figure 4 experienced only two very small mergers at $z \simeq 3$

and grew almost entirely by accretion. This object ends up as a late type spiral galaxy with a very small bulge. The galaxy on the right formed by the merger of several fragments, including a major merger at $z \simeq 0.3$. This galaxy ends up as an elliptical. The asterisks at high z represent the Lyman-break object that satisfies the Steidel *et al.* selection criteria. The spiral galaxy is the descendant of a single fairly massive Lyman-break object; the elliptical harbours the descendants of two less massive Lyman-break objects which merged at relatively recent epochs.

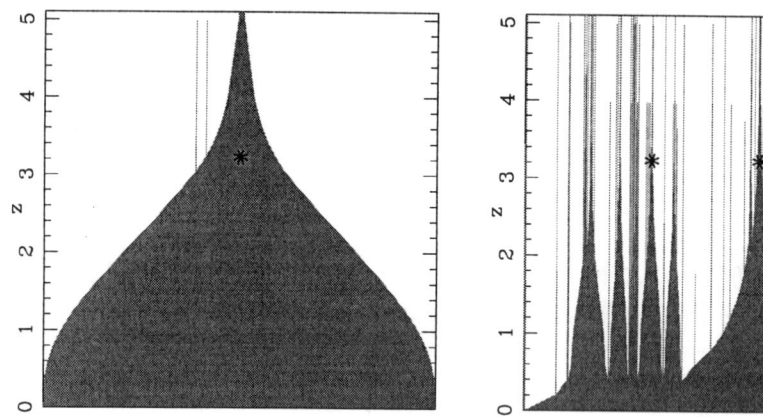

Figure 4: Star formation histories of two present day galaxies that contained a Lyman-break progenitor (marked by the crosses) satisfying the selection criteria of Steidel *et al*[22]. Redshift decreases downward in these plots and the width of the shaded region is proportional to the mass in stars at each epoch. The galaxy on the left is a late-type spiral; the galaxy on the right is an elliptical.

4 Conclusions

Theoretical studies of galaxy formation, based on numerical simulations and semi-analytic techniques are an essential complement to observational studies of the high redshift universe. Such modelling is required in order to establish the connection between different types of data and their relation to the physics of galaxy formation in a cosmological setting. Although some of the physical processes involved, particularly those associated with star formation, are very complex and poorly understood, progress can be made by complementing a physically based description with heuristic rules to describe star formation.

Semi-analytic models now exist in which the various physical processes are characterised by a minimum of free parameters, all of which are fixed by reference to a small subset of the data for local galaxies. The models so specified can then be used to explore the expected properties of the high redshift universe in different cosmologies. We have illustrated the predictive power of these semi-analytic models by comparing our published predictions for the redshift distribution of galaxies of $B \sim 24$ mag with recent data from Cowie *et al.* [34] (Figure 1). The excellent agreement between them is the most striking demonstration so far of the virtues of this approach.

In general, the best understood aspects of galaxy formation are those related to their dark matter component. The abundance, merging history and internal structure of galactic halos are all reasonably well established in a variety of cosmological models of hierarchical clustering. Some understanding also exists of the physical basis of observable properties such as the general shape of the galaxy luminosity function, the slope and scatter of the Tully-Fisher relation, the general features of the colour-magnitude diagram, the gross morphological properties of galaxies in different environments, and the counts of faint galaxies as a function of magnitude, redshift and morphology. All of these properties can be explained, at least at some level, within a broad class of CDM cosmologies.

Several fundamental properties of the galaxy population remain poorly understood. Examples include the faint end slope of the field luminosity function which is predicted to be significantly steeper than the standard estimate [37]. None of the existing models can simultaneously match the zero-point of the Tully-Fisher relation and the overall amplitude of the galaxy luminosity function, a problem which can be traced back to an overabundance of dark matter halos predicted in all CDM cosmologies. While the small scatter in the observed colour-magnitude relation for cluster ellipticals does not seem incompatible with hierarchical clustering, none of the models published to date can account for the measured slope in this relation.

In spite of the unsolved problems just mentioned, semi-analytic modelling remains a powerful tool to interpret the exciting new data on the high redshift universe discussed at this conference. As an example, we presented in this article results from new calculations which attempt to identify the evolutionary status of the Lyman-break galaxies at $z \simeq 3.5$ recently discovered by Steidel *et al.* [22]. Perhaps surprisingly, we found that the abundance and global properties of these objects is almost exactly what is predicted by our fiducial model of galaxy formation based on the standard CDM cosmology. Although the

54

predicted abundance is, in fact, quite sensitive to certain model assumptions such as the IMF and the amplitude of mass fluctuations, in general, these galaxies are among the first objects in which appreciable star formation is taking place. Within a broad class of CDM models, it appears that the Steidel *et al.* objects signal the onset of significant galaxy formation. These objects evolve into the population of normal galaxies seen today. Our models seem to imply that the long awaited discovery of the early phases of normal galaxy evolution has now taken place.

References

1. N. Katz and J.E. Gunn 1991, ApJ, 377, 365
2. J.F. Navarro and S.D.M. White, 1993, MNRAS, 265,271
3. M. Steinmetz and E. Muller, 1995, MNRAS, 276, 549
4. A.E. Evrard, F. Summers and M. Davis 1994, ApJ, 422,11
5. R. Cen & J.P. Ostriker 1993, ApJ, 417, 415
6. S.D.M. White and C.S. Frenk, 1991, ApJ, 379,52
7. S. Cole, 1991, ApJ, 367,45
8. R.G. Bower, 1991, MNRAS, 248,332
9. J.R. Bond, S. Cole, G.P. Efstathiou and N., Kaiser, 1991, ApJ, 379,440
10. G. Kauffmann and S.D.M. White 1993 MNRAS 261,921
11. G. Kauffmann, S.D.M. White and B. Guiderdoni, 1993 MNRAS, 264,201
12. G. Kauffmann, B. Guiderdoni and S.D.M. White 1994, MNRAS, 267,981
13. G. Kauffmann 1995, MNRAS, 274, 153
14. G. Kauffmann 1996, MNRAS, 281, 487
15. S. Cole, A. Aragon-Salamanca, C.S. Frenk, J.F. Navarro, S. Zepf, 1994, MNRAS, 271,744
16. C. Lacey, B. Rocca-Volmerange and J. Silk 1993, ApJ, 402, 153
17. J. Heyl, S. Cole, C.S. Frenk and J.F. Navarro 1995, MNRAS, 274,755
18. C.M. Baugh, S. Cole and C.S. Frenk, 1996, MNRAS 282,L27
19. C.M. Baugh, S. Cole and C.S. Frenk, 1996, MNRAS in press
20. C.S. Frenk, S.D.M. White, G. Efstathiou and M. Davis 1985, Nature, 317, 595
21. C. Steidel, M. Pettini and D. Hamilton 1995, AJ, 110, 2519
22. C. Steidel, M. Giavalisco, M. Pettini, M. Dickinson and K. Adelberger 1996 ApJ, 462L, 17
23. P.J.E. Peebles 1980 "The Large Scale Structure of the Universe" Princeton University Press, Princeton NJ
24. W.H. Press and P. Schechter 1974, ApJ, 187, 452
25. C.S. Frenk, S.D.M White, M. Davis, G.P. Efstathiou, 1988, ApJ 327,507

26. C.G. Lacey and S. Cole, 1993, MNRAS, 262,627

27. J.F. Navarro, C.S. Frenk and S.D.M. White, 1995, MNRAS, 275, 56

28. J.F. Navarro, C.S. Frenk and S.D.M. White, 1996, ApJ, 462, 563

29. J. Barnes and L. Hernquist, 1996, ApJ, 471, 115

30. J. C. Mihos and L. Hernquist ApJ, 464, 641

31. A. Dressler 1980, ApJ, 236, 351

32. R.G. Bower, J.R. Lucey and R.S. Ellis 1992, MNRAS, 254, 601

33. K. Glazebrook, R.S. Ellis, M. Colless, T. Broadhurst, J. Allington-Smith, N. Tanvir 1995, MNRAS 273, 157

34. L.L. Cowie, A. Songaila, E.M. Hu and J.G. Cohen 1996, AJ, 112, 839

35. S.J. Lilly, L. Tresse, F. Hammer, D. Crampton, O. Le Fevre 1995, ApJ, 445, 108

36. R.S. Ellis, M. Colless, T. Broadhurst, J. Heyl, K. Glazebrook 1996, MNRAS, 280, 235

37. J. Loveday, B.A. Peterson, G. Efstathiou and S.J. Maddox 1992 ApJ, 390, 338

QUANTITATIVE MORPHOLOGY OF FAINT FIELD GALAXIES

R.G. ABRAHAM

Institute of Astronomy, University of Cambridge
Madingley Road, Cambridge, CB3 OHA UK

W. FREEDMAN

Carnegie Observatories
813 Santa Barbara Street, Pasadena, CA 91101-1292 USA

B.F. MADORE

Infrared Processing and Analysis Center
California Institute of Technology
Pasadena, CA 91125 USA

The morphological mix of faint galaxies to $I_{814} = 25$ mag is presented, based upon data from the Hubble Space Telescope *Medium Deep Survey* (MDS) and the *Hubble Deep Field* (HDF). Because high redshift galaxies are observed in the rest wavelength ultraviolet, where little is known about local galaxy morphology, we have determined number counts using a quantitative morphological classification system which has been calibrated using numerical simulations of high-z galaxies. We have used images from the Las Campanas 2.5m du Pont telescope and the Ultraviolet Imaging Telescope (UIT) to test the pixel-by-pixel K-correction technique used in these simulations, which accounts for structural changes introduced by cosmological bandshifting. Our analysis shows that the majority of peculiar/irregular/merging galaxies visible in the Hubble Deep Field are not simply bandshifted spiral galaxies, although some very late-type spirals will be misclassified as irregulars at high redshift. The steeply rising number count-magnitude relation for irregular/peculiar/merging galaxies reported earlier using data from the MDS continues to at least $I_{814} = 25$ mag in the HDF. Beyond $I_{814} = 22$ mag the number counts of spiral galaxies rise more steeply than no-evolution predictions, while the counts of elliptical/S0 galaxies only slightly exceed such predictions, and may begin to turn over beyond $I_{814} = 24$ mag.

1 Quantitative Morphology

Recent work from the Medium Deep Survey [9,8,6,5,3] has shown that much of the rapidly evolving faint blue galaxy population [4,10,11] is comprised of morphologically peculiar systems. These systems may be luminous counterparts to local irregular galaxies, tidally disturbed systems, or perhaps members of an entirely new class of objects with no local counterpart. Distinguishing between these possibilities using conventional classification techniques is problematic, because cosmological bandshifting continuously changes the appearance of "benchmark" galaxies upon which classification systems are ultimately based. Furthermore, the subjective nature of visual classification makes comparisons between different groups susceptible to potentially large systematic errors [13].

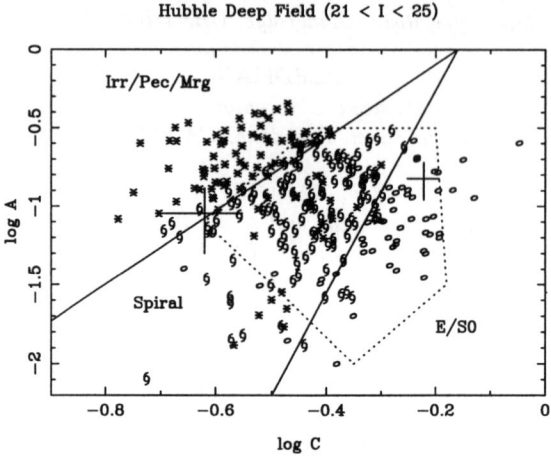

Figure 1: Asymmetry vs. central concentration for galaxies in the Hubble Deep Field. Visual morphological classifications by Ellis are keyed to the plot symbols: E/S0's are shown as ellipses, spirals earlier than Sd are shown as spirals, and irregulars/peculiars/mergers are shown as asterisks. The sectors subdivide the diagram into regions where each of these morphological types dominates. Representative error bars are shown. The dotted polygon shows the "convex hull" enclosing the artificially redshifted local sample of objects described in the text.

To circumvent these difficulties, we have developed a quantitative morphological classification system (based on measurements of central concentration, C, and asymmetry, A) that is well-suited to the investigation of faint galaxies [1,3]. Figure 1 shows the distribution on the $A - C$ diagnostic diagram of I_{814}-band (F814W filter) galaxies with $21 < I_{814} < 25$ mag in the Hubble Deep Field (taken from Abraham *et al.* 1996a). Visual classifications by Ellis are indicated by different symbols. Objects that have been *visually* classified as early-type systems (E/S0), late-type systems (spirals earlier than Sd), and irregular/peculiar/merging systems are well-separated on this diagram.

An important advantage of basing morphological classifications upon position in the $A - C$ diagram is that bandshifting effects can be accounted for directly by numerical simulation. We have developed a technique for artificially redshifting local galaxy CCD images by assigning separate spectral

NGC1566   NGC1365

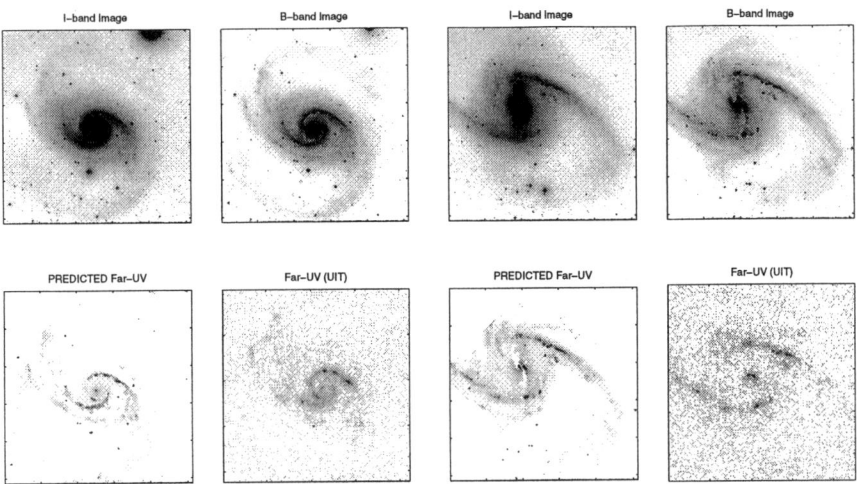

Figure 2: Comparison between the far-UV morphology "predicted" on the basis of optical colors and the observed morphology from the Ultraviolet Imaging Telescope (UIT). Optical images were obtained from the 2.5m du Pont telescope on Las Campanas, and no noise has been added in order to match the signal-to-noise characteristics of the UIT.

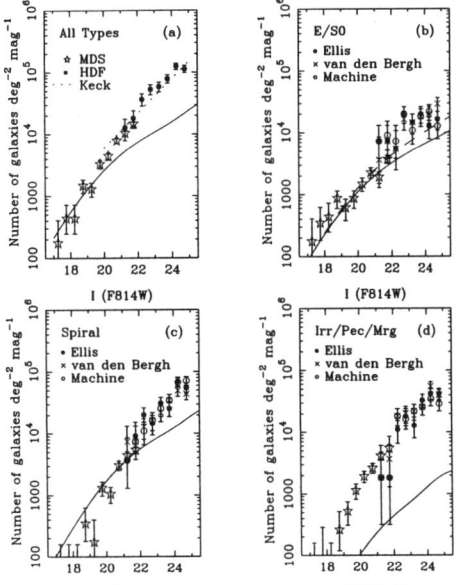

Figure 3: The number-magnitude relations for morphologically segregated samples of galaxies from the HDF and MDS (from Abraham *et al.* 1996a). Open circles indicate counts obtained from automated classifications determined from Figure 1, closed circles indicate the results from the visual classifications of Ellis, and crosses indicate the results from the visual classifications of van den Bergh. The MDS counts are indicated by the stars on each panel. The no-evolution $\Omega = 1$ curves from Glazebrook *et al.* (1995), extrapolated to $I = 25$ mag, are superposed. The dashed line on the E/S0 diagram shows the effect of assuming $\Omega = 0.1$. The dotted line in panel (a) shows the I-band number counts determined by Smail *et al.* (1995) from two deep fields imaged with the Keck telescope.

energy distribution to *individual pixels*. Spectral energy distributions for each pixel are determined by using optical color to interpolate between template spectra corresponding to local S0, Sab, Sbc, Scd, Sdm, and starbursting galaxies (kindly supplied by A. Aragón-Salamanca). Figure 2 illustrates the power of this technique by showing the excellent agreement (for two representative galaxies in our calibration sample) between "predicted" and observed far-UV morphologies, even for the extreme case [a] in which the I_{814}-band morphology is extrapolated to the far-UV. The artificial redshifting procedure is used to construct a "bandshifting vector" which traces the movement of local galaxies in the $A - C$ diagram as the result of cosmological bandshifting (in the absence of evolution, and without explicitly accounting for the effects of dust obscuration).

2 The morphological mix at $21 < I_{814} < 25$ mag

Artificially redshifted images of the 103 objects in the Frei *et al.* (1996) sample of CCD images [b] have been used to subdivide the $A - C$ diagram into sectors within which early-type, late-type, and peculiar/irregular/merger systems dominate at redshifts up to $z = 2.5$. The number count-magnitude diagram resulting from this analysis is shown in Figure 3, along with the results obtained using independent visual classifications by Ellis and van den Bergh. Also shown are the no-evolution predictions for each type, constructed as described in Glazebrook *et al.* (1995) and Abraham *et al.* (1996a,b), by adopting Schechter luminosity functions (LFs) with parameters given by Loveday *et al.* (1992), and a high normalization $\phi_* = 0.03h^3$ Mpc^{-3}. The predicted counts for the elliptical galaxies are based on a flat slope ($\alpha = -1$) for the faint-end of the LF, rather than the turn-over originally found by Loveday *et al.* Beyond $I_{814} = 22$ mag the spiral counts now show a significant excess over the no-evolution predictions. A weaker trend is seen for the spheroidal systems (whose counts are only marginally above the no-evolution prediction) and there is some evidence of a turn-over in the last magnitude interval. The large

---

[a] The great majority of objects at $I_{814} < 25$ mag on the Hubble Deep Field are likely to be at redshifts < 2, so I_{814} filter images of distant objects should be compared with U-band (~ 3600Å) local galaxy data. Images in U-band have much less drastic morphological variations (relative to optical data) than is the case with images in the far-UV (~ 1500Å) (which correspond to $z \sim 4$).

[b] It is emphasized that the Frei *et al.* sample is not not intended to represent a fair volume-limited sample of the local universe; comparatively few very-late-type systems ($T > 7$), and no examples of mergers, are included. However the objects in the Frei catalog *are* intended to be representative of "generic" early-type galaxies and late-type spirals earlier than Sdm systems.

fraction of irregular/peculiar/merger systems detected at faint magnitudes in the HDF suggests that beyond $I_{814} \sim 24$ mag the conventional Hubble system no longer provides an adequate description of the structural characteristics for a significant fraction of galactic (and possibly proto-galactic) systems. Our comparison with artificially redshifted optical images of nearby spiral galaxies rules out the possibility that the bulk of these irregular/peculiar/merger systems are simply misclassified bandshifted spirals.

Acknowledgments

Much of the material in this paper is the result of work done by the official and honorary members of the extragalactic component of the "Cambridge HDF Team". We thank collaborators Richard Ellis, Nial Tanvir, Basilio Santiago, Karl Glazebrook, and Sidney van den Bergh for many stimulating discussions (and more hours of work) that they have contributed to this project. We also congratulate Bob Williams and the staff of STScI on the superb job that has been done in preparing the HDF data, and thank Ted Stecher and the UIT team for their enormous efforts in undertaking the UIT program.

References

1. Abraham, R.G., Valdes, F., Yee, H.K.C. & van den Bergh, S. 1994, ApJ, 432, 75
2. Abraham, R. G, Tanvir, N. R., Santiago, B. X. , Ellis, R. S., Glazebrook, K. & van den Bergh, S. 1996a, MNRAS, 279, L47
3. Abraham, R.G., van den Bergh, S., Glazebrook, K., Ellis, R.S., Santiago, B. X., Surma, P., & Griffiths, R. 1996b, ApJSupp (in press).
4. Broadhurst, T., Ellis, R., & Shanks, T. 1988, MNRAS, 235, 827
5. Cowie, L. L., Hu. E. M., & Songaila, 1995, AJ, 110, 1576.
6. Driver, S.P., Windhorst, R.A. & Griffiths, R.E. 1995, ApJ, 453, 48.
7. Frei, Z., Guhathakurta, P., Gunn, J. E. 1996, AJ, 111, 174
8. Glazebrook, K., Ellis, R., Santiago, B. & Griffiths, R. 1995, MNRAS, 175, L19.
9. Griffiths, R.F. *et al.* 1994, ApJ, 437, 67.
10. Koo, D. C., & Kron, R. 1992, Ann. Rev. 30, 613.
11. Lilly S. J., Tresse, L., Hammer, F., Crampton, D., Le Fevre, O 1995, ApJ, 455, 108
12. Loveday, J., Peterson, B. A., Efstathiou, G., Maddox, S. J. 1992, ApJ, 390, 338
13. Naim, A., *et al.* 1995 , MNRAS, 274, 1107

THE ULTRAVIOLET MORPHOLOGY OF GALAXIES

ROBERT W. O'CONNELL, PAMELA MARCUM

University of Virginia,

Charlottesville, VA 22903-0818, USA

Optical band images of distant (z $\gtrsim$ 0.5) galaxies, such as those of the Hubble Deep Field, record light from the rest-frame vacuum ultraviolet (λ < 3000 Å). Because the appearance of a galaxy is a very strong function of wavelength, and especially so in the UV, evolutionary studies of distant galaxies can be seriously influenced by a "morphological k-correction" effect. We use images obtained by the Ultraviolet Imaging Telescope during the *Astro* missions to explore the extent of this effect and intercompare far-UV with optical morphologies for various types of galaxies.

The morphology of a distant galaxy is often our first index to its evolutionary state. Unfortunately, there is a host of technical complications which afflict the apparent morphologies of distant galaxies. These can conspire to make the familiar look unfamiliar and vice-versa.

Two of these difficulties are well known: *(i)* reduced spatial resolution; and *(ii)* the rapid fading of surface brightness with redshift, $SB \sim (1+z)^{-n}$, where n = 3–5, depending on the method used to characterize surface brightness. These two effects are relatively simple to model, and their consequences for ground-based and HST imaging have recently received much attention[1,2,3,4,5]. Here, we wish to focus on a third difficulty, which can be called the "morphological k correction"—i.e. the fact that the appearance of galaxies is a strong function of wavelength. The photons recorded on exposures of distant galaxies in optical bands, e.g. in the Hubble Deep Field, often originate in the vacuum ultraviolet in the restframe. It has been known for many years that the appearance of galaxies in the UV can be very different than at the optical wavelengths for which the classical morphological typing systems were established[6,7]. In the UV, hot stars are emphasized and cool ones are suppressed. This provides an important opportunity to distinguish stellar populations, but it also means that morphological analyses must take into account a strong UV/optical k-correction.

It has been hard to quantify this effect because few vacuum-UV images of nearby galaxies were available before 1990. Since then, several hundred objects have been observed in the UV by HST, the FOCA balloon experiment[8], and the Ultraviolet Imaging Telescope[9] (UIT) on the two *Astro* missions. Maos et al.[10] have recently published an atlas of 110 nearby galaxy nuclei observed at 2300 Å in the 22″ field of the HST Faint Object Camera. We[11] are preparing

NGC 4631 (Sc)

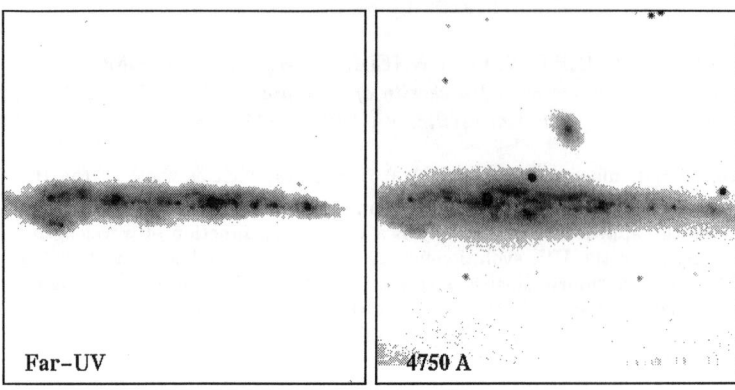

Far–UV   4750 A

Figure 1: Dust does not dominate the UV appearance of normal disk galaxies, even in edge-on systems like this one.

a UV/optical atlas which compares the UV and optical morphologies of 27 nearby galaxies using UIT data mainly from the *Astro-1* mission. A later paper will add 45 galaxies observed during *Astro-2*. UIT's field of view is 40' in diameter and produced typical resolution of 3" FWHM; the two principal (solar-blind) imaging bands were centered at 1500 Å ("far-UV") and 2500 Å ("mid-UV"). UIT's characteristics are well suited to the study of morphologies of nearby galaxies. Some of this data has already been used to simulate the combined effects of the k-correction, reduced resolution, and reduced surface brightness on observations of distant galaxies[5][12].

Here we show a few of the comparisons from the UIT Atlas and briefly mention the effects they illustrate. Data for a number of spirals shows that dust in normal disks does not strongly affect UV morphologies (Fig. 1). However, spirals do shift to later Hubble types in the UV (Fig. 2). The k-correction combined with reduced resolution and bias toward high SB regions can turn a single normal galaxy into an irregular object with "interacting" companions (Fig. 3). Disturbed dust layers can have drastic effects on UV light (e.g. M82, not shown); for instance, an interacting pair of galaxies can turn into an isolated galaxy (Fig. 4). Barred galaxies often become unbarred (e.g. M83[6]). The general trend found is from normal to abnormal. To put this another way, the incidence of abnormal galaxies in the restframe UV appears to be larger than it really is.

M81 (Sb)

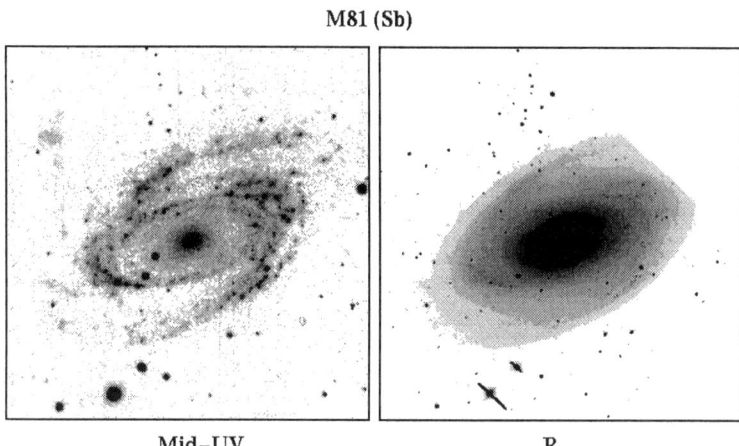

Mid–UV   R

Figure 2: Hubble types for spirals become later in the UV. In the far-UV (not shown), M81 becomes a nearly "empty" ring.

M101 (Sc) Far–UV

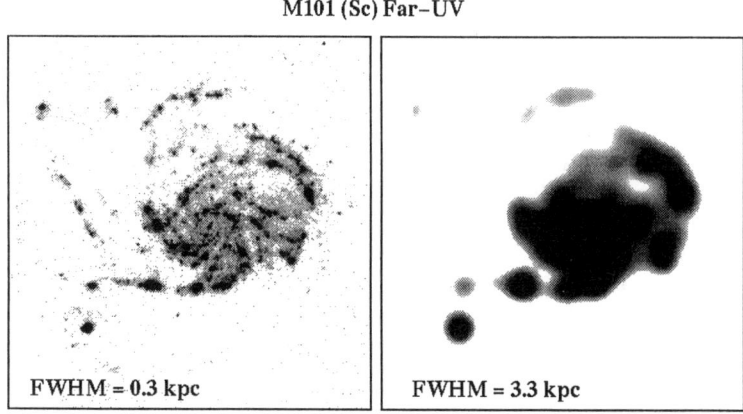

Figure 3: Effects of k-correction and reduced resolution, corresponding to observations at $z \sim 1$ with $0.4''$ FWHM. Bright associations have become "companions".

M51 (Sbc)

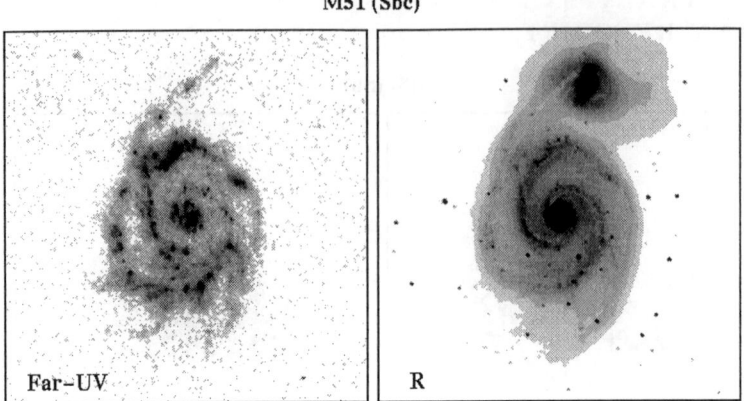

Figure 4: Case of a disappearing perturber.

References

1. S.P. Driver, R.A. Windhorst, and R.E. Griffiths, ApJ, 453, 48, 1995.
2. S.J. Lilly, O. Le Fevre, D. Crampton, F. Hammer, and L. Tresse, ApJ, 455, 50, 1995.
3. R.G. Abraham, N.R. Tanvir, B.X. Santiago, R.S. Ellis, K. Glazebrook, and S. van den Bergh, MNRAS, in press.
4. A. Dressler, this conference.
5. M. Giavalisco, J. Livio, R.C. Bohlin, F.D. Macchetto, and T.P. Stecher, AJ, 112, 369, 1996.
6. R.C. Bohlin, R.H. Cornett, J.K. Hill, A.M. Smith, and T.P. Stecher, ApJ, 274, L53, 1983.
7. C.R. King and R. S. Ellis ApJ, 288, 456, 1985.
8. A. Blecha, M. Golay, D. Huguenin, D. Reichen, and D. Bersier, A&A, 233, L9, 1990.
9. T. P. Stecher et al. ApJ, 395, L1, 1992.
10. D. Maoz, A.V. Filippenko, L.C. Ho, F.D. Macchetto, H.-W. Rix, and D.P. Schneider, ApJS, in press.
11. P. Marcum, R.W. O'Connell, W. Freedman, B. Madore, G. Bothun, K.P. Cheng, P.M. Hintzen, E.P. Smith, R.C. Bohlin, S.G. Neff, M.S. Roberts, A.M. Smith, and T.P. Stecher, in preparation for ApJS.
12. R.C. Bohlin, R.H. Cornett, J.K. Hill, R.S. Hill, W.B. Landsman, R.W. O'Connell, S.G. Neff, A.M. Smith, and T.P. Stecher, ApJ, 368, 12, 1991.

CONSTRAINTS ON GALAXY FORMATION FROM DEEP GALAXY REDSHIFT SURVEYS AND QUASAR ABSORPTION LINE STUDIES

LENNOX L. COWIE

Institute for Astronomy, University of Hawaii, 2680 Woodlawn Dr., Honolulu, HI 96822, USA

Magnitude-limited galaxy redshift surveys are now providing large samples of galaxies to beyond $z = 2$, while color-selected and emission-line-selected samples are finding galaxies to $z = 4.7$. A broad picture is emerging of galaxy formation peaking in the $z = 1$ range, which ties in with the metallicity and density evolution seen in the quasar absorption lines. We still have no direct information beyond $z = 5$, but the ionization of the IGM at this redshift argues for activity prior to this time. The metallicities of around 0.01 solar which appear to be relatively ubiquitous in quasar absorption lines beyond $z = 2$, even in very low column density clouds, could be a relict of this period.

1 Introduction

A consistent picture of galaxy formation and evolution is emerging from combining faint galaxy properties obtained from deep galaxy surveys with what has been learned from quasar absorption line studies. At $z > 2$ a large fraction of the baryon density of the universe appears in quasar absorption line clouds stretching over neutral hydrogen column densities from 10^{12} cm^{-2} to 10^{22} cm^{-2}. This material has largely vanished by the present time, presumably converting into stars. Studies of the metallicity and ionization of the baryonic material in absorbing clouds provide evidence for small amounts of early enrichment followed by a much more rapid rise below redshifts of about 2. At the same time, nearly complete spectroscopic surveys are yielding information on the star-formation history of galaxies, with evidence of substantial evolution in the properties of individual galaxies out to redshifts of one, where there are many massive galaxies which are dominated by the light of massive star formation. These also show that the total star-formation rates in the universe were highest near $z = 0.5 - 1$ and were lower at higher redshifts, although subgalactic or near-galactic sized star-forming galaxies exist at least to redshift 4.7 [refs. 1,2] and presumably extend beyond $z = 5$.

2   Early Metals?

One of the first results to emerge from the HIRES spectrograph[3] on the Keck telescope was the detection of many weak C IV lines corresponding to relatively low column density Lyman alpha forest lines.[4,5] This provided confirmation of previous suggestions that there might be fairly regular metal enrichment in such systems.[6,7] Currently, as is summarized in Fig. 1, we know that nearly all clouds with $N(H I) \gg 10^{15}$ cm^{-2} and a large fraction of those with $N(H I) \gg 3 \times 10^{14}$ cm^{-2} are detected in C IV.[8] The issue of whether yet lower column density clouds may be chemically unenriched remains open since current sensitivity limits would not detect such clouds in C IV at the expected column densities.

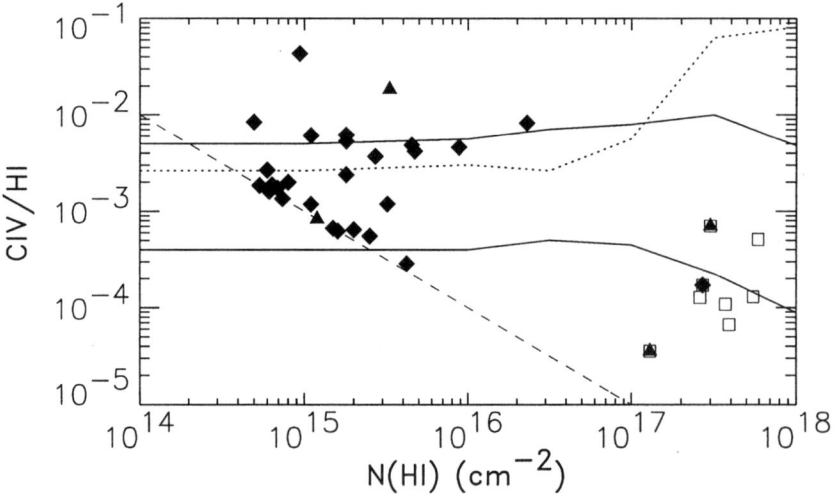

Figure 1: C IV versus H I column density for all systems with $N(H I) > 5 \times 10^{14}$ cm^{-2} at $3.135 < z < 3.60$ toward Q1422+231(diamonds) and for $N(H I) \geq 1.5 \times 10^{15}$ cm^{-2} at $z > 2.95$ toward Q0014+813 (triangles). Also shown (open squares) is C IV/H I in all Lyman limit systems (open boxes) toward eight quasars.[9] The dashed line shows the typical $2\,\sigma$ detection limit for C IV in Q1422+231. The solid lines show model calculations[10] of C IV/H I for $\Gamma = 10^{-2.7}$ (lower), $\Gamma = 10^{-1.7}$ (upper) and the dotted line the model for $\Gamma = 10^{-0.7}$, and a metallicity of 10^{-2} solar, and illustrate that the difference between the higher and lower column density systems is not solely a radiative transfer effect but must arise from a higher ionization parameter or higher metallicity in the weaker clouds.

Remarkably, the metallicity in the forest clouds of around 0.01 solar is similar to that seen in partial Lyman limit system clouds with $N(H I) = 10^{17}$

cm^{-2} (Fig. 1) and in the $z > 2$ damped Lyα systems with N(H I)=10^{21} cm^{-2} [ref. 11]. There is also some evidence that the forest clouds show enhancement of the alpha process elements versus the Fe process elements,[8] as may also be the case in the damped Lyα systems,[11] paralleling the abundances in the low metallicity stars in the Galactic halo.

This relatively ubiquitous enrichment clearly has powerful implications for understanding the early stages of heavy element formation. Essentially it requires that along any line-of-sight through a forest cloud we must see a relatively uniform (at least to order of magnitude) enhancement, or alternatively expressed, that the covering factor of metal-enriched portions of the clouds must be near unity. On smaller scales there may, of course, be lower metallicity pockets or unenriched regions.

We may, of course, be seeing directly the early stages of star formation in the halos of forming galaxies but it is also possible, perhaps even likely, that we may be seeing relicts left over from sub-galactic star formation in the dark ages at $z > 5$. If this is true, this metallicity and the fact that the IGM is ionized at $z = 5$ represent the two pieces of information we have about this period. Even in the $z = 5$ era [ref. 12] it is not easy to understand the uniformity since, based on simple cooling arguments, substantial star formation is only expected in near Galactic-sized clumps.

3 Galaxy Redshift Distributions

Currently, large and near-complete redshift samples exist to $B = 24.5$, $I = 23$ and $K = 20$ as is summarized in Fig. 2 in the left-hand panels. These surveys extend to about $z = 2$ and show that there is a very substantial evolution in the properties of the galaxies even by redshift one. By $z = 1$ there are many large galaxies whose colors and emission-line properties show that they are dominated by massive star formation.[13-15]

This may be most cleanly seen in the K-selected surveys, which do not have the bias against early-type galaxies at $z = 1 - 4$ which is present even in red optical samples. Fig. 3 shows the rest equivalent width of the [O II] 3727 line versus absolute K magnitude in various redshift slices. This figure shows an interesting and (at least for a simple-minded CDM theorist) somewhat surprising effect. Galaxies whose light is dominated by massive star formation typically have [O II] equivalent widths in excess of 25Å [ref. 16], and we can see from Fig. 3 that there are very few such galaxies at any K luminosity at the present time. However at $z = 0.2$ there are small galaxies (the blue dwarfs) which fall into the category of rapid star formers, and as we move to higher redshift more and more massive galaxies appear to be [O II] luminous until at

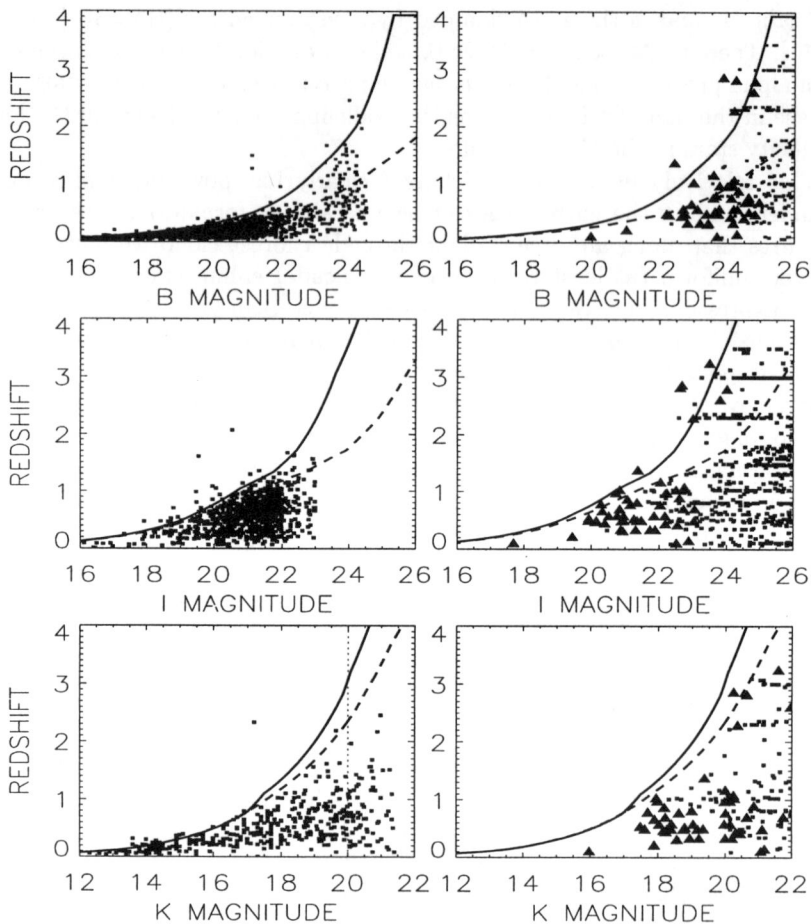

Figure 2: The left-hand panels show the redshift-magnitude diagrams for near-complete spectroscopic samples in the K, I(Kron-Cousins) and B bands. The data are taken from the Autofib sample of Ellis et al.[14] and the Hawaii survey[13] for the B sample, the CFRS survey[15] and the Hawaii survey for the I band and the Glazebrook[18] sample and the Hawaii samples[13,19] for the K band. For the K band we have extended the sample beyond the completeness limit of $K = 20$ shown by the dotted line. The right-hand panels show the spectroscopically identified objects given in Cohen et al.[20] and Steidel et al.[21] in the Hubble Deep Field as triangles. The remaining objects are shown at color-estimated redshifts using a six-color estimator with the four Hubble Deep Field colors and two IR colors (J and $H + K$). This color estimator[15] agrees well with all the measured spectroscopic redshifts. The solid and dashed lines show an unevolving Im (solid) and Sb (dashed) galaxy with $M_K = -25.8$, $M_I = -23.5$ and $M_B = -22$ for $H_0 = 50$ km s^{-1}/Mpc and $q_0 = 0.5$.

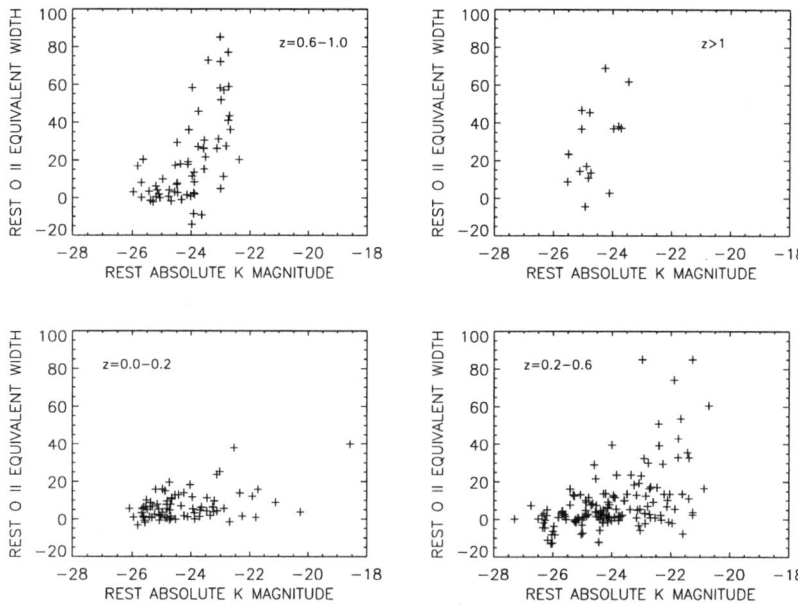

Figure 3: Rest-frame [O II] equivalent width versus absolute rest K magnitude and redshift for the $K < 20$ sample. In the lowest redshift interval (lower left panel) very few galaxies have strong [O II] lines or are undergoing rapid star formation (EW([O II]) $\gtrsim 25$Å). At higher redshifts, progressively more massive galaxies are undergoing rapid formation, until at $z > 1$ the locus of rapidly forming galaxies reaches a luminosity near $M_{K*} \sim -25$. The absolute magnitudes are calculated for $q_0 = 0.5$ and $H_0 = 50$ km s^{-1}/Mpc.

$z > 1$ we see many near L* galaxies ($M_{K*} = -25.1$) in this category. There are very few super L* galaxies at any redshift as can be seen from Fig. 2. Rather, galaxies seem to regulate at this value in their earlier stages. An $M_I = -23.5$ galaxy has an AB magnitude of -23 and would be produced by a star formation rate of about 100 $M_\odot$/yr [ref. 17]. This could easily form a massive galaxy if it persisted until $z = 1$.

In order to push this to higher redshifts we can construct color-estimated redshifts from the Hubble Deep Field (HDF) as was discussed by a number of groups at the conference and in the literature.[22, 23] The color estimates are not straightforward, particularly at the faint end where galaxies generally become very blue, and use of only the four Hubble colors can make this very uncertain. The estimates can be made somewhat more robust by extend-

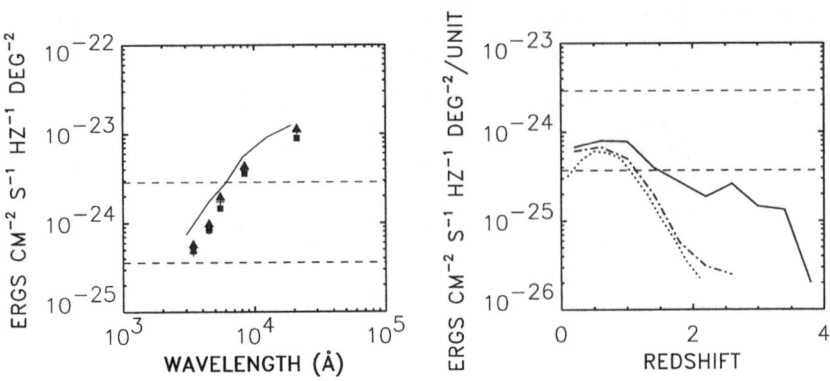

Figure 4: The left panel shows the extragalactic background light (EBL) contribution of the $I < 26$ galaxy sample in the HDF as a function of wavelength (solid line). Also shown is the same quantity computed for a $K < 22$ sample (triangles), an $I < 24.5$ sample (crosses) and a $B < 25.5$ sample (boxes) in the Hawaii survey fields.[25] The dashed lines show the expected UV surface brightness required to form the local metal density with the range reflecting the uncertainty in the metal density estimates.[17, 26] The right-hand panel shows the evolution of the EBL per unit redshift at a wavelength of $3000(1+z)$Å versus redshift in the $I < 23$ spectroscopic sample of the Hawaii fields (dotted line), in the $I < 23$ color estimated sample in the HDF (dot-dash line), and in the $I < 26$ color estimated sample in the HDF (solid line).

ing the wavelength coverage into the near IR, and we have used additional J and $H + K$ colors to make a six-color estimate. These may be found at **http://www.ifa.hawaii.edu/~cowie/hdf.html**. The right-hand panels of Fig. 3 combine these color estimators with spectroscopic data on the HDF to expand the magnitude-redshift relation. This may be compared with the gravitational lensing estimates discussed by Mellier in this volume.[24]

With the exception of a scattering of AGN which lie above the curves, we may see that even at faint magnitudes and high redshifts galaxies remain at or below L*.

4 The history of galaxy formation

The integrated extragalactic background light (EBL) from the galaxies in the HDF is shown over the 3000Å–20000Å range in the left panel of Fig. 4 where it is compared with the integrated EBL in B-, I-, and K-selected samples from the Hawaii Galaxy Survey fields. The extra depth of the HDF does not greatly increase the EBL, although the shape is slightly bluer, since the number counts strongly converge in the red light density, while in the blue they only logarithmically diverge.

The ultraviolet EBL is a direct measure of the metal density production which may be compared with the local value of the metal density in the universe, whose rather uncertain range is shown by the dashed lines.[17, 26] When sliced by redshift it therefore produces a direct history of the star formation in the universe. This method has a considerable advantage over the method of constructing the UV light density as a function of redshift, which was discussed by a number of speakers at the conference.[27] The EBL technique measures the integrated production of metals either in total or in a given redshift interval independent of cosmology. By contrast, the UV luminosity density measures the production rate and is sensitive to the cosmology adopted. It must also be integrated back to study the total production.

As has been known for some years there is enough ultraviolet light to account for most of the current galaxies in relatively recent formation. The spectroscopic or color-estimated redshift distributions allow us to expand this result by slicing in the redshift direction. In the right-hand panel we show the EBL per unit redshift at a wavelength of $3000(1+z)$Å as a function of redshift. The dotted line shows this for the $I < 23$ sample in the Hawaii Survey Fields using full spectroscopic data, while the dashed-dotted line shows the $I < 23$ sample for the HDF using the color-estimated redshifts. The agreement is satisfactory. The solid line then shows the $I < 26$ sample from the HDF. As can be seen by comparing the $I < 23$ sample with the $I < 26$ sample, the $I < 23$ sample maps the ultraviolet EBL well to just beyond $z = 1$ so that the $I < 26$ sample should be good to around $z = 3$. The fall-off in the $I < 26$ curve is therefore real and most star formation has taken place at $z \lesssim 1.5$. As can be seen from the thin dashed lines, the total amount of star formation agrees with the present metal density range. At the present redshift the larger area Hawaii Survey suggests that the star formation rate has begun to turn down.

From Fig. 4 we can see that most of the star formation occurs after $z = 1.5$ though it is probably beginning to die out by $z = 0$. As we have seen in section 3, below $z = 1$ most of this star formation is in relatively evolved galaxies.[13] The redshift range for the formation matches that in which the baryons in the quasar absorption line systems vanish and is also broadly consistent with the general history of chemical evolution and star formation in our own Galaxy.[17, 28]

74

Acknowledgments

This work was primarily supported by grants AR-06377.06-94A, GO-05399.01-93A and GO-05922.01-94A from STScI. I would like to thank Richard Ellis for supplying the Autofib data in tabular format and Esther Hu for assistance.

References

1. E.M. Hu, R.G. McMahon and E. Egami, ApJ, 459, L53, 1996.
2. E.M. Hu and R.G. McMahon, Nature, 382, 231, 1996.
3. S.S. Vogt, et al., Proc. SPIE, 2198, 362, 1994.
4. L.L. Cowie, A. Songaila, T.-S. Kim and E.M. Hu, AJ, 109, 1522, 1995.
5. D. Tytler, et al., in *QSO Absorption Lines,* ed. G. Meylan, (Springer, Heidelberg), p. 289, 1995.
6. D.M. Meyer and D.G. York, ApJ, 315, L5, 1987.
7. D. Tytler, ApJ, 321, 49, 1987.
8. A. Songaila and L.L. Cowie, AJ, 112, 335, 1996.
9. A. Songaila, in preparation, 1996.
10. J. Bergeron and G. Stasińska, A&A, 169, 1, 1986.
11. L. Lu, these proceedings.
12. M.J. Rees, these proceedings.
13. L.L. Cowie, A. Songaila, E.M. Hu and J.G. Cohen, AJ, in press, 1996.
14. R.S. Ellis, et al., MNRAS, 280, 235, 1996.
15. S.J. Lilly, F. Hammer, O. LeFèvre and D. Crampton ApJ, 455, 75, 1995.
16. R.C. Kennicutt, Jr., ApJ, 388, 310, 1992.
17. L.L. Cowie, in *The Post-Recombination Universe,* eds. N. Kaiser and A.N. Lasenby, (Kluwer, Dordrecht), p. 1, 1988.
18. K. Glazebrook, et al., MNRAS, 266, 169, 1995.
19. A. Songaila, L.L Cowie, E.M. Hu, and J.P. Gardner, ApJS, 94, 461, 1994.
20. J.G. Cohen, L.L. Cowie, D.W. Hogg, A. Songaila, R. Blandford, E.M. Hu and P. Shopbell, ApJ, in press, 1996.
21. C.C. Steidel, et al., AJ, 112, 352, 1996.
22. S.D.J. Gwyn and F.D.A. Hartwick, ApJ, 468, L77, 1996.
23. K.M. Lanzetta, A. Yahil and A. Fernández-Soto, Nature, 381, 759, 1996.
24. Y. Mellier, these proceedings.
25. L.L. Cowie, et al., ApJ, 434, 114, 1994.
26. A. Songaila, L.L. Cowie, and S.J. Lilly, ApJ, 348, 371, 1990.
27. P. Madau, H.C. Ferguson, M.E. Dickinson, M. Giavalisco, C.C. Steidel and A. Fruchter, MNRAS, submitted.
28. M. Fall, these proceedings.

GALAXY COUNTS, COLOURS AND REDSHIFTS IN THE HUBBLE AND HERSCHEL DEEP FIELDS

T. SHANKS, N. METCALFE, A. CAMPOS, R. FONG, J.P. GARDNER

Department of Physics, Univ. of Durham, South Road,
Durham, DH1 3LE, England

Here we present the deepest, ground- and space-based, galaxy counts so far obtained. The space-based data consists of galaxy counts from the Hubble Deep Field to $\approx 29^m$ in the B,R,I bands [1]. The ground-based data consists of an $\approx$ 50hr CCD observation [2] at the 4.2m William Herschel Telescope to $B = 28.^m2$ and an $\approx$ 30hr IRCAM3 observation with the 3.8m UK Infrared Telescope to $K = 22.^m75$ [4]. By modelling the counts and colours of galaxies at the faintest detection limits, we show that in both open and flat cosmologies, the faint blue galaxies are likely to lie at high redshift ($z \approx 2$). In an open cosmology, we find that models where galaxies evolve to have an increased star-formation rate at high redshift are consistent with the observational data. In a flat cosmology, such models require an extra population of galaxies which are only seen at high redshift and then rapidly fade or disappear.

1 Galaxy Count Data

We have measured the galaxy counts in the Deep Field [1] recently observed by the Hubble Space Telescope (HST) to $B = 29^m$, $R = 28.^m5$ and $I = 28^m$. The exposure times here were $\approx$ 25 hrs in each band. The ground-based observations comprise a 50hr B CCD exposure to $B = 28.^m2$ made on a random field with the William Herschel Telescope (WHT) [2,3]. In this field we have also obtained a 30hr IRCAM3 exposure with UKIRT to $K = 22.^m75$ [4].

We show in Fig. 1 our derived B, I and K galaxy number counts from both datasets, together with those from other work. The HST B counts extend one magnitude deeper than the WHT data and they are in good agreement where they overlap. The HST I counts extend about 10× deeper than previous data, because of the higher HST resolution and the fainter background sky. The ground-based K counts now also appear to be reasonably well defined in the range $15^m < K < 24^m$.

2 Count Models

As noted previously [6,2,5], if the B count models are normalised at $B \approx 18^m$ rather than $B \approx 15^m$ then non-evolving models give a reasonable representation of the B band counts and redshift distributions in the range $18^m < B < 22.^m5$. This high normalisation has recently received new support from HST

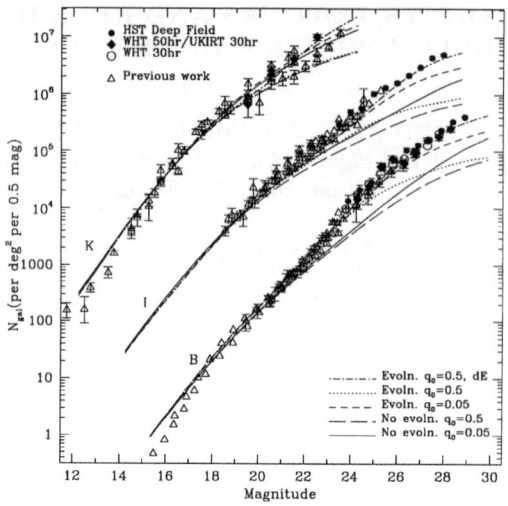

Figure 1: The B and I (×10) galaxy counts from the Hubble Deep Field compared to previous counts and various models. Also shown are deep K counts (×100) from ground-based data. The evolutionary models are Bruzual & Charlot[13] exponential models with $\tau = 2.5$ Gyr, for E/S0/Sab galaxies and $\tau = 9$ Gyr for Sbc/Scd/Sdm galaxies. The fit of the $q_0 = 0.5$ model is improved when an extra high redshift galaxy population (dE) with constant SFR at $z > 1$ and rapidly fading at $z < 1$ is invoked. The K galaxy counts are well fitted by non-evolving models.

galaxy counts subdivided by morphology where non-evolving models with the high normalisation give an excellent fit to both spiral and early type galaxy counts with $17^m < I < 22^m$ [7,8]. Also the high normalisation allows non-evolving models with $0.05 < q_0 < 0.5$ to fit the K counts for $15^m < K < 24^m$.

Simple evolutionary models where galaxy star-formation rates rise exponentially with look-back time, are also known to fit the B counts in the range $18^m < B < 25^m$ [6,2,5]. But previous faint galaxy redshift surveys at $B < 24^m$ presented a problem for such models, as they predict a high redshift tail which was unobserved in these surveys [9]. However, these surveys were usually only $\approx 60\%$ complete and the galaxies with unidentified redshifts were frequently blue and could be at high redshift. Recently Cowie et al [10,11] have used the Keck 10m telescope to make a new $22.^m5 < B < 24^m$ galaxy redshift survey with $> 80\%$ completeness and have detected such a high redshift galaxy component from a Keck galaxy redshift survey to $B = 24^m$, supporting the basic viability of these models. An extended high redshift tail is also consistent with the low galaxy clustering amplitude observed at $23^m < B < 28^m$ [12].

In respect of the evolutionary models, a further new ingredient is that

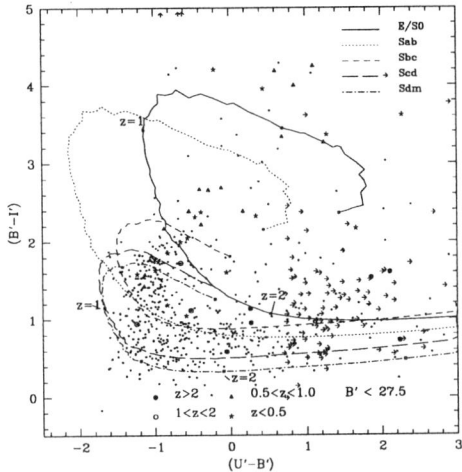

Figure 2: Dots represent the $U' - B' : B' - I'$ colours of $B' < 27.^m5$ galaxies in the Hubble Deep Field. The predicted tracks are for our $q_o = 0.05$ models. The colours of 45 brighter galaxies with Keck spectroscopic redshifts are also shown and these agree well with the predicted colours for these galaxies. It can also be seen that $U' - B' < 0$ is predicted to correspond to galaxies with $z < 2$ and $U' - B' > 0$ to galaxies with $z > 2$.

spiral galaxies with exponentially increasing star-formation rates, of time scale $\tau \approx 9Gyr$, are now thought to be able to sustain large amounts of B band evolution [13,14]. These spiral evolutionary models also have the advantage that evolving the steeper luminosity function of late-type galaxies [3,15] makes it easier to increase the B number counts using lower redshift galaxies, particularly if they contain even small amounts of dust [16,14].

In the low q_o case, this spiral dominated model then produces a reasonable fit to the optical counts to $B \approx 27^m$ and $I \approx 26^m$ (see Fig. 1). In the $q_o = 0.5$ case, the spiral luminosity evolution model only fits the optical data to $B \approx 25^m$ and $I \approx 23.^m5$ and then more seriously underestimates the counts at fainter magnitudes. Thus, the HST data confirms [17,6,2,3] that if $q_o = 0.5$, that there is not enough spatial volume at high redshifts to allow simple luminosity evolution models to fit the counts at $B > 25^m$.

To improve the fit of the $q_o = 0.5$ model, we consider a model with an extra population of high redshift galaxies which have a constant star-formation rate from the formation epoch till z=1 ($\approx 4Gyr$ after formation with our assumed H_o). This 'dE' model is in the spirit of previously proposed 'disappearing dwarf' count models [18] and it gives a good fit to the faint B,I,K counts and

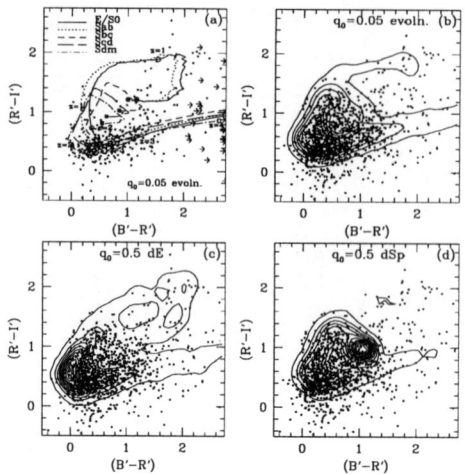

Figure 3: (a) The $q_o = 0.05$ evolutionary models' $B' - R' : R' - I'$ tracks with redshift. The dots indicate the colours of galaxies with $R' < 27.^m5$ and $U' - B' > 0$ which are predicted to have $z > 2$. (b) Dots represent the $B' - R' : R' - I'$ colours of $R' < 28^m$ galaxies in the Hubble Deep Field. The contours represent the prediction for the $q_o = 0.05$ evolutionary model, which are consistent with the data. (c) As Fig. 3b for the $q_o = 0.5$ 'disappearing dwarf' (dE) model, which are also in reasonable agreement with the data. (d) As Fig. 3b for the $q_o = 0.5$ low redshift dwarf (dSp) model, which is rejected by the data.

the Keck n(z) data.

3   Model Constraints from Galaxy Colours and Redshifts

We next test these models against the faint galaxy colour distributions in the Hubble Deep Field. We see in Fig. 2 that our predicted $U' - B' : B' - I'$ model tracks compare well with the observed colours for $B' < 27.^m5$ galaxies. (Primed letters here denote the natural HST magnitude system). In particular, the redshifting of the Lyman α forest/break [19] through the U' band causes the model $U' - B'$ colours to move sharply redwards at $B' - I' \approx 0.^m7$ and the same effect is clearly seen in the data.

The above predictions show that, for the majority of faint galaxies, $U' - B' < 0$ is predicted to correspond to $z < 2$ galaxies and $U' - B' > 0$ corresponds to $z > 2$ galaxies. We find that the proportion of galaxies with $U' - B' > 0$ rises to $47 \pm 7\%$ of the total at $27^m < B < 28^m$, indicating that the redshift distribution may peak at $z \approx 2$. This fraction is matched very well by both the $q_o = 0.05$ model which predicts 47% with $U' - B' > 0$ at the same limit and the $q_o = 0.5$, disappearing dwarf model which predicts 43%. We have also considered another $q_o = 0.5$ model which assumes an extra population of low redshift dwarf spirals (dSp) which evolve more slowly according to our standard exponential model for spiral luminosity evolution [13]. However, this model predicts too few high redshift $(U' - B' > 0)$ galaxies (28%) for compatibility with the faint HST data.

These conclusions are confirmed by consideration of the $B' - R' : R' - I'$ colour-colour plot in Fig. 3. Fig. 3a shows the predicted tracks of the galaxy types with redshift. Also plotted are the galaxies with $U' - B' > 0$ and $R' < 27.^m5$; these are expected to have $z > 2$ by the above arguments and it can be seen that their position on the $B' - R' : R' - I'$ tracks is entirely consistent with their lying in this redshift range. Figs. 3b,c,d then show the HST data at $R' < 28^m$ compared to the predicted galaxy number contours for the various models. Both the $q_o = 0.05$ and the $q_o = 0.5$ dE model contours give a reasonable fit to the data which seems to peak at $B' - R' \approx 0.3$, $R' - I' \approx 0.3$, corresponding to $z \approx 2$ for all galaxy types. However, the $q_o = 0.5$ dSp model contours peak away from this point at $B' - R' \approx 1$, $R' - I' \approx 1$ which corresponds to $z \approx 0.5$ for the dwarf spiral galaxies and we again conclude that the galaxy redshift distribution in this model is skewed to too low redshifts to be compatible with the colour data.

References

1. R.E. Williams et al., AJ, in press.
2. N. Metcalfe, T. Shanks & R. Fong, in prep.
3. N. Metcalfe, T. Shanks, R. Fong & N. Roche, MNRAS, 273, 257, 1995.
4. J. Gardner, T. Shanks, N. Metcalfe, A. Campos & R. Fong, in prep.
5. N. Metcalfe, T. Shanks, R. Fong & L.R. Jones, MNRAS, 249, 481, 1991.
6. D.C. Koo & R.G. Kron, ARA&A, 30, 613, 1992.
7. K. Glazebrook, R.S. Ellis, B. Santiago & R.E. Griffiths, MNRAS, 275, L19, 1995.
8. S.P. Driver, R.A. Windhorst, E.J. Ostrander, W.C. Keel, R.E. Griffiths & K.U. Ratnatunga, AJ, 449, L23, 1995.
9. K. Glazebrook, R.S. Ellis, M. Colless, T.J. Broadhurst, J.R. Allington-Smith & N.R. Tanvir, MNRAS, 273, 157, 1995.
10. L.L. Cowie, E.M. Hu & A. Songaila, Nature, 377, 603, 1995.
11. L.L. Cowie, A. Songaila, E.M. Hu & J.G. Cohen, AJ, in press.
12. N. Roche, T. Shanks, N. Metcalfe & R. Fong, MNRAS, 263, 360, 1993.
13. A.G. Bruzual & S. Charlot, ApJ, 405, 538, 1993.
14. A. Campos & T. Shanks, MNRAS, submitted.
15. T. Shanks in The Galactic and Extragalactic Background Radiations, eds. S. Bowyer & C. Leinert (Kluwer, Dordrecht, 1990).
16. C. Gronwall & D.C. Koo, ApJ, 440, L1, 1995.
17. Y. Yoshii & F. Takahara, ApJ, 326, 1, 1988.
18. A. Babul & M.J. Rees, MNRAS, 255, 346, 1992.
19. P. Madau, ApJ, 441, 18, 1995.

HST OBSERVATIONS OF FIELD GALAXIES TO $z \sim 1$: CLUES TO GALAXY EVOLUTION

O. LE FÈVRE[1], R.S. ELLIS[2], S.J. LILLY[2,3], B. ABRAHAM[2], J. BRINCHMAN, D. SCHADE[3], T.J., BROADHURST[4], M. COLLESS[5], D. CRAMPTON[6], K. GLAZEBROOK[7], F. HAMMER[1], L. TRESSE[2]

[1] *DAEC, Observatoire de Paris-Meudon, France*
[2] *Institute for Astronomy, Cambridge, UK*
[3] *Department of Astronomy, University of Toronto, Canada*
[4] *Astronomy Dept., University of California, Berkeley, USA*
[5] *Mount Stromlo and Siding Spring Observatories, Australia*
[6] *Dominion Astrophysical Observatory, Victoria, Canada*
[7] *Anglo Australian Observatory, Australia*

The Hubble Space Telescope has a unique role to play to understand galaxy evolution. Following the great progress in deep ground-based redshift surveys, the morphological information provided by HST is of considerable importance to evaluate the evolution in the mix of galaxy types, in disc properties, or the role of interactions and mergers. We discuss here the current status of deep redshift surveys up to $z \sim 1.2$, and the on-going follow-up imaging survey conducted with the WFPC2 on HST. Two redshift surveys, the Canada-France Redshift Survey and the LDSS/Autofib survey, have been combined to create a unique data set for a detailed imaging survey with HST. First results on the evolution of the morphological mix and the merger/interaction rate are presented.

1 Introduction

Considerable progress has been made in understanding the evolution of field galaxies. Several deep redshift surveys with well controlled selection criteria and observational biases have allowed to quantify in a comprehensive way the evolution of various populations of galaxies out to redshifts ~ 1.2 from samples of several hundred galaxies (Lilly et al., 1995, Ellis et al., 1996, Cowie et al., 1996). There is now ample evidence that the evolution of galaxies is indeed taking place over one half the current age of the universe. The rest frame B band luminosity function has been determined as a function of redshift up to $z \sim 1$ and shows unambiguously signs of evolution. The mean comoving luminosity density of the field galaxy population at 2800Å has declined by an order of magnitude since a redshift ~ 1 (Lilly et al., 1996a).

At a redshift ~ 1, 1 arcsecond is 8.5 h^{-1}kpc (H_0=50, q_0=0.5). This can be directly compared to the very best seeing obtained from ground based observatories, reaching 0.5 arcseconds FWHM or better at sites like the Mauna Kea. Ground based imaging therefore has great difficulties to identify the morphology of galaxies at such large distances. It is possible to retrieve some infomation on the disc to bulge ratio from the imaging data of the Canada-France Redshift Survey (Schade et al., 1996). However, much has to be gained from the unprecedented spatial resolution provided by the Hubble Space Telescope. In this paper, the recent progress in our understanding of galaxy evolution from $z \sim 1$ is illustrated. The first results of a systematic HST/WFPC2 imaging survey of galaxies with known redshifts in the CFRS and LDSS samples are presented. The contribution from mergers to the evolution of the galaxy population is evaluated. We conclude with prospects for the future.

2   The evolution of galaxies from $z \sim 1$

The recent progress in understanding the evolution of galaxies is linked to the progress in multi-object spectrographs which have allowed to collect unprecedently large samples of faint galaxies with securely measured redshifts (e.g. Le Fèvre et al., 1994). This has open the era of *statistical* investigations at large redshifts, working on sufficiently large samples so that the average properties of the field galaxy population can be described with good accuracy. The success of deep redshift surveys such as the Canada-France Redshift Survey rests on the careful multi-step approach, from the well controlled deep CCD photometry and objet selection from purely magnitude limited samples (Lilly et al., 1995a), to the assignment of the probability for a redshift measurement to be in error (Le Fèvre et al., 1995). In some ways, deep redshift surveys are less affected by unknow or uncontrolled bias effects than some of the first generation local wide-angle redshift surveys.

Galaxy evolution from $z \sim 1$ is now a secure observational fact. Much of the evolutionary properties of the galaxies are summarized in building the luminosity function (LF), with good accuracy (large samples). The LF derived from the CFRS shows a strong evolution for the blue selected galaxies, while the LF of "red" galaxies does not evolve much (Lilly et al., 1995b). This agrees with the surveys of Ellis et al. (1996) and Cowie et al. (1996), which also identify the strongest evolution in the star-forming population. The mean rest frame luminosity density in the CFRS is seen to decrease by an order of magnitude since a redshift of 1 (Lilly et al., 1996). Explaining the changes observed in the LF is by no means an easy task, and identifying the role played by pure luminosity evolution, vs. number density evolution, or their relative contributions, requires quantitative measurements, like the mean central surface brightness of discs, or the merger/interaction rate, vs. redshift, best obtained from HST imaging.

The clustering properties have also been shown to evolve. The CFRS has

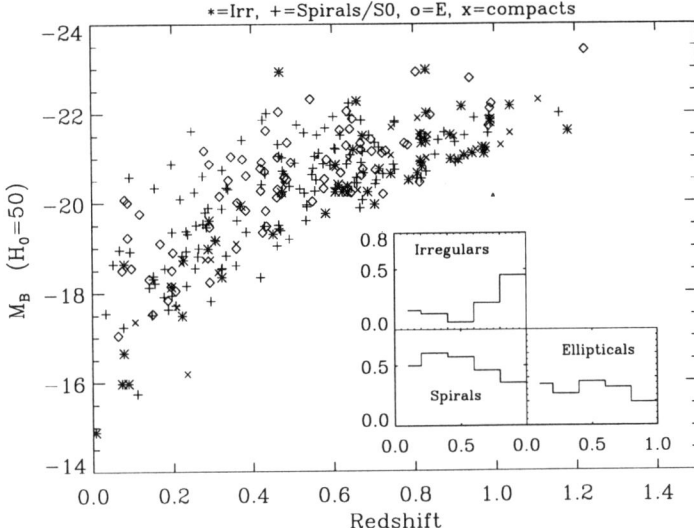

Figure 1: Absolute blue magnitudes ($H_0=0$, $\Omega = 0.1$) and redshifts for 320 field galaxies from the combined CFRS+LDSS HST follow-up. Morphological classifications in the scheme [E/S0: Spiral: Late-type/Irr Compact] are indicated. The inset histograms show the relative fractions of different types as a function of redshift and indicate little evolution for massive regular E/S0 and spiral galaxies, but a marked increase with redshift for later type galaxies.

allowed to compute the spatial correlation function out to $z \sim 1$ (Le Fèvre et al., 1996): at $z \sim 0.6$, the correlation amplitude of L_* galaxies is 10 times lower than at $z \sim 0$ (Le Fèvre et al., 1996). The evolution has been confirmed at lower redshift $z \sim 0.3$ (Shepherd et al., 1996), and up to $z \sim 1.5$ from the smaller sample of the Hawaii Deep Survey (Carlberg et al., 1996).

3 The contribution from HST

With the second version of the Wide Field and Planetary Camera, the sensitivity and the pixel scale of 0.1 arcsec/pix are very well suited to identify the morphological type of galaxies at redshifts up to ~1. In 2h, S/N=1 is reached for $\mu_{I,AB} = 25.4$ mag/arcsec^2, compared to a central surface brightness $\mu_{I,AB} = 22.5$ at z=0.6, after scaling from local galaxy discs (Freeman's law) *without* luminosity evolution. This allows to successfuly detect discs and identify morphological signatures, in a similar way to what can be done from the ground on galaxies in the Coma cluster. This has been shown spectacularly from the Hubble Deep Field images, described elsewhere in this volume.

The morphology of distant galaxies provides information on three aspects of the evolution of the galaxies. First, quantitative fitting of the two dimensional light profiles allows a separation of the bulge and disc components of normal spiral galaxies. We can see whether the disks of normal galaxies evolve mainly in surface brightness or scale length as discriminator between different scenarios for how discs are built up over time. The surface brightness of discs as a function of inclination angle tells us whether disks are optically thick at high redshifts and constrains the size of any population of low surface brightness galaxies. Second, comparison of images in two wavebands located above and well below the 4000Å region of the spectrum can isolate regions of recent star formation, or other activity, which will generaly be much bluer than other regions. This can tell us about the nature of star formation in high redshift galaxies, either in star-burst galaxies or in more quiescent systems. Finally, the incidence of mergers and interactions can be determined and the effects of this activity on the galaxies studied. The high spatial resolution offered by the HST makes it the only facility that can address these questions.

The availability of HST images for faint field galaxies has emphasised the important role that small late type galaxies must play in recent evolutionary history. Using the Medium Deep Survey, Glazebrook et al. (1995) and Abraham et al. (1996), found that galaxy number-magnitude counts for regular ellipticals and spirals are broadly compatible with no evolution expectations to beyond the limits of the ground-based redshift surveys. However, they found an increasing proportion of small irregular galaxies at faint limits, suggesting these galaxies are responsible for the faint blue excess. Schade et al. (1995) have shown that the disks of regular spirals have surface brightnesses about 1 mag brighter than the local "Freeman" value.

4   HST imaging of a combined sample of CFRS–LDSS galaxies

The interpretation of morphological data requires knowledge of the redshifts of the galaxies. The redshift determines not only the cosmic epoch and the rest frame wavelength of the observations, but also the spatial extent, luminosity and surface brightness, which all have strong dependancies on redshift. Field samples exhibit a wide range of redshifts at any given apparent magnitude, making redshift information essential.

Our combined CFRS and LDSS surveys currently consist of some 200 B-selected galaxies ($B \leq 24$, LDSS), 700 I-selected galaxies ($I \leq 22$, CFRS), and 100 K-selected galaxies ($K \leq 20$, LDSS). The redshift range is $0.1 < z < 1.3$, and a median redshift $< z > = 0.6$, corresponding to a look-back time of

50% of the age of the universe. The CFRS survey corresponds most closely to a selection in the rest frame visual waveband at $0.4 < z < 1.0$, and this sample therefore allows the most direct and unambiguous comparison with the local galaxy population. On the other hand, the B-selected selection is weighted at high redshifts towards the star-forming objects that are undergoing the strongest evolution, whilst the K-band selection is based primarily on the properties of the oldest, least evolving stellar populations within each galaxy.

The HST observations have been caried out in cycles 4, 5 and 6. Images have been acquired in I (F814W) and B (F450W), for a minimum of 7200 and 6900 sec. exposure times respectively. A total of $\sim$ 300 galaxies have been observed, and the data processed, while the data for another $\sim$ 70 galaxies is being processed.

Galaxy types have been visualy estimated separately by three of us (RSE, OLF, SJL). The methods used, including automated computer classification, will be described in Brinchman et al (1996, in preparation). Figure 1 shows the evolution in the type mix with redshift, for galaxies in the broad categories ellipticals, spirals, irregulars/peculiars. It is immediately evident that the fraction of irregular/peculiar galaxies is strongly increasing with redshift. This substantiates the claim made from the deep HST counts that the higher fraction of irregular/star forming galaxies observed at fainter magnitudes (Abraham et al., 1996), is coming from galaxies at high redshift, and not from a relatively nearby population of dwarf galaxies.

5   Contribution of merging to the evolution

The effect of mergers of galaxies has long been recognised to be a possibly important process in the evolution of galaxies. Detailled simulations have shown that mergers can affect the morphology of galaxies as well as temporarilly boost the star formation rate, from either major mergers of galaxies with comparable mass, or even from the merger of dwarf galaxies with a more massive galaxy (Mihos&Hernquist, 1994a,b; Hernquist&Mihos, 1995; Mihos, 1995). While we know that *some* luminosity evolution must have occured in field galaxies from $z \sim 1$, it is not clear at all if luminosity changes is the sole responsible in the evolution observed, and if number density evolution through mergers is playing a role. One difficulty is that it might be very rare to observe galaxies in the process of merging. The index for the evolution of the fraction of galaxy pairs P_f, has been estimated to be in the range $1.2 < m < 3$, with $P_f \propto (1 + z)^m$ (Burkey et al., 1994, Carlberg et al., 1994, Yee&Ellingson, 1995, Patton et al., 1996, Neuschaefer et al., 1996). Most of the samples used in these studies face either the lack of redshift measurements for the galaxies studied, small samples, or relatively low redshift.

To identify the role played by interactions and mergers in our sample, we have investigated the environement of the galaxies with know redshifts, within a fixed radius of 20 h^{-1}kpc around each galaxy. As the morphological signatures of major or minor mergers may disappear relatively quickly ($\sim 10^8$ years, Mihos and Hernquist, 1994; Liu and Kennicutt, 1995), it might prove unreasonable to try to look for merger remnants. It seems more appropriate to compute the rate of "upcoming" and "on-going" mergers. We have defined "major mergers" as systems for which at least one galaxy is identified within 20 kpc of the galaxy with known redshift, and the magnitude difference between the two galaxies is less than 1.5 magnitude. This includes systems which should be merging within less than 0.5Gy or so (upcoming mergers), or systems in the process of merging, with evident double nuclei, strong assymetries, wisps, and tails. In addition, we have computed the number of companion galaxies within a radius 20 h^{-1} kpc of each galaxy with redshift, with a magnitude difference larger than 1.5 mag from the primary galaxy, and with an absolute magnitude $M_B \leq -17.5$, corresponding to the depth of the HST imaging data at all redshifts, and estimated assuming the same k-correction as for the primary galaxy. Both these numbers have been corrected from the expected background/foreground contamination, estimated from the deep counts measured in the HDF (Abraham et al., 1996). In addition, we have developped an algorithm based on the Lee statistic to identify upcoming and on-going major mergers in an automated way (Abraham et al., in prep.).

It is striking that more than half of the major mergers identified in our sample are at $z > 0.8$, while the median redshift of the sample is z=0.55. Figure 2 shows the major mergers identified for $z > 0.8$. Figure 3 plots the evolution in the number of major mergers vs. redshift. A strong increase in the number of major mergers is observed for $z > 0.8$. The Lee statistic agrees with the visual classification, as shown in figure 3. as the number of galaxies with large Lee numbers is clearly larger at $z > 0.8$. In addition, the average number of faint companions around a bright galaxy is also showing a strong increase for $z > 0.8$. These preliminary results seem to indicate that interactions/mergers are playing an increasingly important role at $z > 0.8$. A detailled analysis will be presented elsewhere (Le Fèvre et al., in prep.)

6 Prospects

HST imaging will remain a crucial tool in building up our understanding of galaxy evolution. The availability of large samples is certainly the main limitation to overcome in the following years, and concerted ground-based redshift

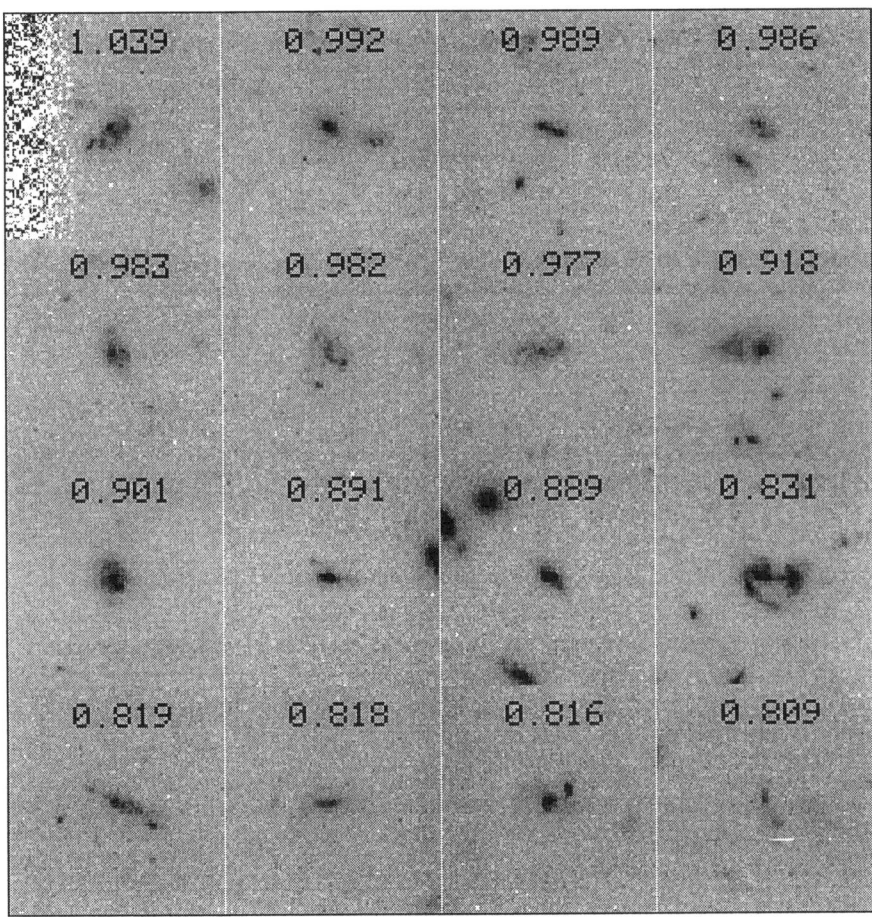

Figure 2: 16 of the 18 major mergers identified in the subsamples of 220 CFRS galaxies in the HST CFRS-LDSS, imaging survey, at $z > 0.8$. This corresponds to half of the total number of major mergers in the sample. The redshift of each galaxy is indicated at the top of each 10×10 arcsec^2 image.

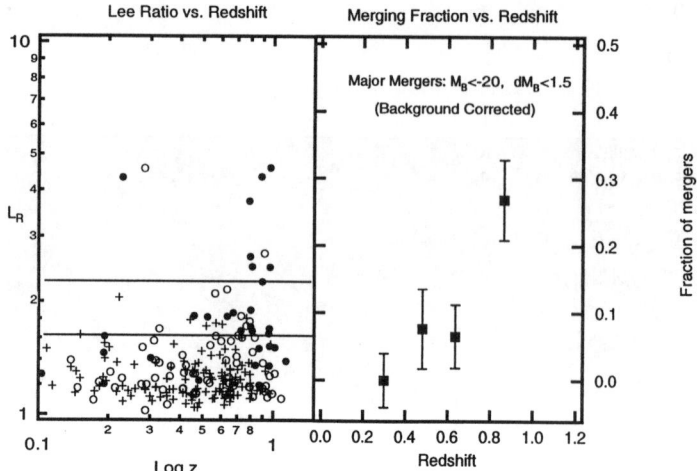

Figure 3: (*Left*) the "Lee-ratio" (objective measure of the lumpiness of an image within an isophote) plotted agaisnt redshift for all sources in the CFRS/LDSS combined catalog. Very high values indicate strong evidence for a "major merger" (solid symbols are classified by eye as mergers); (*right*) the fraction of galaxies brighter than $M_B = -20$ with a projected companion within $20h_{50}^{-1}$ kpc and within 1.5 mag of the primary as a function of redshift, corrected for foreground contamination.

surveys and HST imaging efforts are to be further developped. Although current samples are reaching several hundred galaxies with both redshifts and HST images, the necessity to isolate samples of galaxies with various types, luminosities, and environments, calls for samples with many thousand of galaxies. This will be a challenge to both ground based instrumentation and the HST narrow field of view.

The proposed Visible Infra Red Multi Object Spectrograph (VIRMOS) for the ESO-VLT (Le Fèvre et al., 1996) will enable to assemble a sample of 1.5×10^5 galaxies with measured redshifts in the range $0 < z < 3.5$ or more. This will be possible with an imaging spectrograph working in the 0.5 to 1.8 μm domain, and multi-slit capabilities allowing to observed up to 800 galaxies simultaneously. The imaging follow-up, at high spatial resolution, of such surveys in areas of several square degrees, will be pushing the limits of the HST. A wide field of view, such as currently envisaged for the NGST, will be necessary to ensure a comprehensive approach to better understand galaxy formation and evolution.

References

1. Abraham, R.G., Tanvir, N.R., Santiago, B., Ellis, R.S., Glazebrook, K., van den Bergh, S., 1996, MNRAS, 279, L47
2. Burkey. J., et al., 1994, ApJ, 429, L13
3. Carlberg, R., Cowie, L.L., Songaila, A., Hu, E., astro-ph/9605024
4. Carlberg, R., Pritchet, C., Infante, L., 1994, ApJ, 435, 540
5. Cowie, L.L., Hu, E., Songaila, A., Cohen, J., 1996, AJ, in press
6. Ellis, R.S., Colless, M., Broadhurst, T.J., Heyl, J.S., Glazebrook, K., 1996, MNRAS, 280, 235
7. Glazebrook, K., Ellis, R.S., Santiago, B., Griffiths, R.E., 1995, MNRAS
8. Hernquist, L., Mihos, J.C., 1995, ApJ, 448, 41
9. Le Fèvre, O., Crampton, D., Felenbok, P., Monnet, G., 1994, A&A, 282, 325
10. Le Fèvre, O., Crampton, D., Lilly, S.J., Hammer, F., Tresse, L., 1995, Ap.J., 455, 60
11. Le Fèvre, O., Hudon, D., Lilly, S.J., Crampton, D., Hammer, F., Tresse, L., 1996, Ap.J., 461, 534
12. Le Fèvre, O., Vettolani, Vettolani, P., Cuby, J.G., Cappacioli, M., Maccagni, Mancini, D., D., Mazure, A., Picat, J.P., proc. ESO conf. "Early Universe wit the VLT", Springer, in press
13. Lilly, S.J., Le Fèvre, O., Crampton, D., Hammer, F., Tresse, L., 1995, Ap.J., 455, 50
14. Lilly, S.J., Tresse, L., Hammer, F., Crampton, D., Le Fèvre, O., 1995, Ap.J., 455, 108
15. Lilly, S.J., Le Fèvre, O., Hammer, F., Crampton, D., Ap.J., 1996, 460, L1
16. Lui, C.T., Kennicutt, R.C., 1995, ApJ, 450, 547
17. Mihos, J.C., 1995, ApJ, 438, L75
18. Mihos, J.C., Hernquist, L., 1994, ApJ, 425, L13
19. Mihos, J.C., Hernquist, L., 1994, ApJ, 431, L9
20. Neuschaefer, L.W., et al., ApJ, astro-ph/9609090
21. Patton, D.R., Pritchet, C.J., Yee, H.K.C., Ellingson, E., Carlberg, R.G., ApJ, astro-ph/9608016
22. Schade, D., Lilly, S.J., Le Fèvre, O., Hammer, F., Crampton, D., 1996, Ap.J., 464, 79
23. Schade, D., Lilly, S.J., Crampton, D., Le Fèvre, O., Hammer, F., 1995, Ap.J., 451, L1
24. Shepherd et al., 1996, ApJ, astro-ph/9601014
25. Yee, H.K.C., Ellingson, E., 1995, ApJ, 454, 32

GALAXIES AT z > 4.5

ESTHER M. HU

Institute for Astronomy, University of Hawaii, 2680 Woodlawn Dr.,
Honolulu, HI 96822, USA

RICHARD G. MCMAHON

Institute of Astronomy, Madingley Road, Cambridge CB3 0HA, UK

EIICHI EGAMI

Max-Planck-Institut für extraterrestriche Physik, Postfach 1603,
85740 Garching bei München, GERMANY

We present the results of ground-based and HST imaging studies targeted on $z > 4.5$ quasar fields. High-redshift galaxies identified in deep narrow-band Lyα images of the fields surrounding the quasars BR1202–0725 ($z = 4.694$) and BR2237–0607 ($z = 4.558$) have been confirmed with follow-up spectroscopy using the LRIS spectrograph on the Keck 10m telescope. These high-redshift galaxies are typically only moderately luminous objects and their sub-L^* magnitudes imply star formation rates of only a few solar masses per year. For BR1202–0725 tip-tilt imaging in Lyα at the UH 2.2m telescope and deep narrow-band IR imaging in [O II] at CFHT reveals a complex and structured system which may be merging to produce the host galaxy of the quasar. For BR2237–0607 the Lyα-emitting galaxies show the strong Lyα-emission and weak continuum expected for objects before substantial chemical enrichment and consequent dust formation and extinction has occurred. These objects appear to be galaxies in their first outburst of star formation.

1 Introduction

It is clear that some form of activity took place prior to $z \sim 5$ to generate the ionization of the intergalactic medium[1] and also that some quasar absorption-line systems have small amounts of metals in place[2] by $z = 4–5$, indicating that some galaxy formation has initiated at or prior to these redshifts. These early epochs may be our best chance to see galaxies in the first stages prior to the formation of substantial dust and metals, and hence may be the best place for Lyα emission-line searches. While earlier Lyα searches were unsuccessful,[3,4] improved sensitivities – particularly with the Keck 10m telescope – are now enabling us to detect Lyα emitters at $z \sim 4.5 - 5$, at least around the known $z > 4.5$ quasars, and there have been a flurry of recent results.[5-7,9] Parallel searches for color-selected objects are also taking place, tuned to these redshift ranges, which are designed to turn up more evolved objects.[8-10]

We discuss here several high-redshift galaxies which have been identified in deep narrow-band Lyα imaging of the fields surrounding the quasars BR1202–

0725 and BR2237–0607, and studied with followup spectroscopy using the LRIS spectrograph on Keck. These objects appear to follow very closely our expectations for early galaxies although, consistent with objects at lower redshifts, they are not particularly luminous.

2   The companions to BR1202–0725

Both BR1202–0725 and BR2237–0607 were identified in the APM quasar survey[11,12] and show $z > 4$ damped Lyα (DLA) systems in their spectra.[13] The damped system at $z = 4.383$ [refs. 2,13] seen in BR1202–0725 is the highest redshift DLA known. Narrow-band imaging[5] by Hu, McMahon, and Egami at the wavelength of redshifted Lyα for BR1202–0725 identified emission which coincides with the continuum light from a faint galaxy seen in Hubble Space Telescope WFPC2 images 2.6″ NW of the quasar, and pointing radially towards it. This object was also identified as a high-redshift galaxy by Fontana *et al.* using color-break techniques[9] and by Petitjean *et al.* using integral field spectroscopy.[6] From high-resolution tip-tilt imaging (FWHM 0.45″) of this system at the UH 2.2m telescope in the Lyα line we have now identified a second faint Lyα emission component coincident with a fainter galaxy 3″ SW of the quasar, also pointing radially towards the quasar in the *HST* image (Fig. 1). In these higher resolution images it is possible to see that the peak of the Lyα emission coincides with the two continuum structures seen in the *HST* images, but that there is also a diffuse component extending out towards the illuminating quasar. The second emission system near the quasar at $z = 4.7$ is confirmed in multi-slit LRIS spectra taken on these objects with the Keck 10m telescope. C IV emission is not detected in these spectra. However, [O II] emission is also seen in these objects in narrow-band IR filter imaging at CFHT made using the UH 1024^2 IR camera, QUIRC. The morphology of these nearby systems suggests that we may be seeing here the merging of these subsystems and possibly witnessing the formation of the quasar's host galaxy. Near these emission-line structures there is excess emission at millimeter and submillimeter wavelengths[14,15] and most recently, molecular CO transitions have been detected in close proximity to the quasar.[16]

3   Field Galaxies around BR2237–0607 — Early Star Formation?

For BR2237–0607 based on deep narrow-band Lyα imaging followed by spectroscopy using LRIS at Keck, we identify[7] two Lyα emitting galaxies at $z = 4.55$. In contrast to the BR1202–0725 emission-line objects, these galaxies have separations $> 100″$ (~ 700 kpc) from the quasar, which is unlikely to have a significant role in exciting them. Observed equivalent widths for the Lyα line exceed ∼1000 Å, in each case. While one of these objects is quite compact, and has no detectable continuum in a 1-hr I-band exposure at Keck, and thus might be an AGN, the second has a diffuse extended structure in both continuum and emission, and is most likely a star-forming galaxy. Line fluxes are ∼ 5 × 10^{-17} ergs cm^{-2} s^{-1}. These objects may represent the very earliest stages of galaxy formation prior to the formation of a significant stellar population and the destruction of the Lyα line by dust.

4   Discussion

The objects in the BR2237–0607 field are currently the only galaxies, well separated from quasars, that are known at these high redshifts, and their properties are therefore of considerable interest if only in targeting future searches. The rest frame equivalent widths of the systems are around > 130 Å and > 240 Å, which are marginally consistent with stellar excitation for an initial mass function dominated by massive stars.[19] The Lyα may arise from a combination of internal and external ionization. If the observed emission were primarily due to stars, and there were no internal scattering and extinction, then the luminosity of 3×10^{42} h^{-2} erg s^{-1} (q_0=0.5) would correspond to a star formation rate in solar masses ($M_\odot$) per year of ∼ 3 h^{-2} $M_\odot$ yr^{-1}, where we use Kennicutt's (ref. 21) relation between Hα luminosity and star formation rate (SFR) of SFR = $L(H\alpha) \times 8.9 \times 10^{-42}$ erg s^{-1} $M_\odot$ yr^{-1}, and assume a ratio of Lyα to Hα (8.7) that applies for Case B recombination.[20] Given the Hubble time at this redshift of 7×10^8 h^{-1} yr (q_0=0.5) the integrated amount of star formation is small compared to that of a 'normal' galaxy with 6×10^{10} h^{-1} $M_\odot$ of stars (a so-called L^* galaxy).

Because of the targeted nature of the search it is hard to estimate from the present data whether such objects may be common in the general field or whether they are preferentially found around quasars. Observations, currently in progress, of additional quasars and blank field regions should answer this question.

References

1. M.J. Rees, these proceedings.
2. L. Lu, et al., ApJ, 457, 1, 1996.
3. C.J. Pritchet, PASP, 106, 1052, 1994.
4. D. Thompson, S. Djorgovski and J. Trauger, AJ, 110, 963, 1995.
5. E.M. Hu, R.G. McMahon and E. Egami, ApJ, 459, L53, 1996.

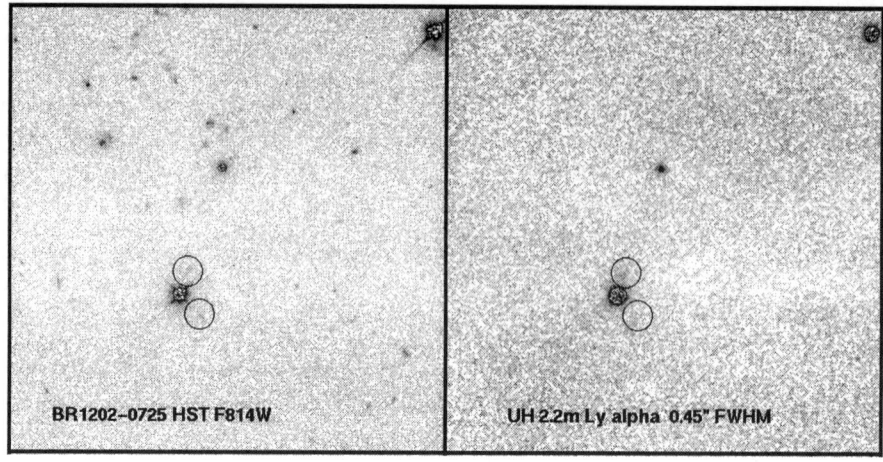

Figure 1: The left-hand panel shows a deep *HST* WFPC2 image of the BR1202–0725 field with faint neighboring continuum structures circled (3″ diameter). The right-hand panel shows the corresponding Lyα image taken in a 4-hr integration through an 80 Å-wide narrow-band filter at the f/31 tip-tilt secondary of the UH 2.2m telescope. Diffuse Lyα emission was detected coincident with both NW and SW companion galaxies to the quasar.

6. P. Petitjean, *et al.*, Nature, 380, 411, 1996.

7. E.M. Hu and R.G. McMahon, Nature, 382, 231, 1996.

8. E. Egami, PhD thesis, Univ. of Hawaii, 1995.

9. A. Fontana, *et al.*, MNRAS, 279, L27, 1996.

10. S. D'Odorico *et al.*, these proceedings.

11. M.J. Irwin, R.G. McMahon and C. Hazard, in *The Space Distribution of Quasars*, ed. D. Crampton, 117, 1996.

12. L.J. Storrie-Lombardi, R.G. McMahon, M.J. Irwin and C. Hazard, ApJS in press, [astro-ph/9604021].

13. L.J. Storrie-Lombardi, M.J. Irwin and R.G. McMahon, MNRAS in press, [astro-ph/9608146].

14. R.G. McMahon, *et al.*, MNRAS, 267, 9L, 1994.

15. K.G. Isaak, *et al.*, MNRAS, 267, L28, 1994.

16. A. Omont *et al.*, Nature, 382, 428, 1996.

17. L.L. Cowie, A. Songaila, E.M. Hu and J.G. Cohen, AJ, in press.

18. L.L. Cowie, E.M. Hu and A. Songaila, Nature, 377, 603, 1995.

19. S. Charlot and S.M. Fall, ApJ, 415, 580, 1993.

20. M. Brocklehurst, MNRAS, 153, 471, 1971.

21. R.C. Kennicutt, Jr., ApJ, 272, 54, 1983.

INTERNAL KINEMATICS OF A SAMPLE OF GALAXIES AT z =0.25-0.45

L. SIMARD, C.J. PRITCHET

Department of Physics and Astronomy, University of Victoria
P.O. Box 3055, Victoria, Canada, V8W 3P6
E−mail: (simard,pritchet)@uvastro.phys.uvic.ca

We present the results of a limited survey of the internal kinematics of intermediate redshift (z = 0.25−0.45) field galaxies. Using the Canada-France-Hawaii Telescope, spatially-resolved spectra of the [O II] $\lambda\lambda$ 3726−3729Å doublet emission line have been obtained for 22 galaxies. 25% of the field galaxies in the sample have [OII] kinematics unrelated to rotation. The [OII] emission is confined to the nucleus in these galaxies. Anomalous kinematics is related to minor merger events. Kinematically normal galaxies in the sample are ∼1.5−2.0 mag brighter than expected from their rotation velocity and the Tully-Fisher (TF) relation. The main source of uncertainty is the large scatter in the local TF relation for late-type galaxies. Luminosity-dependent luminosity evolution neatly reconciles the lack of evolution seen in other works with the results of our survey.

1 Introduction

Internal kinematics directly measures a fundamental property of galaxies: mass. It can therefore be a powerful probe of the nature of intermediate redshift galaxies. The idea is simple: if intermediate redshift galaxies are as massive as "normal" spirals such as the Milky Way (instead of being dwarf galaxies), we would observe rotation velocities of ∼ 200 km/s. On the other hand, if they are really lower mass objects that have been boosted in luminosity by ∼10× as suggested in the luminosity-dependent luminosity evolution scenario (Broadhurst *et al.* 1988), then the TF relation predicts that their rotation velocities will be ∼ 100 km/s. This approach is direct as it is not affected by uncertainties in models based on local luminosity functions.

2 Observations

Objects with 0.25 < z < 0.45 were selected from the CNOC cluster survey database (Yee *et al.* 1996). Rest-frame [OII] equivalent widths W_{3727} were between 20 and 50 Å. Target galaxies may be [OII] strong relative to local early-type (Sb and earlier) spirals, but they have the same W_{3727}'s as many local late-type (Sc and later) galaxies. That range of W_{3727} is also representative of the excess galaxy population at intermediate redshifts (Broadhurst *et al.*

1988, Broadhurst *et al.* 1992). *Elongated* objects were preferentially selected to minimize sin i effects.

Spatially-resolved observations of the [O II] $\lambda\lambda$ 3726−3729 Å doublet emission line were obtained with the Multi −Object Spectrograph (MOS) and the Subarcsecond Imaging Spectrograph (SIS) at the Canada−France−Hawaii 3.6-m Telescope (CFHT) in July-August 1994. Typical total integration time per galaxy was 7200 seconds split in 2-3 exposures for cosmic ray removal. 20 field and 2 cluster galaxies were observed.

3   Synthetic Rotation Curve Fitting

The [O II] flux levels were very low (typical S/N ratio *per pixel* $\sim$ 2−3), and it was important to choose a method, such as synthetic rotation curve fitting, which used all the pixels simultaneously to statistically find the best parameter values and their respective uncertainties. The parameters of the fitting model were the disk projected rotation velocity V_{rot} sin i, the [OII] exponential disk scale length r_d, the [OII] total line flux and the [OII] doublet ratio I_{3726}/I_{3729}.

The synthetic rotation curves were constructed using an [OII] emission distributed in a thin exponential disk with a flat rotation curve. These [OII] disks were convolved with a point−spread−function (PSF) extracted from direct images, and the result of placing a slit in front of the image was computed. The flux passing through the synthetic slit was convolved with the spectrograph's instrumental profile extracted from comparison arc lines.

The best fitting parameter values and their error distributions were found using the Metropolis algorithm (Saha and Williams 1994) which Monte−Carlo samples parameter space with a sampling density proportional to the likelihood.

4   Results

4.1   [OII] Morphologies

7 galaxies (5 field and 2 cluster galaxies) have [OII] emission confined to their nucleus, so their [OII] gas kinematics is decoupled from their rotation. All but one have close companions. This suggests that enhanced star formation activity may be the result of merger events. Some of these galaxies also appear to be of early-types. Kinematically anomalous field galaxies make up 25% of the field sample. This is similar to the fraction of blue-nucleated galaxies observed in HST images of galaxies at z $\sim$ 0.6 (Schade *et al.* 1995).

4.2 An Intermediate Redshift TF plot

Figure 1 shows kinematical evidence for luminosity evolution at intermediate redshifts. The local H_α −B band TF relation for all morphological types (open symbols) was defined using data taken from Mathewson *et al.* (1992). V_{rot} sin *i*'s of kinematically normal CFHT galaxies are plotted as solid circles. The upper long dashed line is an unweighted linear fit to all the local morphological types. This linear fit was then shifted by $\Delta M_{B_0} = -1.0$ mag (middle dashed line) and $\Delta M_{B_0} = -2.0$ mag (lower dashed line) to produce fiducial lines representing various degrees of luminosity evolution.

Galaxies in the sample are ∼1.5−2.0 mag brighter than expected from their rotation velocity and the local TF relation.

5 Discussion

In order to accurately measure magnitude offsets from the TF relation, each galaxy in our CFHT sample (or any other sample for that matter) should be compared to the local TF relation for galaxies with similar [OII] emission line strengths. However, the existing body of local TF data is insufficient to ascertain the dependence of local galaxy luminosity on the strengths of emission lines indicative of star formation.

The luminosity-dependent luminosity evolution scenario neatly reconciles the various amounts of luminosity evolution seen in surface brightness and internal kinematics studies. At the low end of the galaxy mass spectrum, compact narrow emission line galaxies have linewidths 2−3× smaller than expected from the TF relation of normal spiral galaxies (Koo *et al.* 1995). At the high mass end, large ($r_d \geq 3.0$ kpc) and bright ($M_B \leq -20.7$) galaxies have typical rotation velocities of 200 km/s, and their increase in B luminosity with respect to the local TF relation was less than 0.6 mag (Vogt *et al.* 1996).

Our CFHT sample occupies a niche in size and mass right in between the above two Keck samples. Compared to the Vogt *et al.* sample, CFHT galaxies are typically fainter ($\Delta M_B \simeq +1.0$), intrinsically smaller ($r_d < 2.0$ kpc) and less massive ($\simeq 100$ km/s). They were however more massive than the Koo *et al.* galaxies. If mass is taken as an indicator of the luminosity all the galaxies would have had in a quiescent phase, then all three internal kinematics studies can be understood with *mass*-dependent luminosity evolution.

The B-band surface brightness $\mu_0(B)$ of field disk galaxies undergoes a strong evolution over the range $0.1 < z < 0.6$ (Schade *et al.* 1996). At z ∼ 0.5, $\Delta\mu_0(B) = -1.1$ mag. This is consistent with or slightly less than the evolution seen in our CFHT sample, and it is certainly more than the amount

98

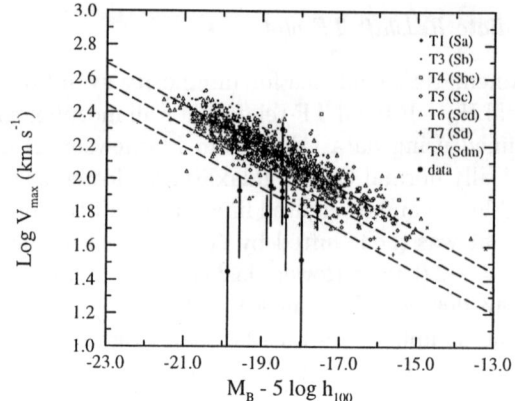

Figure 1: Kinematical evidence for luminosity evolution. M_B is the k-corrected B-band absolute magnitude, and V_{max} is the disk rotation velocity.

of evolution seen in the Keck sample of Vogt *et al.* Figure 1 of Schade *et al.* hints that smaller galaxies evolve more drastically than large galaxies (see their highest redshift bin where the log $r_d - M_B$ relation clearly curves "down"), and a large number of galaxies at (log $r_d = 0.8$, $M_B = -21$) show little or no evolution on the Schade diagram as observed in the Keck sample.

Acknowledgments

We gratefully acknowledge support from NSERC of Canada.

References

1. T.J. Broadhurst *et al.*, MNRAS, 235, 827, 1988.
2. T.J. Broadhurst *et al.*, Nature, 355, 55, 1992.
3. D.C. Koo *et al.*, ApJ, 440, L49, 1995.
4. D.S. Mathewson *et al.*, ApJS, 81, 41, 1992.
5. P. Saha and T.B. Williams, AJ, 107, 1295, 1994.
6. D. Schade *et al.*, ApJ, 451, L1, 1995.
7. D. Schade *et al.*, ApJ, 465, L103, 1996.
8. N.P. Vogt *et al.*, ApJ, 465, L15, 1996.
9. H.K.C. Yee *et al.*, ApJS, 102, 269, 1996.

OLD GALAXIES AT HIGH REDSHIFT

J.S. DUNLOP

Institute for Astronomy, Department of Physics & Astronomy, University of Edinburgh, Royal Observatory, Edinburgh EH9 3HJ, UK

The most passive galaxies at high redshift are unlikely to be identified by either narrow-band emission-line searches, or by Lyman limit searches (both techniques which have been highlighted at this meeting) simply because such selection methods rely on the presence of a strong ultraviolet component. Selection on the basis of extreme radio power has also proved to yield optically active objects with the majority of high-redshift objects studied to date displaying complex elongated optical/UV morphologies, relatively blue optical-ultraviolet continuum colours, and strong emission lines. These features, coupled with the failure to detect any spectral signatures of old stars at $z > 1$, has led to the suggestion that these galaxies are being observed close to or even during a general epoch of formation. However, we have recently demonstrated that radio selection at significantly fainter (mJy) flux densities can be used to identify apparently passively evolving elliptical galaxies at high redshift. Deep Keck spectra have now been obtained for two such objects yielding absorption line redshifts $z \simeq 1.5$; 53W091 at $z = 1.552$ (Dunlop *et al.* 1996)[1] and most recently 53W069 at $z = 1.432$. The ultraviolet SEDs of these galaxies indicate minimum ages > 3 Gyr while, as stressed in this article, the strength of the reddenning-independent ultraviolet spectral breaks actually indicate a greater minimum age of 5 Gyr for both objects assuming solar metallicity. Since the spectra comprise the integrated light of each galaxy to radii greater than r_e, I argue that it is difficult to justify the adoption of significantly super-solar metallicity in interpreting these data. It thus seems hard to escape the conclusion that $\Omega_0 < 1$ and that, irrespective of the adopted cosmology, at least some massive ellipitical galaxies were formed at high redshift ($z > 5$).

1   Background: locating passively evolving galaxies at high z

The recent discovery of a substantial population of star-forming galaxies at ($3.0 < z < 3.5$) has revolutionised the study of radio quiet galaxies at high redshift, as evidenced by a number of contributions at this meeting[2]. However a selection method which depends on a Lyman continuum break superimposed on an otherwise blue far-UV continuum can shed little light on the evolutionary state of the most passively evolving systems which exist at a given epoch. This is unfortunate since, given the ease with which a relatively small starburst can mask the true properties of an underlying galaxy, it is the reddest/most-passive systems at any redshift which are of greatest interest for constraining the first epoch of galaxy formation and indeed the age of the Universe.

Radio-based selection has long provided an alternative and effective method of locating high redshift galaxies which, at least in principle, should not be so directly biassed towards star-forming sources. Indeed, if anything it should be biassed towards the precursors of old elliptical galaxies since at low-redshifts it is well-established that the hosts of powerful radio sources are elliptical galaxies with well-evolved stellar populations. Despite this, identification of high-redshift objects on the basis of extreme radio power has also yielded optically active objects with the majority of high-redshift radio galaxies studied to date displaying complex elongated optical/UV morphologies, relatively blue optical-ultraviolet continuum colours, and strong emission lines[3]. However, the fact that the optical-ultraviolet properties of high-redshift radio galaxies are known to correlate with radio power[4] suggests that any radio-based search for 'normal' elliptical galaxies at high redshift should be confined to milli-Jansky flux levels. Accordingly, over the past few years we have investigated the properties of weak radio galaxies with $S_{1.4GHz} > 1$ mJy from the Leiden Berkeley Deep Survey, and have isolated a sample of 10 extremely red objects that have $R - K > 5$ and $z_{est} > 1$ for intensive spectroscopic study.

2 Keck Spectroscopy

While a red $R-K$ colour can be taken as indicative of an old stellar population, deep optical spectroscopy is vital for the reliable dating of these objects for four reasons. First, a spectroscopic redshift is required. Second, it is necessary to show that the red colour of the object arises from a lack of young stars rather than, for example, from a dust-reddened active nucleus. Third, the shape of the rest-frame ultraviolet spectrum of a galaxy is extremely sensitive to the age of the stellar population[5]. Fourth, for high-redshift galaxies it should be possible to use evolutionary synthesis models to derive relatively robust age estimates from ultra-violet SEDs because, for the potential age range of interest (*i.e.* age < 8 Gyr for $z > 1$) the ultraviolet SED is completely dominated by stars near the main-sequence turnoff point on the HR diagram[6] (*i.e.* disagreements over, for example, the strength and colour of the AGB or HB are unimportant).

We have now obtained deep optical spectra of two red mJy radio galaxies using LRIS on the Keck telescope. Our redshift determination and spectral dating of the first of these (53W091; $z = 1.552$) have been published[1] and will be described in more detail elsewhere[7]. In brief, the ultraviolet SED of this source is, as illustrated in Figure 1, very similar to those of low-redshift ellipticals such as M32, and essentially identical to that of an F6V star. Both these comparisons suggest an age of $\simeq 3.5$ Gyr, a result confirmed by spectral synthesis modelling. We have recently (June 1996) obtained a deep LRIS spectrum of a second red mJy radio galaxy, 53W069. This object also lies at $z \simeq 1.5$ ($z = 1.432$) and has an ultraviolet SED which is in fact slightly redder,

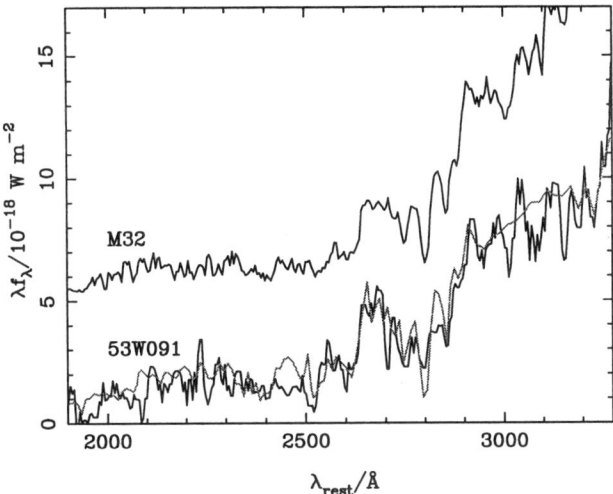

Figure 1: The rest-frame spectrum of 53W091 compared with an instantaneous starburst at an age of 3.5 Gyr (grey line) and a transposed IUE spectrum of the elliptical M32.

indicating an age of $\simeq 4.5$ Gyr[8].

3 Galaxy ages from ultraviolet spectral breaks

While the large number of stellar absorption features detected in the spectra of both these objects proves that their UV light is dominated by stars, dating on the basis of the overall shape of their UV SED is susceptible to distortion either by dust reddening or by low-level direct/indirect AGN contamination. However the strengths of the spectral breaks at 2640Å and 2900Å being relatively immune from such complications should yield more robust age estimates. Certainly the strength of these breaks in the IUE spectra of stars is well studied[9], and despite misgivings over the understanding of the relevant opacities, evolutionary synthesis models based on both observed and theoretical stellar spectra do indeed seem to produce reasonably consistent results from both breaks. As indicated in Table 1, this analysis indicates that in both galaxies $\simeq 5$ Gyr has elapsed since the last era of significant star-formation activity.

The most attractive way to reconcile an Einstein-de Sitter Universe with such large ages at $z \simeq 1.5$ is to assume that the strong breaks in both 53W069 and 53W091 are due to high metallicity rather than age. However, the metal-

Table 1: Minimum ages deduced from the break strengths in the rest-frame spectra of 53W091 and 53W069 using 3 alternative models of galaxy spectral evolution. The evolution of the 2640Å break in Worthey's model is anomolously rapid (*e.g.* it yields a MS turnoff age of < 3 Gyr for the sun). The other ages in the table are consistent with 5 Gyr.

Ages (Gyr)		Model	Average Age
2640Å	2900Å		
(1.8)	4.6	Worthey (1994)[11]	4.6
4.0	6.5	Jimenez *et al.* (1996)[12]	5.3
6.2	4.6	Bruzual & Charlot (1993)[13]	5.4

licity dependence of these breaks does not appear to be strong[1,9], and in any case the *mean* metallicities of comparably massive giant ellipticals at low redshift are at most only mildly super-solar when averaged out to $r \simeq r_e$ ($\simeq 0.5$ arcsec at $z \simeq 1.5$ and thus within the LRIS slit)[10]. Taken at face value an age > 5 Gyr at $z \simeq 1.5$ implies $\Omega_0 < 0.2$ for $H_0 > 55 \mathrm{kms}^{-1}\mathrm{Mpc}^{-1}$ unless $\Lambda > 0$.

Acknowledgments

I gratefully acknowledge my collaborators in this work: John Peacock, Hy Spinrad, Arjun Dey, Rogier Windhorst, Raul Jimenez and Daniel Stern.

References

1. J.S. Dunlop *et al.*, 1996, Nature, 381, 581.
2. M. Giavalisco, 1997, in this book.
3. P. McCarthy, 1993, ARA&A, 31, 639.
4. J.S. Dunlop & J.A. Peacock, 1993, MNRAS, 263, 936.
5. R.W. O'Connell, 1988, in *Towards Understanding Galaxies at large Redshift*, 177, eds R.G. Kron & A. Renzini, Kluwer.
6. G. Magris C. & G. Bruzual A., 1993, ApJ, 417, 102.
7. H. Spinrad *et al.*, ApJ, 1996, in press.
8. A. Dey, 1997, in this book.
9. M.N. Fanelli *et al.*, 1992, ApJS, 82, 197.
10. N. Arimoto, 1996, in *Fresh Views of Elliptical Galaxies*, ASP Conf. Ser. Vol. 86, 239.
11. G. Worthey, 1994, ApJS, 95, 107.
12. R. Jimenez *et al.*, MNRAS, in preparation.
13. G. Bruzual A. & S. Charlot, 1993, ApJ, 405, 538.

THE HST/LAS CAMPANAS EXTRAGALACTIC BACKGROUND LIGHT PROGRAM

R.A. BERNSTEIN, B.F. MADORE

California Institute of Technology, MS 105-24, Pasadena, CA 91125, USA

W.L. FREEDMAN

Carnegie Observatories, 813 Santa Barbara St. Pasadena, CA 91101, USA

A measurement of the Extragalactic Background Light (EBL) from unresolved sources will help constrain both galaxy evolution and cosmological models; it can also place limits on the existence of populations of objects below the detection threshold of current redshift and number count surveys. We discuss the status of a program to measure the mean flux and spatial fluctuations (power spectrum) of the EBL in three wide optical passbands centered around 3000Å, 5000Å, and 8000Å, using data obtained contemporaneously from the WFPC2 aboard HST and from Las Campanas Observatory in Chile.

1 Introduction

The Extragalactic Background Light (EBL) is the integrated light from all extragalactic sources, both resolved and unresolved. With the new 10m–class telescopes and HST, the limits of resolved–source detection are being extended to ever fainter levels; however, a measurement of the EBL remains an invaluable complement to the source–count approach. Populations of low surface brightness objects, as well as the bulk of the luminosity function at high redshifts, are easily missed in both surface brightness limited galaxy counts and redshift surveys, but they can be identified through their contribution to the total background flux.

In addition to including the light from unresolvable sources (and thus providing an independent check on the completeness of galaxy counts and redshift surveys in general), the EBL is immune to surface brightness selection effects, which can potentially bias the results of magnitude limited redshift surveys and galaxy counts. The EBL also avoids the "counting" ambiguities which arise from the non–uniform surface brightness distribution of individual galaxies in their rest–frame UV emission. This can cause counting errors even among the detected populations: single objects can be erroneously split into several and flux is lost from the regions in those objects where the surface brightness drops below the sky noise. Finally, photometry and identification of faint galaxies becomes uncertain near detection limits. A measurement of the mean flux of the EBL avoids these inherent difficulties.

Several long–standing questions of cosmology and galaxy evolution can be addressed with a *multiwavelength* measurement of the mean level of the EBL, *i.e.*, by a measurement of the broad–band spectrum of the EBL. Because the Lyman limit shifts through the bandpass of observation as a function of redshift, objects in different redshift ranges dominate the EBL at different wavelengths. For example, at 3000Å, the EBL is dominated by the UV flux of galaxies at low z and can therefore be used to constrain the star formation rate at $z < 2$. At 8000Å, the total volume of space at high z has the greatest influence on the EBL, so the flux at longer observed wavelengths is sensitive to the cosmological model. In an ongoing project to measure the EBL at 3000, 5000, and 8000Å, we hope to isolate these effects, and also constrain the relative contribution of low surface brightness and high redshift objects that are beyond the detection limits of galaxy counts and redshift surveys.

2   Predictions and previous attempts.

A lower limit to the mean flux of the EBL can be derived by integrating the flux from observed galaxies. An estimate of the mean EBL comes from extrapolating the observed galaxy counts beyond the detection limits in isophotal magnitude and surface brightness. However, the extrapolation must be designed in such a way that it successfully includes populations missed in the direct counting.

By making assumptions about the population of galaxies as a function of redshift, one can create a diverse range of models which fit the observed number counts and redshift distributions; by including galaxies which are difficult to observe, due, for example, to rapidly fading starbursts or low surface brightness, the range of acceptable models becomes even greater. Such models do, however, predict different mean levels and colors for the EBL, so that a measurement of the EBL provides powerful constraints on models of galaxy populations and evolution (see Väisänen[3] for a recent summary).

Previous attempts to measure the EBL have lead to the upper limits indicated in Figure 1 (see Matilla[2] for a review). The lower and upper limits plotted there demonstrate that the required sensitivity to measure the EBL in the range 3000–8000Å is roughly 10^{-9}ergs/s/cm^2/sr/Å, which corresponds to a value of roughly 28 AB mag/arcsec^2 at 5000Å.

3   Measuring the EBL

To measure the EBL, one must not only measure the absolute flux of the sky to better than 28 mag/arcsec^2, but also identify all other foreground components to that same accuracy. With the advent of new detectors, new telescopes, and new techniques this level of accuracy can be now achieved. In the optical, diffuse foreground components include airglow, zodiacal light, and Galactic scattered light. Airglow is not an issue for HST, and so will not be discussed here. Zodiacal light is sunlight scattered off of dust in our solar system and is well described by Mie theory.[1] Thus the color and features of the source (*i.e.*, the Sun) are well preserved in the zodiacal light with very little (and very measurable) deviation. The zodiacal light in a particular field varies strongly with the line of sight through the ecliptic plane, which is a function of the Earth's orbital position. Therefore, it is crucial that zodiacal light be measured at the same time (orbital position) that the total background flux is measured. Diffuse galactic contributions from scattered starlight can be effectively minimized by judicious selection of the field to be studied. Great care has been taken to minimize galactic effects by chosing the fields in minima of 100μm emission in the IRAS maps. The relative fluxes of these foreground components are also indicated in Figure 1.

From HST, the airglow contribution is avoided entirely , and so zodiacal light becomes the dominant foreground component to be removed. We are making a direct measurement of this component by identifying the solar Fraunhoffer lines in the spectrum of the zodiacal light. The equivalent widths intrinsic to the Fraunhoffer lines in the solar spectrum are well determined; by measuring (from the ground or from space) the apparent equivalent widths of those same features in the zodiacal light, we can determine the absolute contribution of the zodiacal light to the background in a given field at the time of observation. Currently, spectra with the required resolution and sensitivity can only be obtained from the ground, where one can use long–slit spectroscopy with CCD detectors to maximize integration time, area (and thus, signal and sensitivity) and spectral resolution.

While spectra are optimal for measuring the zodiacal light, the EBL itself is better measured using broad–band images. The increased collecting area not only improves the signal–to–noise ratio purely by virtue of the number of pixels involved — a 1024^2 CCD image provides a statistical accuracy of 2% of the expected mean EBL — but also assures that a more representative area of the Universe is being sampled. Also, the spatial resolution allows us to identify and remove stars, consider the resolvable objects separately from and in addition to the diffuse background, and measure the spatial fluctuations of the EBL in addition to the mean level.

In order to measure the EBL, we then require two things: (1) the absolute calibration of the total flux of the night sky as measured from HST and (2) a measurement of the zodiacal light to 1% accuracy in the exact field and at the exact time of the HST observations. The requisite observations have been conducted over 3 years, first with imaging and simultaneous spectroscopy from the 2.5m du Pont and 1m Swope telescopes at Las Campanas Observatory, and, in the final year, with HST images and spectra taken concurrently with the ground–based observations. In order to use the WFPC2 images for this project, we have developed a technique to improve the zero–point calibration due to dark current by roughly a factor of 5. The observations were made only in the shadow of the Earth, and care was also taken to avoid off–axis scattering of light from bright stars and the Moon.

With the data in hand, we are now in a position to make the first decisive measurement of the mean flux of the EBL in three broad passbands centered on 3000, 5000, and 8000Å. Such a measurement will be a unique and discriminating probe of galaxy populations and evolution.

References

1. Ch. Leinert, H.J. Staude, Ap&SS, 116, 415, 1985.
2. K. Mattila in *Proc IAU 139*, ed. S. Bowyer and Ch. Leinert (Dordrecht: Dluwer Academic Publisher,1990)
3. P. Väisänen, preprint, 1996.

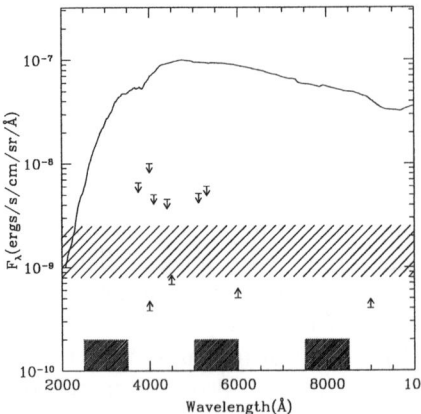

Figure 1: The expected EBL from models (hatched region) is shown,together with upper limits from previous attempts to measure the EBL and lower limits from direct integration of the flux from galaxy counts. The flux of the zodiacal light at the time of our observations (solid line), and the HST band–passes used (shaded regions) are also shown.

A VLA SURVEY OF THE HUBBLE DEEP FIELD

K.I. KELLERMANN AND E.B. FOMALONT
National Radio Astronomy Observatory,
Charlottesville, VA 22903, USA

E.A. RICHARDS
University of Virginia and NRAO,
Charlottesville, VA 22903, USA

R.A. WINDHORST
Arizona State University,
Tempe, AZ 85207, USA

R.B. PARTRIDGE
Haverford College,
Haverford, PA 19041, USA

We have used the VLA at 8.4 GHz (3.6 cm) to image the Hubble Deep Field (HDF) and surounding flanking fields. This paper reports on the initial results from observations made in February and March 1996.

1 The HDF Radio Image

The HST observations of the HDF are described by Williams et al. 1996. The radio observations were made using the VLA with a resolution of $3''$ between February 5 and March 12, 1996. The observations were spread out over six separate days each of length 10 hours. The field center for the radio observations was $\alpha = 12^h36^m49^s$, $\delta = +62°12'58''$ (epoch J2000). The full width at half power (FWHM) of the VLA is $312''$ and radio sources were detected in a field of view of diameter $480''$. Observations were alternated between the HDF and the calibrator source 1217+585. The flux density scale was fixed by observation of the source 3C286.

The rms noise after 50 hours of integration was 2.8 μJy. There are 18 radio sources above 12 μJy listed in Table 1 which gives the integrated image flux density and sky flux density (with errors), corrected for the primary beam attenuation, the right ascension and the declination (with errors) of the centroid of each source, the angular size estimate or limit (13 of the 18 sources are unresolved), the redshift (Cohen et al 1996), and our estimate of the I-magnitude of the identification, as measured from the HDF and HFF images. The question mark by the three radio sources indicates a possible identification with an

Name	Flux Density (μJy)				Right Ascension (J2000)		Declination (J2000)		Size "	z	I
	Image		Sky								
3634+1212	40	3	90	6	12 36 34.46	0.03	62 12 12.8	0.2	< 2		19
3634+1240	40	3	82	5	12 36 34.49	0.03	62 12 40.9	0.2	< 2	1.215	22
3640+1011	30	9	324	93	12 36 40.68	0.11	62 10 11.1	0.6	4		25?
3642+1331	80	10	100	13	12 36 42.08	0.02	62 13 31.4	0.2	3		
*3644+1249	13	3	15	3	12 36 44.02	0.07	62 12 49.9	0.6	< 2	0.556	21
*3644+1133	458	3	783	5	12 36 44.38	0.02	62 11 33.0	0.2	< 2	1.013	20
3646+1448	36	13	81	30	12 36 46.04	0.06	62 14 48.7	0.4	3		> 25
*3646+1404	152	3	206	4	12 36 46.33	0.02	62 14 04.7	0.2	< 2	0.960	21
*3649+1313	22	2	22	2	12 36 49.61	0.10	62 13 13.8	0.6	< 5	0.475	21
*3651+1221	18	3	20	3	12 36 51.67	0.06	62 12 21.1	0.4	< 2	0.299	24
3652+1444	121	3	256	6	12 36 52.88	0.02	62 14 44.1	0.2	< 2	0.372	22
3653+1139	15	3	23	5	12 36 53.35	0.06	62 11 39.7	0.4	< 2		23
*3655+1311	12	3	14	3	12 36 55.47	0.11	62 13 11.7	0.6	< 3		23
3701+1147	19	2	41	4	12 37 01.59	0.12	62 11 47.2	0.5	< 4		24?
3708+1055	13	3	146	31	12 37 08.24	0.10	62 10 55.8	0.5	< 3		21
3708+1245	15	2	49	8	12 37 08.70	0.08	62 12 45.0	0.6	< 3		24?
3721+1129	100	3	+	+	12 37 21.25	0.02	62 11 29.7	0.2	< 2		23
3725+1128	48	13	+	+	12 37 25.96	0.05	62 11 28.3	0.2	3		23

+ Outside primary beam. Correction factor to the sky flux density from the image
flux density is uncertain but greater than 20.
* Contained within the HDF

Table 1: RADIO SOURCES IN THE HUBBLE DEEP AND FLANKING FIELDS

object which is a less than two arcseconds away from the radio position.

2 The Optical Identifications

The optical identification of the weak radio sources with faint galaxies requires an alignment of the radio and optical grids to better than $1''$. However, the guide star positions used by HST to determine its a priori registration can be up to $2''$ in error, and these can produce offsets in each of the WFPC II frames from the radio FK5 frame by over $1''$. A more accurate registration with respect to the FK5 frame can be made by a comparison with an astrometric quality Palomar image and by using these high quality identifications of radio sources. For the HDF image, we estimate that the a priori positional accuracy is better than $0''.5$; for some of the flanking fields the position error may be larger than $1''$. Nevertheless, most radio sources can be unambiguously identified even with the present astrometric precision.

The radio positions given in Table 1 is tied to the inertial quasar reference frame (by using the quasar 1217+585 as the primary position calibrator) to an accuracy better than $0''.02$; however, the noise in the images limits the positional accuracy of most sources to $0''.3$ in each coordinate. These sources can be used to determine the most accurate registration of the Hubble Deep Field using the best identifications, and further radio observations at higher resolution should produce an alignment of the radio and optical grids to better than $0''.1$.

The source HDF 3649+1313 is near the center of the Hubble field where the radio sensitivity is maximum. The peak emission is identified with a distorted galaxy and emission extends to the north and west to other distorted galaxies in this probable interacting system. Source HDF 3644+1133 is intriguing. It is one of the brightest sources in the radio field and is identified with a typical red elliptical galaxy. The main radio component is less than 2'' in diameter, but there is also low level radio emission extending north and south from the peak. The northern spur ends near a peculiar looking blue chain galaxy; the southern spur curves west to a small asymmetric object (on the flanking field). Higher resolution radio observations are needed to determine if the emission peaks are connected, or just merely blended together in this relatively low resolution radio image.

Twelve radio sources lie in the flanking fields where the limiting magnitude is about I=25 mag. Eight sources are reliably identified, three have uncertain identifications and one appears to be an empty optical field. As with the HDF, identifications are associated with elliptical galaxies and early-type spiral galaxies and many identified sources appear to be interacting with a nearby galaxy, or to be in small groups.

Source HFF 3634+1212 appears to be an interacting system; source HFF 3701+1147 lies at the end of an arc which eminates from the nearby 22-mag galaxy (this feature is weak in the flanking field image and must be confirmed); one source (HFF 3721+1129) may be associated with a ring-shaped galaxy.

For the seven sources with measured redshifts, the monochromatic radio luminosity at 8 GHz is 10^{21-25} W Hz^{-1} which is comparable to that of nearby FRI radio galaxies identified with strong radio sources.

The three uncertain identifications in the flanking fields (HFF 3640+1011, HFF 3701+1147, and HFF 3708+1245) are displaced more than an arcsecond from the radio position. Their status should be resolved with better radio-optical astrometry. The empty field, source HFF 3646+1448, is associated with one of the few slightly extended radio sources. No optical object above 25 mag is seen in the flanking field at the position of the peak of the radio source or within 2'' of it.

3 Discussion and Conclusions

The first radio image covering the Hubble Deep Field and surrounding flanking fields shows that most of the identifications are with galaxies in the magnitude range 20 to 24 mag. The most common types are red elliptical galaxies and blue early-type spirals, in agreement with the results from other deep surveys.

It is interesting that all of the radio sources in the HDF above our flux density limit of 12 μJy are identified with galaxies substantially brighter than the HDF detection limit of 29 mag. Further observations with the VLA at a resolution of 0.''4 are planned to improve the sensitivity and angular resolution, and to obtain radio spectral information. We expect that all of the radio sources found in the HDF with improved sensitivity will have optical counterparts brighter than magnitude 26 or 27. Failure to identify any radio sources down to the HDF limit of magnitude 29 will by itself be of great interest.

These radio studies, together with additional optical observations of the identified galaxies should further enrich our understanding of both the process of star formation at high redshift and the distinction between radio-loud and radio quiet early type spirals.

Acknowledgments

This work was made possible through the foresight of R. Williams and the staff of the STScI in obtaining and making available the HDF images. We thank David Hogg for the use of the redshifts and magnitudes of galaxies in the Hubble Deep Field. Part of this work was supported by NASA through grant number AR-06337.02-94A to the Space Telescope Science Institute, which is operated by tthe Association of Universities for Research in Astronomy, Inc., under NASA contract NAS5-222655 and by a National Science Foundation grant AST-9320049 to Haverford College. EAR gratefully acknowledges the support of a Sigma Xi Grant-in-AID-of Research. The National Radio Astronomy Observatory is a facility of the National Science Foundatation which is operated by Associated Universities Inc., under a Cooperative Agreement with the National Science Foundation.

References

Cohen, J. G. et al. 1996, ApJ, submitted.

Fomalont, E. B., Partridge, R. B., Lowenthal, J. D., & Windhorst, R. A. 1993, ApJ, 404, 8; 1993, ApJ, 405, 498.

Williams, R. E. et al. 1996, AJ, in press.

CONSEQUENCES OF A NON-DETECTION OF FORMING GALAXIES BY AN INFRARED SURVEY

F. MANNUCCI

CAISMI–CNR, Largo E.Fermi 5, 50125 Firenze, Italia

D. THOMPSON, S.V.W. BECKWITH

Max-Planck-Institut für Astronomie, Königstuhl 17, 69117 Heidelberg, Germany

We present the results of a narrowband search for forming galaxies at high redshifts. Given the coverage of 276 square minutes of arc, this is the most extended search for extragalactic emission line objects at near infrared wavelengths. Despite of flux limits down to 1.4×10^{-16} erg cm^{-2} s^{-1} and of a total comoving volume surveyed of 1.4×10^5 Mpc^3 (for $H_0 = 50$ km s^{-1} Mpc^{-1}, $\Omega_0 = 1$), no such population was detected. We show how this null detection can be used to derive upper limits to a) the comoving volume density of this population and b) to the metal production density at redshift between 1 and 4.

1 Introduction

One of the most important issues of galaxy formation is understanding when the majority of the stars of early-type galaxies formed because identification of the first generations of stars and measurement of their ages are of great importance for models of the universe and galaxy formation.

A genuine population of forming galaxies has been discovered by Steidel et al. (1996, ApJ 462, L17). Our complementary approach consists in looking for redshifted optical emission lines in the near-IR. The main advantage of this approach with respect to the optical searches is that these lines are emitted at much longer rest-frame wavelengths and, therefore, suffer considerably less from extinction by dust. As an example, while the high-redshift galaxies detected by Steidel et al. (1996) appear to have fairly significant star formation rates, they generally have weak or absent Lyα emission.

2 Observations and data reduction

We obtained deep images of selected fields through narrow and broad band filters and looked for objects which are relatively brighter in the narrow filter, thus indicating a substantial flux in an emission line. We chose the fields to contain objects with known redshifts which put optical emission lines in the passbands of the narrow filters. If there is any tendency toward clustering, these objects will pinpoint regions of overdensity.

A total of 30 image pairs were obtained at the Calar Alto 3.5m and ESO/MPI 2.2m telescopes with exposure times of 1-2 hours through the narrowband filter and 15-30 minutes through the broadband filter. Plots of the (broad−narrow) color vs. narrowband magnitude were constructed for each of the 30 image pairs. Objects with relatively strong emission lines stand away from the locus of the remaining objects. There was only a single strong candidate identified in the survey data, but this is a complex object and will be discussed in Beckwith et al. (1997, in preparation).

This survey searches about two orders of magnitude more volume in the universe than similar infrared surveys previously completed. More details about observations, data reduction and results can be found in Thompson et al. (1996, AJ, in press).

3   Limits on the comoving volume density of forming galaxies

If the detected object is not a forming galaxy, we can put upper limits on the comoving volume density of such objects versus their star formation rates. To calculate the volume sampled by the survey, we include the five strong lines: $H\alpha$, $H\beta$, [OIII]λ5007, [OII]λ3727, and Lyα. Each of these would appear in the narrowband filters at different redshifts. The survey limits plotted in figure 1 assume no obscuration by dust, and therefore represent lower limits to the true star formation rate. The arrow in fig. 1 shows the effect that an extinction of $E_{(B-V)} = 0.3$ would have on the plotted survey limits.

These upper limits are compared to the expected density and luminosity of the forming galaxy population, taking into account various star formation histories and mass evolution. The model are described in Mannucci and Beckwith (1995, ApJ 442, 569) and estimate the density of young galaxies which are necessary to produce the local population of elliptical galaxies. We consider three classes of models: the first and simplest model (labelled *constant* in fig. 1) assumes no evolution in the mass function of elliptical galaxies and a constant star formation rate for each galaxy during a certain period of time. The boundaries for each model in fig. 1 correspond to the maximum and minimum reasonable span of time, i.e., to the minimum and maximum expected brightness, respectively. The other two models correspond to mild evolution of the mass function of the elliptical galaxies: in the *burst* model there are more massive but less numerous objects, and vice versa for the *hierarchical* model.

Figure 1 shows survey limits and model expectations in the redshift bin $2.0 < z < 3.5$. In this redshift range, sampled mainly by the Hα line in the K band and by the Hβ and [OIII] lines in the H and K bands, objects with unobscured star formation rates equal to $100 \, M_\odot \, yr^{-1}$ would be readily detected.

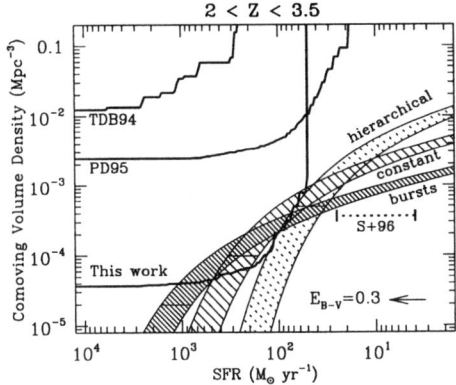

Figure 1: Upper limits on the volume density of forming galaxies in the redshift range $2.0 < z < 3.5$. The *constant, bursts* and *hierarchical* models are discussed in the text. The dashed line marks the comoving density and range of SFR for the population of star-forming galaxies detected by Steidel et al. (1996) at $3.0 < z < 3.5$. The three thick lines are the upper limits to the PG volume density from three surveys (this work, TDB94: Thompson et al. (1994, AJ 107, 1); PD95: Pahre and Djorgovski (1995, ApJ 449, L1)), where the regions to the upper left of these curves are excluded by the surveys. The arrow shows the effect on these limits of an extinction of $E(B - V) = 0.3$.

Figure 1 shows that the data sample enough volume to exclude the *constant* and *burst* models, although the limits only partially overlap the expectations for *hierarchical* models. This means that unobscured young galaxies can be present in this redshift range only if either they are small systems, or have low surface brightness, or are less efficient in emitting lines. Alternatively, the majority of galaxy formation could have occurred at higher redshifts.

4   The comoving metal production density

The same data can be used to compute upper limits to the density of metal production to be compared with other measurements and lower limits. The procedure is shown in fig 2. We assume that both luminosity and SFR are proportional to the mass of a galaxy, and that the mass function for the elliptical galaxies does not evolve strongly with redshift (i.e., we can neglect effects like inflow, outflow or merging). This supplies us with the α and ϕ^* values of a Schechter "SFR function" (number of galaxies with a given SFR vs. SFR) at any redshift. We compute the maximum value of the third parameter SFR*, i.e., the maximum SFR of an L^* galaxy giving no detection in our survey at a given redshift (changing this value corresponds to shifting these SFR function horizontally in fig. 2a) By integrating the resulting SFR function we compute

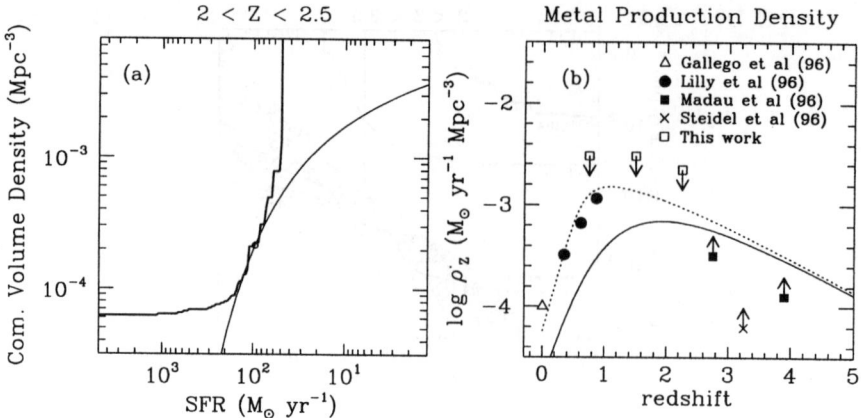

Figure 2: (a): measure of the maximum SFR compatible with our data: the SFR function has $\phi^* = 0.0087h^3 Mpc^-3$ (1/3 of the value for field galaxies given by Ellis et al., 1996, MN 280, 235) and SFR* = 95 M$_\odot$/yr, derived by matching to our survey limits. (b) Upper limits to the metal production density vs. redshift from our survey compared with the models by Pei and Fall (1995, ApJ 454, 69) (solid line: closed and outflow models; dotted line: inflow model) and with other measurements and lower limits (see Madau et al, 1996, for references).

the upper limit to the total SFR density and, using the recipe in Madau et al. (1996, ApJ, in press), the metal production density. We apply this procedure for those redshift ranges in which our survey is sensitive, obtaining the upper limits in fig. 2b.

Even if these points are subject to some uncertainties (discussed in Mannucci et al., 1997, in preparation) due to the various assumptions, it seems that they can put intertesting constraints on the models of galaxy formation

5 Conclusions

Young galaxies with emission lines at infrared wavelengths are rare at the level that can be reached with the current generation of detectors. The failure to detect many young galaxies makes it unlikely that most galaxies had star formation histories with continuous formation starting at any redshift and continuing to about $z \sim 2$. Hierarchical formation, in which galaxies were assembled from many pieces over a long interval are consistent with the results.

It is also possible that physical conditions not included in the models could reduce the observable line flux and weaken the conclusions derived. Two examples: a) dust along the lines of sight; b) young galaxies might be very extended, making the surface brightness too low to see in this survey.

THE UNIVERSE AT $z > 5$: WHEN AND HOW DID THE 'DARK AGE' END?

MARTIN J. REES

Institute of Astronomy,
Madingley Road, Cambridge, CB3 OHA, UK

This paper considers how the first subgalactic structures produced the UV radiation that ionized the intergalactic medium before $z = 5$, and the 'feedback' effects of the UV radiation on structure formation. The relevance of pregalactic activity to heavy element production and the origin of magnetic fields is briefly addressed.

1 Introduction

When the primordial radiation cooled below a few thousand degrees, it shifted into the infrared. The universe then entered a dark age, which continued until the first bound structures formed, releasing gravitational or nuclear energy that lit up the universe again. How long did the 'dark age' last? We know that at least some galaxies and quasars had already formed by a billion years. But how much earlier did structures form, and what were they like?

The density of quasars and large galaxies thins out at observed redshifts, but subgalactic structures may exist even at redshifts exceeding 10. I shall discuss the effects of the earliest stars and supernovae – production of UV radiation, reheating of the IGM, and the production of the first heavy elements – and the implications for observations at ultra-high redshifts.

2 Clustering in hierarchical models

I will focus on the cold dark matter (CDM) model. But this is just a 'template' for some more general deductions, which essentially apply to any 'bottom up' model for structure formation. There is no minimum scale for the aggregation, under gravity, of cold non-baryonic matter.[1-3] However the baryons constitute a gas whose pressure opposes condensation on very small scales. The gas therefore does not 'feel' the very smallest condensations. The baryonic Jeans mass is

$$M_J = 3 \times 10^5 \left(\frac{1+z}{10} \right)^{-\frac{3}{2}} \left(\frac{T_g}{500\text{K}} \right)^{\frac{3}{2}} \frac{\Omega b}{\Omega} M_\odot \qquad (1)$$

On scales larger than this, baryons can condense into bound systems, along with the dark matter.[1-3] During the 'dark age' the gas became even cooler than the microwave background: if it had cooled adiabatically, with no heat input since recombination, its temperature T_g would, at $z = 10$, have been below 5 K. The smallest bound structures, with mass $\sim M_J$, would have virialised at temperature of a few times larger than T_g. Larger masses would virialise at temperatures higher by a further factor $(M/M_J)^{2/3}$. This virial temperature would be reached not solely by adiabatic compression, but also because of a shock: it is unlikely that the gas could contract by more than a factor of 2 in radius before being shocked.

These virialised systems would, however, have a dull existence as stable clouds unless they could lose energy and deflate due to atomic or molecular radiative processes — clouds that couldn't cool would simply remain in equilibrium, being later incorporated in a larger scale of structure as the hierarchy builds up. On the other hand, clouds that can cool will deflate, even go into free-fall collapse, and (perhaps after a disc phase) fragment into smaller pieces.

Three 'cooling regimes' are relevant during successive phases of the cosmogonic process, each being associated with a characteristic temperature.

1. For a H-He plasma the only low-temperature ($< 10^3$ K) cooling comes from molecular hydrogen. This cuts off below a few hundred degrees; above that temperature it allows contraction within the cosmic expansion timescale. The H_2 fraction is never high, and it is in any case not a very efficient coolant (eg Fig 1 of Tegmark et al.[1]) but molecular cooling almost certainly played a role in forming the very first objects that lit up the universe.

2. If H_2 is prevented from forming, then a H-He mixture behaves adiabatically unless T is as high as 8-10 thousand degrees, when excitation of Lyman alpha by the Maxwellian tail of the electrons provides efficient cooling whose rate rises steeply with temperature; gas in this regime contracts almost isothermally.

3. The UV from early stars will photoionize some (and eventually almost all) of the diffuse gas. When this happens, the HI fraction is suppressed to a very low level, so there is is no cooling by collisional excitation of Lyman lines; moreover the energy radiated when a recombination occurs is quickly cancelled by the energy input from a photoionization, so the only net cooling is via bremsstrahlung. The cooling is, in effect, then reduced by a factor of ~ 100 (see, for instance, ref 4). The minimum temperature (below which there is a net heating from the UV) depends on the UV spectrum, and on whether He is doubly ionized: it is in the range 20-40 thousand degrees.

3 The role of molecular hydrogen, and the UV feedback

The role of molecular cooling at early cosmic epochs has been considered by many authors, dating back to the 1960s; recent discussions are due to Tegmark et al.[1] and Haiman et al.[5] This process allows clouds to contract if their temperature exceeds ~ 500 K. The exact efficiency depends on the density , and therefore on the redshift when the first collapse occurs.

But even at high redshifts, H_2 cooling would be quenched if there were a UV background able to dissociate the molecules as fast as they form. Photons of $h\nu > 11.18$ eV can photodissociate H_2, as first calculated by Stecher and Williams.[6] These photons can penetrate a high column density of HI and destroy molecules in virialised and collapsing clouds, even when they are far less intense than the background needed to fully ionize the medium.[5] *Only a small fraction of the UV that ionized the IGM can therefore have been produced in systems where star formation was triggered by molecular cooling. Most must have formed in systems large enough to have been able to cool by atomic line effects.*

There is then a further transition when the medium becomes completely ionized: the UV background gets a boost, because the contributions from remote regions (which dominate in Olbers-type integrals) are less severely attenuated. This means that it can maintain high ionization of a cloud until it has either collapsed to an overdensity exceeding the IGM ratio of ions to neutrals, or until it becomes self-shielding (which happens at more modest overdensities for large clouds). Until that happens the cooling rate will be reduced by the elimination of the (otherwise dominant) 'line' contribution to the cooling.

When this third phase is reached, the thermal properties of the uncollapsed gas will resemble those of the structures responsible for the observed Lyman-forest lines in high-z quasars spectra– these are mainly filaments, draining into virialised systems. Such systems have velocity dispersions of ~ 50 km/sec, and will turn into galaxies of the kind whose descendents are still recognisable.

4 The first stars: some uncertainties

The three uncertainties here are:

(i) What is the IMF of the first stellar population? The high-mass stars are the ones that provide efficient (and relatively prompt) feedback. It plainly makes a big difference whether these are the dominant type of stars, or whether the initial IMF rises steeply towards low masses, so that very many faint stars form before there is a significant feedback.

(ii) The influence of the early stars depends on where their energy is deposited. The UV radiation could, for instance, be mainly absorbed in the gas immediately surrounding the first stars, so that it exerts no feedback on the condensation of further clumps – the total number of massive stars needed to build up the UV background, and the concomitant contamination by heavy elements, would then be greater.

(iii) Quite apart from the uncertainty in the IMF, it is also unclear what fraction of the baryons that fall into a clump would actually be incorporated into stars before being re-ejected. The retained fraction depends on the virial velocity: gas more readily escapes from shallow potential wells. Ejection is even easier in potential wells so shallow that they cannot confine gas at the photoionization temperature.

All these three uncertainties would, for a given fluctuation spectrum, affect the redshift at which molecules were destroyed, and the (smaller) redshift at which full ionization occurred.

5   Heavy elements, magnetic fields, and the oldest stars

If the main UV source is stars, there is inevitably an associated build-up of heavy elements. (In more radical pictures where black holes are involved in the early energy input, this inference doesn't hold, because the energy supply could be gravitational rather than nuclear). The question then arises of how this processed gas would be distributed. Would it be confined in the virialised systems, or could it spread through the entire IGM?

The ubiquity of carbon features in intermediate and high ($N > 3.10^{14}$ cm^{-2}) column density systems (reported by other speakers) implies that heavy elements are broadly enough dispersed to have a large covering factor. These absorption systems may be associated with the subgalactic ($\sim 10^9 M_\odot$) sites of star formation. The nucleosynthesis sites cannot therefore be too sparse if these elements are, within the time available, to diffuse enough so that they are encountered somewhere along every line of sight through a typical high-column-density cloud. The absorption line data tell us the *mean* abundance through the relevant cloud. They are compatible with 99 percent of the material being entirely unprocessed, and the heavy elements being restricted to 1 percent of the material – the early heavy elements need not be thoroughly mixed, but they must have spread sufficiently to have a large 'covering factor' in the intermediate- and high-N clouds.

The first stars are important for another reason: they may generate the first cosmic magnetic fields. Moreover, mass loss (via winds or supernovae permeated by magnetic flux) would disperse magnetic flux along with the heavy elements. This flux, stretched and sheared by bulk motions, can be the 'seed' for the later amplification processes that generate the larger-scale fields pervading disc galaxies.

The efficiency of early mixing is important for the interpretation of stars in our own galaxy that have ultra-low metallicity – lower than the mean metallicity that would have been generated in association with the UV background at $z > 5$. If the heavy elements were efficiently mixed, then these stars would themselves need to have formed before galaxies were assembled. To a first approximation they would thereafter cluster non-dissipatively; they would therefore be distributed in halos (including the halo of our own Galaxy) like the dark matter itself. More careful estimates slightly weaken this inference, This is because the subgalaxies would tend, during the subsequent mergers, to sink via dynamical friction towards the centres of the merged systems. There would nevertheless be a tendency for the most extreme metal-poor stars to have a more extended distribution in our Galactic Halo, and to have a bigger spread of motions.

The number of such stars depends on the early IMF. If this were flatter, there would be fewer low-mass stars formed concurrently with those that produced the UV background. If, on the other hand, the IMF were initially steeper, there could in principle be a lot of very low mass (macho) objects produced at high redshift. These could be distributed like the dark matter. They could provide a few percent of the halo if Ω were 1; a larger proportion in a low-density universe.

6 Summary

There are thus three stages in the build-up of hierarchical structure, characterised by different masses and virial temperatures. They occur at three successive epochs – however, the demarcation is unlikely to be sharp because the range of amplitudes (for gaussian fluctuations) translates into a broad spread of turnaround times for a given mass scale.

These general conclusion are relevant to any model where the initial fluctuations have amplitudes decreasing with scale, so that cosmic structures form 'bottom up'. Such models differ, of course, in the epoch at which 'first light' would have occurred. In PIB models, this may be at $z > 100$; for CDM it is in the range 10-20; for 'mixed dark matter' models the first structures may form still more recently. Molecular cooling tends to be more efficient at high densities, and therefore at large redshifts; but in all cases it determines the scale of the first objects that condense out and contribute the first injection of heat into the universe.

The amount of background UV generated per solar-mass of material in these first objects is very uncertain – it depends on the efficiency of star formation,on whether the IMF favours massive stars (or even supermassive objects or black holes), and on how much of the UV is 'soaked up' by dense gas within the bound objects themselves. But irrespective of all these uncertainties, the

UV background exerts an important feedback on the cosmogonic process, by quenching H_2 cooling, long before photoionizing the entire IGM.

We therefore draw the robust conclusion that the IGM remained predominantly neutral until a sufficient number of objects above $\sim 10^9((1 + z)/10)^{-3/2}M_\odot$ had gone non-linear. Such systems have virial temperatures above 10,000 K – hot enough for HI line emission to permit very efficient cooling. Most of the O-B stars (or accreting black holes) that photoionized the IGM had to form in systems at least as large as this.

Formation of such systems would have continued unimpeded until the universe became, in effect, an HII region. This must have happened before $z = 5$. The only net cooling of a fully photoionized gas comes from bremsstrahlung, which is less effective than the collisionally-excited line emission from gas that is only partly ionized The completion of photionization may therefore signal another pause in the cosmogonic process,[7-8] associated with a further increase in the minimum scale that can collapse, and in the efficiency of cooling.

By the epoch $z = 5$, some structures (albeit perhaps only exceptional ones) must have attained galactic scales. Massive black holes (manifested as quasars) accumulate in the deeper potential wells of these larger systems (see, for instance, ref 9); quasars may dominate the UV background at $z < 4$.

Acknowledgments

I am grateful to Hugh Couchman with whom I first studied this subject more than 10 years ago. I also thank my recent collaborators, especially Zoltan Haiman, Avi Loeb and Max Tegmark.

References

1. M. Tegmark, J.I. Silk, M.J. Rees, A. Blanchard, F. Palla and T. Abel, ApJ, 1996 (in press)
2. J.R. Bond and A.S. Szalay, ApJ, 274, 443, 1983.
3. H.M.P. Couchman and M.J. Rees, MNRAS, 214, 137, 1986.
4. G.P. Efstathiou, MNRAS, 256, 43P, 1992.
5. Z. Haiman, M.J. Rees and A. Loeb, ApJ, 1996 (in press)
6. T.P. Stecher and D.A. Williams, ApJ, 149, L1, 1967.
7. D. Weinberg, L. Hernquist, and M. Katz, ApJ, 1996 (in press)
8. J. Navarro and M. Steinmetz, ApJ, 1996 (in press)
9. M. Haehnelt and M.J. Rees, MNRAS, 263, 168, 1993.

THE TTF SURVEY OF GALAXY POPULATIONS

J. BLAND-HAWTHORN

Anglo-Australian Observatory, Epping, NSW 2121, Australia

D.H. JONES

Mount Stromlo Observatory, Weston, ACT 2611, Australia

The TAURUS Tunable Filter (TTF) affords a new approach to observational cosmology, permitting wide-field monochromatic imaging at R and I bands ($\delta\lambda \approx$ 6 − 60Å FWHM). Here, we describe an emission-line survey currently underway using the TTF on the AAT to detect redshifted Hα over the ranges $z = 0.06 - 0.1$ and $z = 0.22 - 0.26$ in order to search for local counterparts to the numerous faint blue galaxies observed at higher redshift.

1 Introduction

A current problem of observational cosmology is the inability to reconcile the large numbers of faint blue galaxies (FBGs) with the results of spectroscopic surveys. These surveys ($B \lesssim 24$) indicate the FBGs lie at modest redshift ($z \sim 0.3$) rather than the high redshifts originally expected. The excess population may arise from a *separate* population of star-forming galaxies characterised by broad [OII] equivalent widths. No such equivalent population has yet been detected locally ($z \lesssim 0.1$). This is a crucial issue in the interpretation of the excess FBG counts, for if local galaxy samples are incomplete, then an *artificial* excess in the galaxy counts will appear at higher redshift.

2 Survey

The TTF Field Galaxy Survey aims to place stronger constraints on galaxy star formation rates and luminosities both at $z \sim 0.25$ where the FBG excess is observed, and at $z \lesssim 0.1$ where an equivalent local population has long been expected but not observed. We are undertaking a volume-limited survey for Hα line emission in the redshift intervals $0.06 \leq z \leq 0.10$ and $0.22 \leq z \leq 0.26$. (The survey will also be sensitive to redshifted Lyα, [OII], Hβ, and [OIII].)

Fig. 1(a) shows apparent Hα luminosities, $F_{H\alpha}$, as a function of redshift for galaxies with star formation rates (SFRs) between 0.01 and 10 $M_{\odot}$ yr^{-1}. Fig. 1(b) shows the signal-to-noise ratio (SNR) per pixel that we expect to attain using the TTF for galaxies with SFRs between 0.1 and 10 $M_{\odot}$ yr^{-1}, at redshifts of 0.08 (interval A; solid curve) and 0.24 (interval B; dashed curve). Since a typical star forming region in an LMC-type dwarf is about 300 pc in

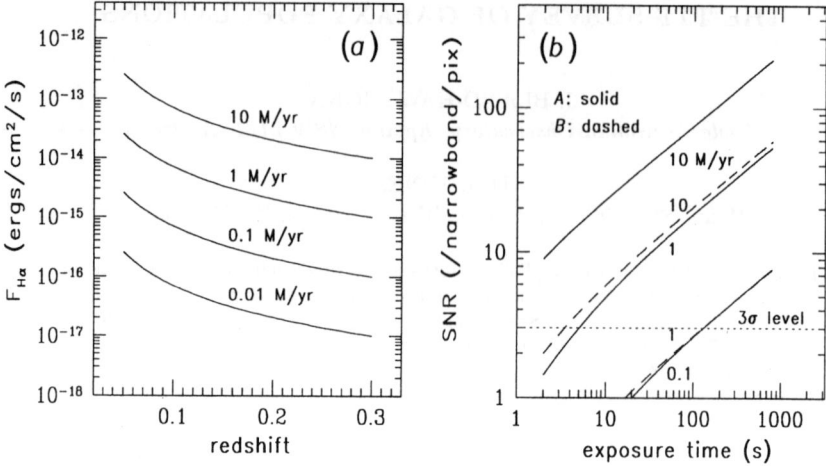

Figure 1: Calculations used to optimize TTF survey coverage in $2''$ seeing ($H_0 = 75$ km s^{-1} Mpc^{-1}, $\Omega_0 = 0.2$, $\Lambda_0 = 0$). We assume 17% overall AAT/TTF efficiency.

diameter, we expect the emission-line regions to be spatially unresolved. SFRs as low as 0.1 $M_\odot$ yr^{-1} at $z = 0.08$ and 1 $M_\odot$ yr^{-1} at $z = 0.24$ can yield a 3σ detection in only 130 s of exposure time. By way of comparison, a low-activity Sb spiral such as the Milky Way has a SFR of ~ 5 $M_\odot$ yr^{-1} while the LMC has a rate of 0.26 M$_\odot$ yr^{-1}. We have settled upon an exposure time of 200 s per slice as a compromise between flux limit and sky coverage. Exposures of this duration allow us to obtain Hα fluxes to $\sim 10^{-16}$ ergs cm^{-2} s^{-1}. We aim to cover 0.56 $\square°$ on the sky (~ 13200 Mpc^3) in eight nights.

3 Scientific Return

The survey promises unprecedented detection of Hα emission in the separate star-forming galaxy population at $z \sim 0.3$, thereby providing a quantifiable and independent confirmation of its extent compared to the [OII] detections of recent spectroscopic surveys. We expect to extend the current limits for detection of distant emission-line galaxies due to our low background advantage and high line/continuum ratio compared to broadband surveys. At $z \sim 0.1$, we are probing for Hα emission from local counterparts to the FBGs, the results of which will provide much stronger constraints on star-formation rates at $z \lesssim 0.1$ than has previously been possible.

THE ABSENCE OF DIFFUSE GAS AROUND LEO I

DAVID V. BOWEN

Royal Observatory, Blackford Hill, Edinburgh, EH9 3HJ, U.K.

ELINE TOLSTOY

ST-ECF, ESO, Karl-Schwarzschild-Straße 2, Garching bei München, Germany

ANDREA FERRARA

Osservatorio Astrofisico di Arcetri, Firenze, Italy

J. CHRIS BLADES

STScI, 3700 San Martin Drive, Baltimore, MD 21218, USA

ELIAS BRINKS

Departmento de Astronomia, Universidad de Guanajuato, Mexico

We outline the results of a search for UV absorption lines from diffuse gas surrounding the dwarf spheroidal (dSph) galaxy, Leo I. Although there is good reason to expect detection of metal enriched, highly ionized gas around the galaxy, none is detected.

1 Why should there be gas around a dSph

Many theories advanced to explain the origin and evolution of dSph galaxies suggest that there might be diffuse gas extending beyond their optical radius; dwarf irregulars may evolve from dwarf ellipticals by accreting gas from the IGM; dSphs may form as a result of interactions between more massive galaxies, (interactions are an excellent method of dispersing interstellar gas away from stars); they may form from gas-rich systems as supernovae drive gas out after an initial burst of star formation; and if the galaxy mass is high enough, gaseous outflows (in the form of mini galactic fountains) will remain bound to the galaxy and subsequently fall to the center, re-igniting star-formation. The only way to search for this diffuse gas is to search for the UV absorption lines that the gas produces in the spectra of background probes.

2 HST observations

We have observed 3 QSOs/AGNs which lie behind Leo I using the GHRS and G140L grating aboard *HST*. The probes are listed in Table 1, along with their V magnitude, redshift, z_{QSO}, separation on the plane of the sky, ρ, and the distance that the line of sight passes from the center of the galaxy, s, assuming

Table 1: QSOs and AGN behind Leo I

QSO probe	Alias	V	z_{QSO}	QSO-galaxy sep's ρ (')	s (kpc)
Q1004+1303	4C+13.41	15.2	0.240	34.0	2.1
Q1008+1319	...	16.3	1.287	60.7	3.7
Q0957+1317	NGC 3080	15.0	0.035	132.4	8.1

a distance to Leo I of 210 kpc. The wavelength region was selected to search for Si IV, C IV, and Si II absorption, but none was found from Leo I. Converting equivalent width limits to column density limits, N, assuming that a significant fraction of the carbon and silicon gas is in the form of C IV and Si IV [so that $N(C) \approx N(C\,IV)$ & $N(Si) \approx N(Si\,IV)$], and adopting a gas phase abundance of ≈ 0.1 dex of solar, we conclude that $\log N(H) < 18$ at separations of $2-8$ kpc from the center of the galaxy, where $N(H)$ is the total hydrogen column density. The lack of low ionization lines from these and other data also suggest that the H I column density is $\log N(H\,I) < 17$.

3   Has all the gas gone?

Our results demonstrate that there are no dense flows of gas in or out of Leo I, from, e.g., concentrated galactic fountains flowing out of the galaxy or dense inflows destined to re-ignite star formation. Similarly, there is no evidence for tidally disrupted gas which might have accompanied the galaxy's formation or evolution. If the galaxy has undergone a period of star formation during which time most of the gas has been ejected, the gas could have merged with the intergalactic medium for it to be no longer detectable. However, our detection limits are insufficient to rule out the existence of a sphere or shell of ionized gas around the dSph, with a mass up to that constituting the entire galaxy. Our models show that dSph galaxies similar to Leo I are not massive enough to have halos which can contribute significantly to the metal line absorption cross-section of QSO absorbers seen at high redshift.

CANDIDATE PRIMEVAL GALAXIES IN THE HUBBLE DEEP FIELD

D.L. CLEMENTS

European Southern Observatory, Karl-Schwarzschild-Strasse 2,
D-85748 Garching-bei-Munchen, Germany

W.J. COUCH

School of Physics, University of New South Wales,
Sydney, NSW 2052, Australia

We present the results of colour-selection of candidate high redshift (2.6 < z < 3.9) galaxies within the Hubble Deep Field based on the Ly-break at 912Å. We find 8 such objects in the region, giving a comoving number density comparable to that of nearby bright galaxies (for a flat $q_0=0.5$, $H_0=100$ kms^{-1}Mpc^{-1} universe). We provide basic data on the properties of these objects, and show that despite their absolute magnitude being significantly brighter than L* (typically $M_B = -22$), they are generally smaller than nearby galaxies. Furthermore, visual inspection of their images shows that they are all highly disturbed systems, with multiple nuclei, tails and plumes, suggesting that they are undergoing merging processes similar to most nearby starburst galaxies. Theoretical models suggest that galaxies form by accumulation of numerous subcomponents, and we suggest that we are seeing this process underway in these objects. It is thus possible that the epoch of galaxy formation might have been discovered.

1 Introduction

The Hubble Deep Field (HDF) [1] is the deepest and highest resolution optical image ever obtained. The provision of four colours (F300W, F450W, F606W and F814W) allows us to use the Lyman-break technique [2] to select candidate high redshift objects. This method uses the presence of the Lyman cutoff at 912Å in the bluest filter, and the assumption of a fairly flat spectral energy distribution (SED) at longer wavelengths, to provide a photometric signature for high redshift objects. For the HDF data the presence of the Lyman cutoff in the F300W filter allows us to select galaxies with 2.6<z<3.9.

2 Results and Discussion

We apply this technique to the HDF data by choosing those objects with F300W - F450W > 2, to select objects with the Lyman cutoff in the F300W filter, and the remaining colours (F450W - F606W and F606W - F814W) between -0.6 and 0.6 to select a fairly flat SED. We are restricted to objects with F450W brighter than 25.7 by the F300W band magnitude limit of 27.7 (all magnitudes referred to here are ST mags). Applying this selection to catalogues of HDF galaxies generated by the SExtractor programme [3] we find 8 sources meeting our colour selection criteria. This corresponds to a number density of 0.0005 $h^3 Mpc^{-3}$ (where h=H_0/100), which compares to the local bright galaxy (L>L*) number density [4] of $0.015 Mpc^{-3}$.

Examination of the properties of these objects shows that they are (i) luminous, with $M_B \sim$-22 (assuming a K correction appropriate for star forming galaxies); (ii) small, with a typical major axis of 4 kpc; (iii) disturbed systems, exhibiting a variety of morphological disturbances including plumes, double nuclei and asymmetric flux distributions.

We conclude from this that these objects may well be sub-galactic units assembling into the galaxies we see today by processes similar to those we see in galactic mergers in the nearby universe. Mergers are accompanied by bursts of star formation locally [5] which would account for the brightness of these objects. This scheme agrees well with current theories of galaxy formation [6]. We might thus have uncovered the early stages of galaxy formation.

More details on these results can be found in our paper [7].

References

1. R.E. Williams et al., AJ, in press, , 1996.
2. C.C. Steidel and D. Hamilton AJ, 104, 941, 1992.
3. I. Smail et al. ApJ, in press, , 1996.
4. I.M. Parkes, C.A. Collins and R.D. Joseph MNRAS, 266, 983, 1994.
5. D.L. Clements et al. MNRAS, 279, 477, 1996.
6. G. Kauffman MNRAS, 274, 161, 1995.
7. D.L. Clements and W.J. Couch MNRAS, 280, L43, 1996.

EXTRACTING PHYSICAL PARAMETERS OF GALAXIES FROM BROADBAND PHOTOMETRY

ANDREW CONNOLLY, ALEX SZALAY

Department of Physics and Astronomy, Johns Hopkins University, Baltimore, MD, 21218, USA

MARK DICKINSON

Space Telescope Science Institute, 3700 San Martin Drive, Baltimore, MD 21218

ROBERT BRUNNER

Department of Physics and Astronomy, Johns Hopkins University, Baltimore, MD, 21218, USA

The Hubble Deep Field provides a unique opportunity to study the evolution of galaxies at very faint magnitudes ($R = 29$). We demonstrate here that by determining an empirical correlation between the magnitudes and colours of galaxies and their distance we can estimate redshifts using broadband photometry. Using the U, B, R and I optical data for the HDF we show that redshifts can be estimated with a dispersion of $\sigma_z = 0.097$ to $z = 1.0$. If, however, near-infrared J, H and K' photometry is incorporated into these fits the dispersion can be reduced by almost a factor of two and the relation extended to $z > 2$. We find that the dn/dz relation for the spectroscopic and photometric redshift samples (derived from the optical and near-infrared data) are statistically identical.

The effectiveness of estimating galaxy redshifts to an accuracy of $\sigma_z < 0.05$ from multicolour broadband magnitudes has been demonstrated using ground-based optical photometry[1]. More recently photometric-redshift techniques have been applied to the Hubble Deep Field (HDF) data[2][3]. Redshifts are, typically, estimated by fitting the observed galaxy colours to those predicted by model spectral energy distributions (SED's). While there has been some success in using this approach there remains the uncertainty that the models do not fully represent the evolution of galaxy colours. These uncertainties can lead to systematic errors in the redshift estimator[4].

Rather than assume a particular model for the SED's and their associated star formation history we take an empirical approach to the problem. We derive a photometric-redshift relation by fitting the multicolor broadband photometry to the spectroscopic redshift for a subset of galaxies (essentially using them as a training set). We then apply this relation to the full data set. In such a way we estimate the SED of the galaxies internally from the data.

Using the HDF optical photometry we fit a second order polynomial to the U, B, R and I magnitudes (see Fig 1a). For low redshift galaxies ($z < 0.5$)

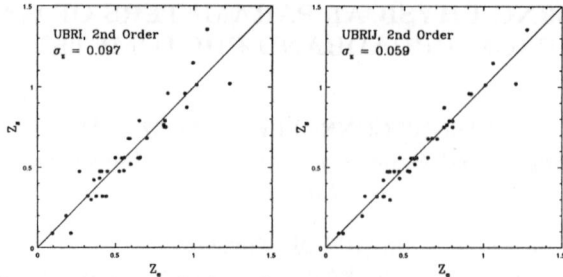

Figure 1. A comparison of the photometric vs spectroscopic redshifts for the HDF. By incorporating the near infrared passbands the dispersion in the photometric-redshift relation is reduced by 40%.

the intrinsic dispersion in the relation is $\sigma_z = 0.03$. As we increase the redshift limit the dispersion in the relation increases. The reason for this is simply that, as the SED is redshifted, the spectral features are sequentially rotated through the adjacent filters. To try and represent this effect with a simple quadratic relation is clearly not feasible. We must, therefore, fit the relation in an iterative fashion (fitting between $0.6 < z < 1.2$ reduces the dispersion to $\sigma_z = 0.055$).

As we progress to higher redshifts the estimation of a photometric-redshift is complicated by the restframe spectral features moving out of the optical bandpasses (at $z \sim 1$). We, therefore, require the use of near-infrared pass-bands (J, H and K'). Using IRIM on the KPNO 4m telescope we [a] obtained J, H and K' near-infrared photometry for the HDF field to K'$\sim$ 23. Simply adding the near-infrared J band to the photometric-redshift relation decreases the dispersion by 40% (from 0.097 to 0.059) for $z < 1.5$ (see Fig 1b). Applying this relation to an $R < 24.5$ sample of galaxies the dn/dz is found to be statistically identical to that derived from the spectroscopic data. Clearly the combination of optical and near-infrared photometry has the potential for estimating galaxy redshifts to a very high accuracy.

1. Connolly, A.J., *et al.*, AJ, 110, 1071
2. Gwyn, S.D.J. & Hartwick, F.D.A., ApJ, in press
3. Mobasher, B., *et al.*, MNRAS, in press
4. Yee, H., 37th Herstmonceux conference on "HST and the High Redshift Universe"

---

[a] Mark Dickinson (PI), Matthew Bershady, Andrew Connolly, Peter Eisenhardt, Richard Elston, Giovanni Fazio, Harry Ferguson, Mauro Giavalisco, David Koo, Tod Lauer, Marc Postman, Adam Stanford, Chuck Steidel, Alex Szalay, Eric Tollestrup, Ned Wright.

STAR-FORMING GALAXIES AT REDSHIFT $z > 3$ IN THE QSO BR 1202-07 FIELD

S. D'ODORICO

European Southern Observatory, K.Schwarzschild St. 2, D-85748 Garching

A. FONTANA, E. GIALLONGO

Osservatorio Astronomico di Roma, I-00040 Monteporzio

S. CRISTIANI

Dipartimento di Astronomia, Universita' di Padova, I-35100 Padova

E. EGAMI

Max-Planck-Institut für Extraterrestrische Physik, Postfach 1603, D-85740 Garching

E.M. HU

IFA, University of Hawaii, Honolulu, HI96822, USA

R.G. MCMAHON

Institute of Astronomy,University of Cambridge,Madingley Road,Cambridge CB30HA, UK

Deep photometry in the B,V,r and I colors down to magnitudes r = 25 has been obtained for a field centered on a $z = 4.7$ QSO. A number of high redshift ($z > 3$)candidates has been selected by comparison with models of star-forming galaxies. The corresponding density is ~ 2.5/arcminute^2. The redshifts of three galaxies at $z = 4.7, 4.8$ and 4.6 have been confirmed spectroscopically.

1 Color selection of high z candidates

In a star forming galaxy with the UV spectrum dominated by hot massive stars , the Lyman continuum break at 91.2 nm is a conspicuous feature which can be used to identify high redshift objects with multi-colour observations. Steidel[1] et al. pioneered this technique by using deep images in three bands, U_n, G and R. A large fraction of the candidates were later confirmed by spectroscopy with LRIS at the Keck , revealing a substantial population of galaxies of this type at redshifts > 2.8(Steidel[2] et al.). Their three band identification technique is not very effective beyond $z \sim 3.4$ because of the increasing effect of the Ly α IG absorption . We have carried out deep observations in a $2.2' x 2.2'$ field centered on the $z = 4.7$ QSO BR 1202-07 in B,V, r Gunn and I colors at the ESO 3.5m NTT. Our data are characterized by the excellent image

Table 1: Colors of candidates with spectroscopic redshifts.

Identification	z	m(r)	B-r	V-r	r-I	I-K
QSO NW comp	4.7	24.3	> 2	1.9	0.2	1.1
ESO #12	4.8	23.7	> 2.7	no V mag	0.7	0.95
ESO #47	4.6:	24.1	> 2.1	1	0.6	no K

quality ($FWHM$ < 0.6 for the r and I images) and depth ($m(r)_{3\sigma}$= 25.6). Details of the observations and the discovery of an high redshift galaxy close to the QSO have been reported elsewhere (Fontana[3] et al).The data have been complemented by K observations down to a magnitude limit of $\sim$ 21.5 at the IFA 2.2m telescope at Manua Kea and by HST data. A catalogue of the colors of around 200 galaxies down to m(r)=25 has now been prepared and compared with a grid of model spectra for star forming galaxies of different redshift, age, time scale of star formation and initial mass function from Bruzual and Charlot (in preparation). Preliminary results were presented by Giallongo[4]. The full catalogue of photometric redshifts is in preparation. The use of 5 color bands, including the infrared, leads to the selection of 12 robust candidates for the redshift interval 3 < z < 4.5 corresponding to a density of 2.5/arcminute^2.

2 Spectroscopy results

A few of the high redshift candidates have been observed with the ESO NTT EMMI spectrograph (7hrs integration) and with the LRIS spectrograph at Keck1 (2 hrs). We have measured the spectra of three of the robust, high z candidates (D'Odorico et al., in preparation). The redshift of the QSO companion is based on a strong emission line which is unambiguously identified with Ly α. The redshift of ESO #12 is based on the Lyman break and on the IG Ly α absorption trough as seen both in the Keck (see review of E.Hu in this volume) and in the ESO spectra. The resdhift of ESO #47 is also based on the Lyman feature identifications in a LRIS spectrum of low S/N.

1. C.S. Steidel *et al* ,AJ, 110, 2519, 1995.
2. C.S. Steidel *et al* ,ApJ, 462, L17, 1996.
3. A.Fontana *et al*, MNRAS, 279, L27, 1996.
4. E. Giallongo *et al*, in *The Early Universe with the VLT*, ed. J.Bergeron (Springer Verlag, 1996)

GLOBAL CHEMICAL EVOLUTION

M.G. EDMUNDS[a], S. PHILLIPPS[b] & S.A. EALES[a]

[a] Department of Physics and Astronomy, University of Wales, College of Cardiff,
PO Box 913, Cardiff CF2 3YB

[b] Astrophysics Group, Department of Physics, University of Bristol, Tyndall
Avenue, Bristol BS8 1TL

We give two weighted estimates of the current abundance of heavy elements in the Universe, and discuss its past evolution with time and redshift. Elementary chemical evolution models show the need to understand just *what* is being observed at a given redshift in quasar absorption spectra. We point out implications for global mm-wave emission from dust in galaxies, and for the general evolution of the star formation rate in galaxies.

Quasar absorption spectra are the result of a cross-section *area weighted* sample of absorbing gas in the Universe, along a line-of-sight. An elementary estimate[1] of the average abundance which would be seen locally (with various assumptions about radial gradients in galaxies, metallicity-luminosity relations and luminosity functions) gives $12 + \log(O/H) = 8.4$, compared to the solar abundance of 8.95. An immediate consequence is that the rate of chemical evolution between distant absorption systems and now may not be as great as is sometimes claimed, since a solar or even higher value has often been assumed as the current value. The spread locally may be expected to be of order 1.5 dex, depending on what the sightline happens to hit. A similar analysis[2] can be made with *mass* weighting, and including the stellar component of galaxies and the hot gas in clusters of galaxies. The result in this case is a nice Copernican value of $12 + \log(O/H) = 8.8$, i.e. the Universe's heavy element abundance is very nearly solar!

How have these mean abundances changed with time? An estimate of the total *mass* of heavy elements formed up to a given epoch can be made from simply estimating the total amount of star formation - because when stars are formed, so inevitably are supernovae progenitors, and hence heavy elements soon afterwards. We assume reasonable bounds on the global star formation rate: either it has remained constant since the start of galaxy formation at redshift z_f, or has not decreased more rapidly than an exponential with a peak rate of 27 times, and a mean rate of 8 times, the current rate. Mapping this into redshift with a cosmological model shows a rather narrow region (only about 0.4 dex) between the two bounds at any redshift. But this is what is *there* not what is *seen*. The global star formation rate bounds we assume are considerably milder than some recent estimates based on counts of damped Lyman-alpha systems [3] or galaxy counts [4], but these other investigations may be sampling only a fast-evolving subset of the global population.

Moving from the global mass of metals (which corresponds to approximately solar composition at the present epoch) to what may be seen observationally at different redshifts requires a chemical evolution model. As a preliminary investigation we use, like other workers [5], a "linear" inflow or outflow model in which the flow of gas is directly proportional to the star formation rate. We can derive [2] a few results for *arbitary* outflows, but arbitary inflows are more difficult. We differ from Pei & Fall [5] in evolving to a *fixed* galaxy mass at the current epoch. The spread in expected gas phase abundances (which might be observed by qso sightlines) is much larger than the spread in *total* abundance at a given redshift z. With inflow models there can be comparatively modest evolution in gas phase abundances with redshift in recent epochs. But what is particularly clear is that we need to be certain whether we are looking at a region in which outflow or inflow has ocurred - any individual observation may be a poor representation of the mean abundance (and by implication the star formation history) either in the Universe, or in the individual galaxy being observed.

It can be argued [6] that the dust mass primarily determines the output of mm-wave radiation from galaxies. Our models can be used to predict dust masses, assuming that the amount of dust follows the mass of heavy elements in the interstellar medium. What is surprising is how *late* ($z \sim 1$) the total dust mass would peak, although individual sources may peak much earlier. The consequences for the general sub-mm background are being assessed. [7]

Acknowledgments

We thank Max Pettini for encouragement, and wish Alec B. well in sorting out the z = 1 to 5 Universe over the next few years!

References

1. S. Phillipps & M.G. Edmunds MNRAS, 281, 362, 1996.
2. M.G. Edmunds & S. Phillips *submitted* MNRAS 1996.
3. K.M. Lanzetta, A.M. Wolf & D.A. Turnshek ApJ, 440, 435, 1995.
4. S.J. Lilly, O. Le Fevre, F. Hammer & D. Crampton ApJ, 406, L1, 1996.
5. Y.C. Pei & S.M. Fall ApJ, 454, 69, 1995.
6. S.A. Eales & M.G. Edmunds MNRAS, 280, 1167, 1996.
7. S.A. Eales & M.G. Edmunds *submitted* MNRAS 1996.

UNRESOLVED BLUE OBJECTS IN THE HUBBLE DEEP FIELD: BURSTING DWARFS?

REBECCA A.W. ELSON

Institute of Astronomy, Madingley Road, Cambridge CB3 0HA, England

1 Summary

We used DAOPHOT to search for point-like objects in the Hubble Deep Field. We fit point-spread functions to all objects down to a detection threshold of 6σ, and used the "sharpness" parameter to distinguish between resolved and point-like sources, by comparing with simulated stars. Details are given in Elson, Santiago & Gilmore (1996). Our search revealed 50 objects with $V_{606} < 29$ which are too blue and faint to be low mass main-sequence stars in our Galaxy. They all have $V_{606} > 26$, and are among the bluest objects in the HDF. Although their colours are consistent with those of white dwarfs, we argue that they are unlikely to be. About half have one or more nearby companion. They are probably distant systems undergoing intense bursts of star formation.

The objects have FWHM~ 0.15 arcsec, which at $z \sim 1$, implies radii of less than ~ 1 kpc. Typically companions appear at an angular separation of about 0.4 arcsec, corresponding to ~ 3.5 kpc. If the objects are at $z \sim 1$ then they would have $M_V \sim -17$, suggesting dwarf galaxies in the process of forming.

Babul & Rees (1992) suggest that the "excess" faint blue galaxies may be dwarf ellipticals undergoing a starburst at $z \approx 1$, and Babul & Ferguson (1996) explore the observational properties of such a population. Figure 1 compares the expected $B_{450} - I_{814}$ distribution with that of our sample. The agreement is qualitatively good.

Figure 2 shows the spatial distribution of the objects in Chip 2. Interestingly, of the 50 objects in the sample, nearly half are in this chip, and there is evidence for clustering.

Preliminary redshifts are available for four of the 50 objects; all four have $z > 2.5$ (J. Lowenthal 1996, private communication; this volume).

References

1. A. Babul and H. Ferguson, ApJ, 458, 100, 1996.
2. A. Babul and M. Rees, MNRAS, 255, 346, 1992.
3. R. Elson, B. Santiago, and G. Gilmore, New Astronomy, in press

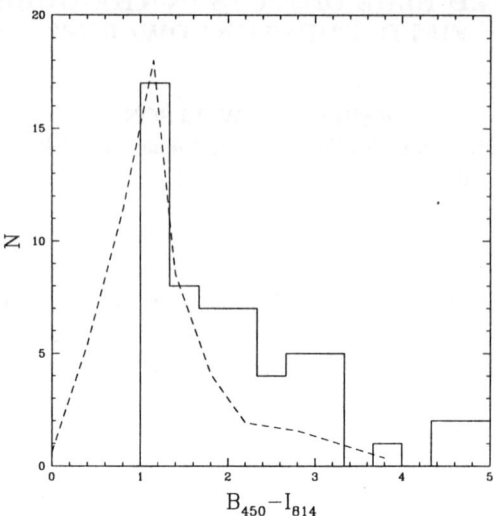

Figure 1: $B_{450} - I_{814}$ distribution for the 50 objects in our sample (solid histogram), compared to the distribution predicted by Babul and Ferguson (1996) for starbursting dwarf ellipticals with $z > 1$ (dashed line).

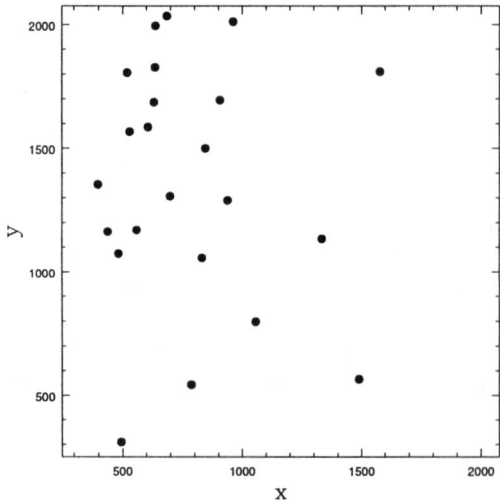

Figure 2: Spatial distribution for the 23 objects in Chip 2.

CONSTRAINTS ON GALAXY MERGERS AT MODERATE REDSHIFTS FROM THEIR OBSERVED LUMINOSITY AND SPATIAL DISTRIBUTION FUNCTIONS

FAN FANG

Department of Astronomy, University of Virginia, Charlottesville, VA 22903, USA

WILLIAM C. SASLAW

Department of Astronomy, University of Virginia, Charlottesville, VA 22903, USA
and
Institute of Astronomy, Madingley Road, Cambridge, CB3 0HA, UK

Galaxy merging is known to be a possible form of evolution which may explain the excess of faint-blue galaxy counts at $z \lesssim 1$. If merging dominates galaxy evolution, it would also have left its signature in the evolution of the galaxy spatial distribution. We report a useful statistical approach which can describe this effect, and relate the galaxy spatial counts-in-cells distribution directly to the evolution of the luminosity function.

We use a master equation to describe the merging evolution of the counts-in-cells distribution function $f(N)$. The equation has the form

$$\Delta f(N,t) = \sum_{N_j>0} [\Gamma(N+N_j, N)\Delta t] f(N+N_j, t) - \sum_{N_j>0}^{(N-1)} [\Gamma(N, N-N_j)\Delta t] f(N,t),$$

where $\Gamma(N+N_j, N)$ describes the probability of N_j mergers during an interval Δt in cells containing $N + N_j$ galaxies originally. It is related to the merging cross section, which is given by a parameterized gravitational focusing model, considering upper limits for the relative speed of galaxy halo merging[1]. A similar master equation describes the merging evolution of galaxy mass, and can be directly related to the evolution of the galaxy luminosity function by assuming a constant (or varying) mass-to-light ratio. The master equations relate different evolution processes through the common merging cross section, so that these evolution processes can constrain each other.

The Autofib Redshift Survey[2] obtained the galaxy luminosity function up to $z \sim 0.7$, which constrains the merging evolution in our model. Figure 1 shows the constrained evolution at $z \sim 0$ (dot-dashed line). We notice that merging cannot fully reproduce the observed local faint-end luminosity function unless these faint galaxies have become differentially fainter by an amount proportional to 10^M (dot-dot-dot-dashed line). This may imply a relative lack of star-forming activity for these galaxies after merging, possibly caused by more mass loss in smaller mergers.

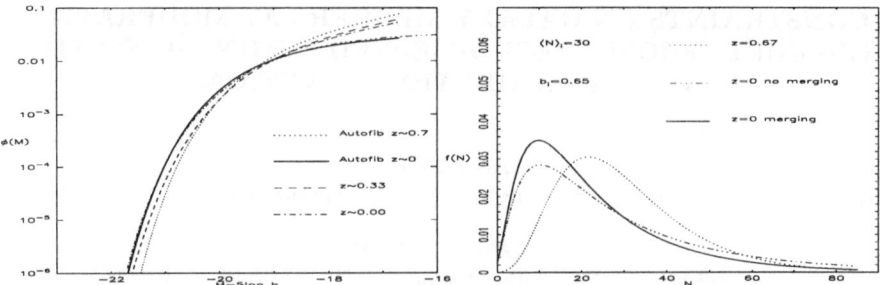

Figure 1: (left) Luminosity functions for merger-driven galaxy luminosity evolution with $M \propto L$. The dot-dot-dot-dashed line represents a form of differential dimming, which forces the faint-end to fit the observation at $z \sim 0$.

Figure 2: (right) The combined evolution with adiabatic galaxy clustering and merging considering a Gaussian distribution for galaxy halo sizes. Starting with the GQED distribution with $\overline{N}_{initial} = 30$, $b_{initial} = 0.65$ (dotted line), the solid line shows the result of this combined evolution which gives $b = 0.74$ at $z = 0$. The dot-dot-dot-dashed line shows the resulting distribution for pure adiabatic clustering which gives $b = 0.8$ at $z = 0$. There is a significant difference between these two distributions in both the values of b and the shape of the distribution function.

The constrained merging leads to a constrained evolution of the counts-in-cells distribution shown in Figure 2, where we assume a Gaussian distribution for the size of galaxy halos (average value is $\sim 10^2 Kpc$ constrained by merging in Figure 1). We also assume galaxy clustering at $z \sim 0.7$ to be described by a gravitational quasi-equilibrium distribution function (GQED)[3]. This function describes[4] the galaxy spatial distribution at $z \sim 0$ with b, the ratio of gravitational potential energy to kinetic energy, to be about 0.8. If galaxies did not merge in the past, the evolution of the galaxy distribution would be adiabatic, described by the cosmic energy equation[5]. Figure 2 indicates significant differences between the evolution with and without merging. The merging evolution is also found to retain the GQED form (solid line), which probably implies that the centers of mass of merged galaxies become dynamically relaxed. We conclude that a counts-in-cells analysis for distant galaxies (when statistical samples become available) would directly indicate the significance of the role of merging in the past galaxy evolution.

References

1. Roos, N. & Norman, C. A. 1979, A&A, 76, 75
2. Ellis, R. S., Colless, M. M., Broadhurst, T., Heyl, J., & Glazebrook, K. 1996, MNRAS, in press
3. Saslaw, W. C. & Fang, F. 1996, ApJ, 460, 16
4. Sheth, R. K., Mo, H. & Saslaw, W. C. 1994, ApJ, 427, 562
5. Saslaw, W. C. 1992, ApJ, 391, 423

LINEAR RECONSTRUCTION OF THE HUBBLE DEEP FIELD

A.S. FRUCHTER

Space Telescope Science Institute, 3700 San Martin Dr., Baltimore, MD 21218, USA, fruchter@stsci.edu

R.N. HOOK

Space Telescope-European Coodinating Facility, ESO, Karl-Schwarzschild-Str. 2, D-85748 Garching, Germany, rhook@eso.org

We have developed a method for the linear reconstruction of an image from under-sampled, dithered data, which has been used to create the distributed, combined Hubble Deep Field images. The algorithm, known as Variable-Pixel Linear Reconstruction (or informally as "drizzling"), preserves photometry and resolution, can weight input images according to the statistical significance of each pixel, and removes the effects of geometric distortion both on image shape and photometry. In this paper, the algorithm and its implementation are described, and measurements of the photometric accuracy and image fidelity are presented. In addition, we present an example of the use of drizzling to combine dithered images in the presence of cosmic rays.

Although the optics of WFPC2 now provide a superb PSF, the detectors at the focal plane severely undersample the image. This problem is most severe on the three WF chips, where the width of a pixel equals the FWHM of the optics in the the near-infrared, and greatly exceeds it in the blue. While much high spatial frequency information in the image is permanently destroyed by smearing with the response of the "fat" pixels, the quality of the image can nevertheless be greatly improved by combining sub-pixel dithered images, as is illustrated on the next page.

In order to combine the dithered images of the Hubble Deep Field, we have developed a new technique for the linear combination of images which we call variable-pixel linear reconstruction, or, informally, "drizzling." Drizzling can be thought of as a continuous set of linear functions that vary smoothly from the optimum linear combination technique – interlacing – to the old-standby, shift-and-add. The degree to which one must depart from interlacing and move towards shift-and-add is determined by the nature of the input data.

Sadly, a lack of space prevents us from further explaining the technique in these proceedings. However, the reader can find the entire contents of our poster at http://www.stsci.edu/fruchter/dither/drizzle.html.

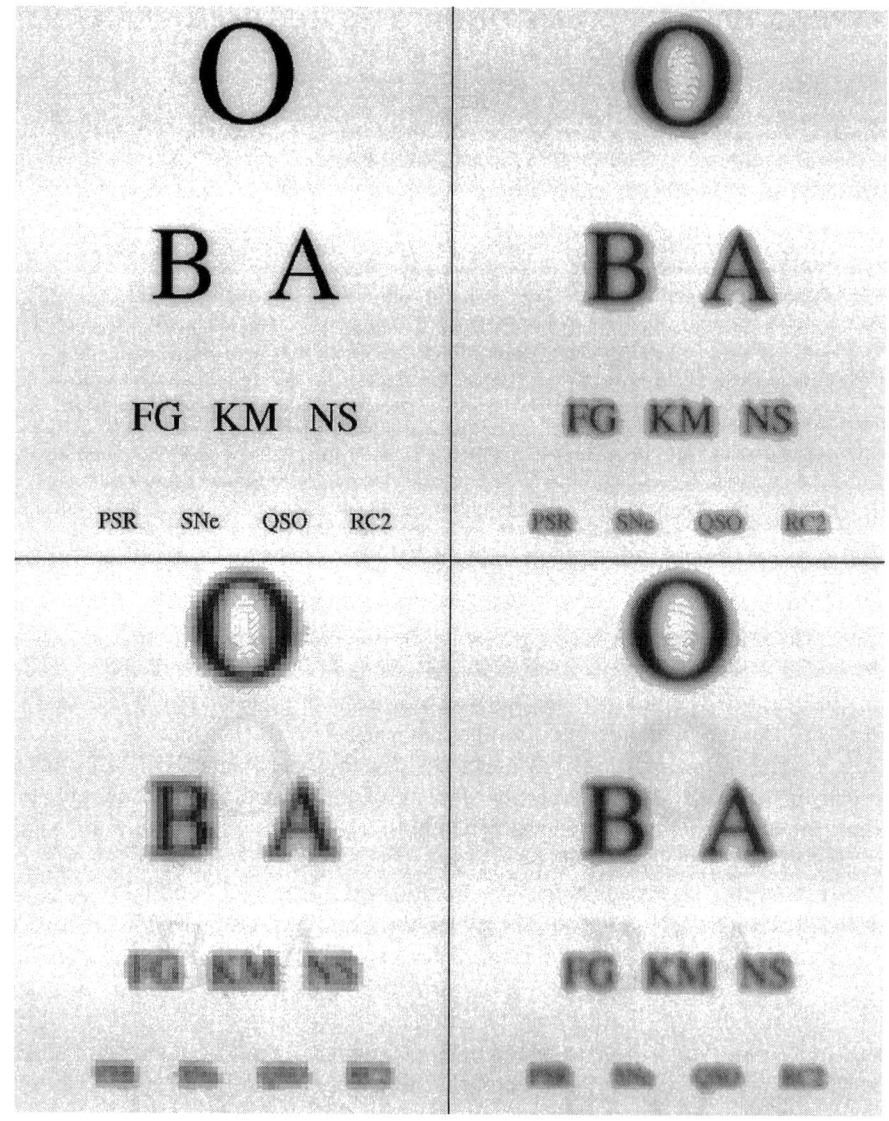

Figure 1: In the upper left corner of this figure, we present the "true image", *i.e.* the image one would see with an infinitely large telescope. The upper right shows the image after convolution with the WFPC PSF. The lower right shows the image after sampling by the WF2 CCD, and the lower right image shows a linear reconstruction of dithered CCD images.

DETECTION OF Hα IN A $z = 1$ STAR-FORMING SPIRAL GALAXY

KARL GLAZEBROOK,

Anglo-Australian Observatory, PO Box 296, Epping, NSW 2121, AUSTRALIA

FROSSIE ECONOMOU

*Joint Astronomy Centre, 660 North A'ohoku Place,
University Park Hilo, HI 96720, USA*

In the last year HST observations by a variety of teams have discovered that an increasing fraction of field galaxies beyond $z = 0.5$ have peculiar or disturbed morphologies. These galaxies usually exhibit blue colours and strong [OII] line emission which would indicate that they are undergoing vigorous star-formation, possibly induced by dynamical processes such as galaxy assembly or tidal interactions. We report the early results of a new program to detect $H\alpha$ in the J band in $z \sim 1$ galaxies. $H\alpha$ luminosity provides a more reliable measurement of actual star-formation rates than [OII] and would allow us to investigate how star-formation processes correlates with physical morphology.

1 OBSERVING BETWEEN THE OH LINES

The bright night sky in the J-band ($J = 15.3\,\mathrm{mags/arcsec^2}$) is a problem for obtaining spectra of extremely faint ($J \sim 19\text{--}20$) $z = 1$ galaxies. However it is believed that most of this background (possibly as much as 95%) derives from narrow OH emission lines.

Our strategy with CGS4 on UKIRT is to go to an intermediate resolution ($R = 2000$) where the OH lines are highly resolved and should contaminate only about 50% of the J-band. We select candidates from a sample with known redshifts (the combined CFRS/LDSS2 field galaxy catalog previously observed with HST) and *choose* the redshifts such that the $H\alpha$ line lies away from any strong OH line. As we are just interested in measuring a single line at a known position the reduced spectral range does not matter. We benefit from from removing the OH background without any throughput losses associated with OH-suppression technology.

2 THE DISTANT GALAXY CFRS 14.0600

We detected $H\alpha$ in the CFRS object 14.0600 ($z = 1.038$) which is classified from our HST data as morphologically peculiar: its large asymmetry would place it at the extreme end of the asymmetry/concentration plane of Abraham

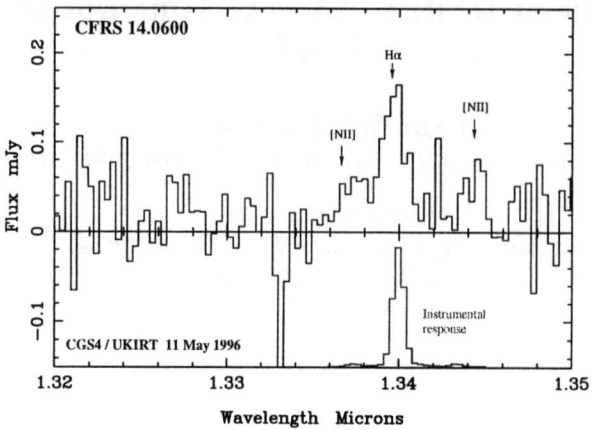

Figure 1: Extracted, combined and fluxed total spectrum from the OBJECT and SKY rows. The profile of an arc line is also shown — the $H\alpha$ line is clearly resolved (with a velocity width of $150\,\mathrm{km\,s^{-1}}$ which is reasonable for this galaxy) and is accompanied by possible [NII] lines at the correct redshift.

et al (1995, ApJS, in press). Closer visual examination suggests it could be a spiral galaxy with very bright blue knots in spiral arms.

The CGS4 data is shown in Figure 1. The continuum is marginally detected — the level is $J \simeq 19.8$ which is reasonable since the CFRS magnitude is $I = 21.0$.

The total $H\alpha$ flux is $3.9 \times 10^{-19}\,\mathrm{W\,m^{-2}}$ giving a luminosity (for $H_0 = 50\,\mathrm{km\,s^{-1}\,Mpc^{-1}}$) of $2.5\text{–}3.9 \times 10^{35}\,\mathrm{W}$ (for $\Omega_0 = 1\text{–}0.1$). When converted into a SFR using the scaling of Kennicutt (1983, ApJ 272, 54) this gives 22–34 $M_\odot\mathrm{yr^{-1}}$ which is enough to make an L^* galaxy in 3–5 Gyr. (This particular galaxy is $L \sim 2L^*$). For comparison the age of the Universe at $z = 1$ is 4–9 Gyr ($H_0 = 50\,\mathrm{km\,s^{-1}\,Mpc^{-1}}$, $\Omega_0 = 1\text{–}0$). Interestingly the [OII] emission in the optical spectrum it is not particularly high — the equivalent width (26Å) is pretty typical of galaxies at these redshifts. The ratio $H\alpha/[\mathrm{OII}] = 1.9$ is close to the median of the local values found by Kennicutt (1992, ApJ 388, 310).

We conclude that this is the first definite evidence that the prevalence of disturbed morphologies seen in high-redshift HST galaxy samples is correlated with major episodes of star-formation in these galaxies. A complete sample is now obviously desirable and this will be the target of our winter observing campaign at UKIRT.

OPTICAL IDENTIFICATIONS OF FAINT RADIO SOURCES

C. GRUPPIONI

Dipartimento di Astronomia dell'Università di Bologna, via Zamboni 33,
I-40126 Bologna, Italy

M. MIGNOLI, G. ZAMORANI

Osservatorio Astronomico di Bologna, via Zamboni 33,
I-40126 Bologna, Italy

Deep imaging and spectroscopy have been carried out searching for optical counter-parts of two complete samples of $S > 0.2$ mJy radiosources in the Marano Field. The samples consist of 63 and 48 sources respectively at 1.4 and 2.4 GHz (43 sources are in common to the two samples). Photometric identifications, based on ESO 3.6-m plates (reaching $m \sim 22.5$ in the U, J and F bands) and on deep CCD exposures taken at ESO NTT (reaching $m \sim 24$–25 in the U, B, V and R bands), show that a significant fraction ($\gtrsim 30\%$) of the identified radio sources occurs in pairs or small groups. This suggests an enhanced radio emission due to interac-tion or merging phenomena. Spectroscopic data with the ESO 3.6-m telescope have been obtained for 25 optical candidates and a reliable redshift determination has been derived for 21 of them. The majority of the identifications resulted to be galaxies, about equally subdivided among early–type galaxies and low redshift starburst and post–starburst galaxies.

1 The Radio Surveys

Deep radio surveys with the Australia Telescope Compact Array (ATCA) have been carried out at 1.4 and 2.4 GHz, with a limiting flux of ~ 0.2 mJy at each frequency, in the *Marano Field* (centered at $\alpha(2000) = 03^h 15^m 09^s, \delta(2000) = -55°13'57''$), for which deep optical and X-ray data are also available. The two radio samples, complete at the $5\sigma_{local}$ level, consist of 63 and 48 sources respectively at 1.4 and 2.4 GHz. The main results of our analysis of the radio data are described by Gruppioni *et al* (1996)[1]. By studying the spec-tral index distribution as a function of flux, we found a significant flattening of the spectral index toward fainter fluxes for the higher frequency selected sample (2.4 GHz), while the median spectral index (α_{med}) was consistent with remaining constant at ~ 0.8 ($f_\nu \propto \nu^{-\alpha}$) for the sample selected at 1.4 GHz. However, at both frequencies a significant number of sources with inverted spectrum do appear at flux densities $\lesssim 2$ mJy. In particular, objects with in-verted spectra constitute $\sim 13\%$ of the total 1.4 GHz sample and $\sim 25\%$ of the total 2.4 GHz one. For this sample this percentage increases to $\sim 40\%$ for $S < 0.6$ mJy.

2 Optical Observations

The entire radio field (0.36 sq. deg.) is covered by 3.6–m plates (U, J and F), reaching $m \sim 22.5$, while CCD exposures, taken at ESO NTT in the U, B, V and R bands and reaching $m \sim 24$–25 in each band, cover $\sim 60\%$ of the radio field.

Spectroscopic data with ESO 3.6–m telescope have been obtained for 25 optical counterparts (with $m_R \lesssim 23$) of our radio sources ($\sim 40\%$ of the complete samples) and a redshift determination has been derived for 21 of them.

The main results from our optical identification analysis are:

i) A significant fraction ($\gtrsim 30\%$) of the radio sources with identification in a CCD frame occurs in pairs or small groups and this suggests an enhanced radio emission due to interaction or merging phenomena.

ii) Of the spectroscopically observed objects, three are classical AGNs: two of them are broad lines QSOs at z=1.663 and 2.166 and one, with narrow lines, is probably a Seyfert 2 at z=0.688. All the other spectroscopically identified objects are galaxies, about equally subdivided among $0.1 \lesssim z \lesssim 0.7$ early–type galaxies (3 with inverted radio spectrum, then probably powered by a low luminosity AGN) and low redshift (z<0.3) starburst and post–starburst galaxies. One object is identified with a galaxy with a tentative redshift of 1.26 on the basis of a strong emission line assumed to be [OII]. The four objects for which the redshift could not be measured from the spectrum do not show emission lines.

iii) Radio sources identified with objects showing enhanced starburst activity do all have steep radio spectra ($\alpha > 0.5$), consistent with the hypothesis of non–thermal radio emission from supernova remnants. Almost all of the spectroscopically identified sources with inverted radio spectrum ($\alpha < 0$) are identified with early–type galaxies at intermediate redshift. No evidence of enhanced starburst activity is seen in any of them. Thus, we conclude that radio emission from these sources is probably powered by an AGN, in agreement with the results of Hammer et al (1995) [2].

References

1. C. Gruppioni, G. Zamorani, H.R. de Ruiter, P. Parma, M. Mignoli and C. Lari, submitted to MNRAS, 1996.
2. F. Hammer, D. Crampton, S.J. Lilly, O. Le Fèvre and T. Kenet, MNRAS, 276, 1085, 1995.

THE REDSHIFT DISTRIBUTION AND LUMINOSITY FUNCTIONS OF THE HUBBLE DEEP FIELD

STEPHEN D. J. GWYN, F. D. A. HARTWICK

Department of Physics and Astronomy, University of Victoria,
Box 3055, Victoria, British Columbia, V8W 3P6, Canada
gwyn@uvastro.phys.uvic.ca, hartwick@uvastro.phys.uvic.ca

Photometric redshifts have been determined for the galaxies in the Hubble Deep Field[1] selected to be brighter than I_{ST}=28 as measured through a 0.2″ aperture. The resulting redshift distribution shows two peaks: one at $z \sim 0.6$ and one at $z \sim 2.2$. Luminosity functions derived from the redshifts show strong evolution as a function of redshift.

The Hubble Deep Field[1] (HDF) optical images are the deepest yet obtained. At this point in time, only a few spectroscopic redshifts have been measured for the brighter galaxies and none for the faintest galaxies in these images. Photometric redshifts[3,4] have been calculated for the galaxies brighter than $I_{ST} = 28$ in the Hubble Deep Field. Here, the redshift distribution and luminosity functions out to $z = 5$ are presented.

The photometric redshift technique proceeds as follows: The photometric data for each galaxy in the field are converted into spectral energy distributions (SED's) and compared to a set of template spectra of all Hubble types and redshifts ranging from $z = 0$ to $z = 5$ compiled from the Bruzual & Charlot[5] galaxy evolution models. The best matching template determines the redshift. When the photometric redshifts are compared to the available spectroscopic redshifts in the HDF[6,7], it was found that the uncertainties in the photometric redshifts at low ($z < 1.5$) and high ($z > 1.5$) redshifts are $\sigma_{pz} \simeq 0.2$ and $\sigma_{pz} \simeq 0.5$ respectively.

The left hand figure shows the redshift distribution thus derived. It shows two peaks: one at $z \sim 0.6$ and one at $z \sim 2.2$. It should emphasized that our selection criterion was essentially one of surface brightness ($I_{ST} < 28$ within a 0.2″ aperture). When allowance is made for this difference in sample selection our results are compatible with that of Yee (these proceedings). Our redshift distribution also shows similarities to the distribution that Mellier (these proceedings) derives from gravitational lensing. The right hand figure shows the evolution of the luminosity function with redshift. Shown for comparison as a solid line is the local luminosity function of Loveday.[8] We wish to re-emphasize that these luminosity functions describe those galaxies with the highest surface brightness and therefore presumably the most active star formation.

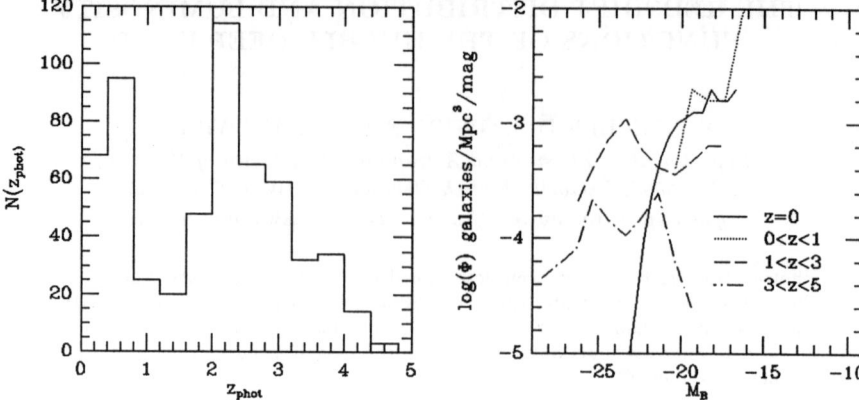

The variations in the luminosity function with redshift can be explained if there are two epochs at which galaxies undergo their first major burst of star formation. The first occurs from $2.5 < z < 5$ (the high redshift peak); at this time the larger galaxies start forming stars. These bright, bursting galaxies populate the bright end of the higher redshift luminosity functions. Babul & Rees[2] postulated that star formation in dwarf galaxies is delayed until $z \sim 1$ by photoionization of their gas by inter-galactic ultraviolet radiation. This second epoch of star formation explains the observed excess at moderate redshifts of faint blue galaxies and the low redshift peak of the histogram. These galaxies end up on the steeply rising faint tail of the lowest redshift luminosity function.

References

1. R.E. Williams, *et al.*, AJ, in press, 1996
2. A. Babul & M.J. Rees MNRAS, 255, 364, 1992.
3. S.D.J. Gwyn, unpublished M.Sc. thesis, 1995
4. S.D.J. Gwyn & F.D.A. Hartwick, [astro-ph/9603149], ApJL, 468, L77, 1996.
5. G.A. Bruzual & S. Charlot, ApJ, 405, 538, 1993.
6. J.G. Cohen, L.L. Cowie, D.W. Hogg, A. Songaila, R. Blanford, & E. M. Hu, to appear in ApJL, [astro-ph/9608121], 1996
7. C.C. Steidel, M. Giavalisco, M. Pettini, M. Dickinson & K.L. Adelberger, to appear in AJ [astro-ph/9604140], 1996
8. J. Loveday, B.A. Peterson, G. Efstathiou & S.J. Maddox ApJ, 390, 338, 1992.

THE TWO-POINT CORRELATION FUNCTION AND MORPHOLOGICAL SEGREGATION IN THE OPTICAL REDSHIFT SURVEY

S. HERMIT

Astronomical Observatory, Juliane Maries Vej 30, 2100 Kbh. Ø, Denmark

B.X. SANTIAGO, O. LAHAV

Institute of Astronomy, Cambridge University, Madingley Road, Cambridge CB3 0HA, United Kingdom

M.A. STRAUSS

Dept. of Astrophysical Sciences, Princeton University, Princeton, NJ 08544, U.S.A.

M. DAVIS

Physics and Astronomy Departments, University of California, Berkeley, CA 94720, U.S.A.

A. DRESSLER

Observatories of the Carnegie Institution of Washington, 813 Santa Barbara Street, Pasadena CA 91101, U.S.A.

J.P. HUCHRA

Center for Astrophysics, 60 Garden Street, Cambridge, MA 02138, U.S.A.

In this brief contribution we report on results for the clustering of galaxies in real and redshift space using the Optical Redshift Survey (ORS).

1   Introduction

It is an important goal of astronomy to understand the formation of galaxies and large scale structure. One of the ways to constrain these models is by quantifying the clustering of the galaxies. In this brief presentation we present results for the two point correlation function in redshift space and real space, $\xi(s)$ and $\xi(r)$, respectively.

We first find $\xi(s)$ for several subsamples of ORS, spanning nearly a factor of 30 in volume. We detect significant variations in $\xi(s)$ among the subsamples covering small volumes. For volumes $> (75h^{-1}\mathrm{Mpc})^3$, however, the ORS subsamples present very similar clustering patterns. Fits of the canonical form $\xi(s) = (\frac{s}{s_0})^{-\gamma_s}$ give best-fit values in the range $1.5 \leq \gamma_s \leq 1.7$ and $6.5 \leq s_0 \leq 8.8h^{-1}$ Mpc for several samples extending to redshifts of 8000 km s^{-1}. We find significant differences in clustering between the diameter-limited and magnitude-limited ORS samples within a radius of 4000 km s^{-1} centered on the Local Group; $\xi(s)$ is larger for the magnitude-limited samples than for diameter-limited ones. We interpret this as an indirect result of morphological segregation coupled with differences in morphological mix.

2 Morphological segregation

It has been suggested by several works that galaxies of different morphological type have different clustering properties. In order to test this we split ORS into morphological subsamples and compute $\xi(s)$ for each subsample. Our results confirm the existence of morphological segregation of galaxies out to scales of $s \sim 10h^{-1}$ Mpc, and indicate that the relative bias factor between early type galaxies and late-types may be weakly dependent on scale. If real, this would suggest non-linear biasing.

3 Clustering in real space

We also compute correlations as a function of radial and projected separations, $\xi(r_p, \pi)$, from which we derive the real space correlation function, $\xi(r)$. We obtain $4.9 \leq r_0 \leq 7.3h^{-1}$ Mpc and $1.5 \leq \gamma_r \leq 1.7$ for various ORS samples. The results obtained in real space confirm those found using $\xi(s)$, i.e. in small volumes, magnitude limited samples show larger clustering than do diameter limited ones. There is no difference when large volumes are considered.

4 Conclusions

For volumes $> (75h^{-1}\mathrm{Mpc})^3$, the ORS subsamples present very similar clustering patterns. We find that magnitude limited samples show stronger clustering than diameter limited ones only in small local volumes. This result is shown to hold in real space as well. Finally, we detect morphological segregation out to scales of $s \sim 10h^{-1}$ Mpc and find evidence that the relative bias factor may be weakly dependent on scale.

See Hermit et al.[?] for the entire paper.

References

1. S. Hermit, B.X. Santiago, O. Lahav, M.A. Strauss, M. Davis, A. Dressler & J.P. Huchra, 1996, MNRAS in press. Preprint astro-ph 9608001

THE REDSHIFT DISTRIBUTION IN THE HUBBLE DEEP FIELD

DAVID W. HOGG, JUDITH G. COHEN, ROGER BLANDFORD, PATRICK SHOPBELL

California Institute of Technology, mail code 130-33, Pasadena CA 91125

LENNOX L. COWIE, ESTHER M. HU, ANTOINETTE SONGAILA

Institute for Astronomy, University of Hawaii

These observations represent the first results of a redshift survey in the Hubble Deep Field[1] (HDF) with LRIS[2] on the Keck Telescope (Cohen, Caltech PI; Cowie, Hawaii PI), extending ongoing surveys[3,4] to an additional field. A total of 140 successful redshifts were acquired in two rectangular strips 2 x 7.3 arcmin^2 centered on the HST field in 1996 January, March and April. One strip, containing K-selected objects, was aligned east-west while the second, visual-selected, was aligned at a position angle of 30 deg. The sample is incomplete; plans exist to complete the sample in a number of photometric bandpasses; but in view of the great interest in the HDF and the many follow up studies in progress, we are making the redshift list available now on the WWW[5]. Some additional redshifts are presented by Illingworth et al and Moustakas et al (both this proceedings).

The median redshift is 0.53. The redshift distribution shows extremely pronounced, statistically significant peaks at redshifts 0.321, 0.457, 0.475, 0.516, 0.559 and 0.680, with rest-frame velocity dispersions 170, 310, 315, 595, 410 and 265 km s^{-1} respectively (no correction for instrumental uncertainty has been made so these are upper limits). These peaks together contain ~ 40 percent of the sample. Their appearance in this sample strongly supports the hypothesis that similar features found in other fields[3,6] are generic.

There is no evidence that the K-selected sample in the E-W strip shows more clustering than the visual-selected objects in the 30 deg strip. Furthermore, from by-eye morphological classification of the entire sample, there is no excess of ellipticals among peak members relative to non-members as might be predicted by the morphology-density relation[7]. We believe that the peaks are robust to selection technique and not a result of a selection bias.

By itself, this sample is too small for reliable measurement of the two-point correlation function. However, there is a 5σ excess correlation in the 500 to 1000 km s^{-1} interval with a correlation scale 600 ± 200 km s^{-1} which corresponds to a correlation length of $\sim 3\, h^{-1}$ Mpc comoving (cf. other determinations[8,9]). There is no evidence (internal to this survey) for evolution in the clustering with redshift, or for periodicity in the peak redshifts[10].

Although in one sense these peaks in velocity space are simply evidence for galaxy–galaxy correlations, when viewed in configuration space the structures ought to have morphologies of spheres, filaments, or walls. They are obviously not conventional clusters because they are too numerous[11], they show no central concentration or increased fraction of elliptical galaxies, and they have low velocity dispersions. The structures extend laterally over at least ~ 6 arcmin or $\sim 2\,h^{-1}$ Mpc, so if they are quasi-spherical, the inferred space density is $\sim 3 \times 10^{-3} h^3$ Mpc^{-3}, $\sim 1/3$ the density of L^* galaxies. Walls are observed locally[12,13,14]; on this basis we speculate that the structures we are observing are actually their high-redshift counterparts. Only follow-up observations can resolve this matter; we have been allocated time to take redshifts in a strip adjacent to a previous field[8] in order to map this large scale structure.

Acknowledgements

We thank the Hubble Deep Field team, led by Bob Williams, for planning, taking, reducing, and making public the HDF images and the Keck Foundation for construction of the Keck Telescope. Thanks to Tomislav Kundić, Keith Matthews, Gerry Neugebauer, Mike Pahre, John Peacock, and Tom Soifer for helpful conversations and to Bev Oke, Jerry Nelson and Gerry Smith for instrument and telescope design and construction.

1. Williams R. E. et al, 1996, AJ, in press
2. Oke J.B., et al, 1995, PASP 107, 3750
3. Cohen J. G. et al, 1996, ApJ, 462, L9
4. Cowie L. L. et al, 1996, AJ, in press
5. http://astro.caltech.edu/~dwh/hdf.html
6. Crampton D. et al, 1995, ApJ, 455, 96
7. Dressler A. 1980, ApJ, 236, 351
8. Carlberg R. G. et al, 1997, in press
9. LeFèvre O. et al, 1996, ApJ, 461, 534
10. Broadhurst T. et al, 1990, Nature, 343, 726
11. Postman M. A. et al, 1996, AJ, 111, 615
12. de Vaucouleurs G. H. 1975, in *Galaxies and the Universe*, ed. Sandage A. et al, pg. 557
13. de Lapparent V., Geller M. & Huchra J. P., 1986, ApJ, 302, L1
14. Landy S. D. et al, 1996, ApJ, 456, L1

K-BAND NUMBER COUNTS AND THE K-BAND HUBBLE DIAGRAM: HOW TO DISENTANGLE GEOMETRICAL AND EVOLUTIONARY EFFECTS

J.C. JACKSON, MARINA DODGSON

Department of Mathematics and Statistics, University of Northumbria at Newcastle,
Ellison Building, Newcastle-upon-Tyne
NE1 8ST, UK

1 Introduction

The observed K-band magnitude-redshift relationship for radio-bright galaxies looks like a classical Hubble diagram; $K(z)$ varies smoothly with z, and has small dispersion at fixed z. Lilly and Longair[1] presented K-magnitudes for a complete sample of 3CR galaxies in the range $0 < z < 1.7$, shown in Figure 1. In their rest frames these objects appear in the infra-red to be giant elliptical galaxies, indistinguishable from their radio-quiet counterparts. However, at high z they are distinctly brighter than would be allowed by $0 \leq \Omega_0 \leq 1$, $\Lambda_0 = 0$ (the broken curves in Figures 1 and 2); this is not just a selection effect[2], but represents passive luminosity evolution, amounting to approximately one magnitude at $z = 1$[1]. With $\Lambda_0 = 0$ and no evolution the best-fit curve corresponds to $\Omega_0 = 6.3$, a figure which is quite unacceptable. In contrast, Figure 2 shows number counts $n(K)$ for a very recent compilation of K-band data[3]. With $\Lambda_0 = 0$ and no evolution the least-bad curve corresponds to $\Omega_0 = 0$, compared with which there are too many galaxies at faint magnitudes. If radio-brightness is just a label, which indicates that 3CR galaxies are at the bright end of the K-band luminosity function but are otherwise representative[4], then passive evolution should have noticeable effects upon $n(K)$[5,6]. Here we determine compromise best-fit figures for the cosmological and evolutionary parameters, which reconcile the conflicting claims of the two diagrams.

2 Quantitave Considerations

We adopt fixed absolute magnitude $M = -24.0$ for the 3CR galaxies, and a Schechter luminosity function with fixed parameters $\phi^* = 0.75 \times 10^{-2}$ Mpc^{-3}, $\alpha = -1.2$, $M^* = -23.6$, ($H_0 = 100$km sec^{-1} Mpc^{-1}). Evolution is modelled by $L^* \propto (1 + z)^\beta$. Figures 1 and 2 define standard deviations $\sigma_1 \sim 0.58$ and $\sigma_2 \sim 0.13$, so that we can construct a joint $\chi^2 = \chi_1^2 + \chi_2^2$, to be minimised. A two-parameter search with $\Lambda_0 = 0$ gives $\Omega_0 = 0.47$, $\beta = 1.49$; Figure 3

150

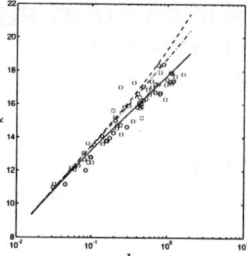

Figure 1: Hubble diagram.

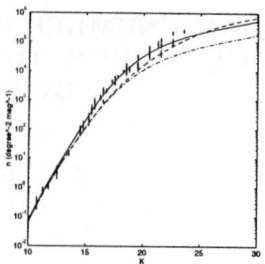

Figure 2: number counts.

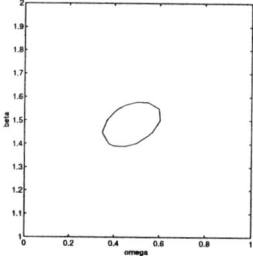

Figure 3: confidence region.

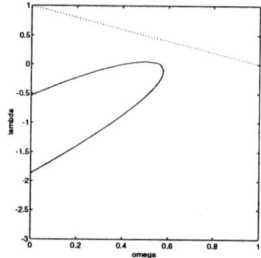

Figure 4: confidence region.

shows the corresponding 95% confidence region. A three-parameter search with $\Omega_0 \geq 0$ gives $\Omega_0 = 0$, $\Lambda_0 = -1.26$, $\beta = 1.58$ (the continuous curves in Figures 1 and 2); Figure 4 is a representation of the corresponding three-dimensional 95% confidence region, with β as the vertical axis, showing a typical horizontal slice, at $\beta = 1.5$. Flat models (the dotted line) are clearly excluded.

References

1. S.J. Lilly and M.S. Longair, MNRAS, 211, 833, 1984.
2. G.J. Hill and S.J. Lilly, ApJ, 367, 1, 1991.
3. N. Metcalfe *et al*, Nature, in press.
4. P.J. McCarthy, ARA&A, 31, 639, 1993.
5. T. Shanks *et al*, in *High Redshift and Primeval Galaxies*, eds. J. Bergeron *et al* (Editions Frontières, Paris, 1987).
6. Y. Yoshii and F. Takahara, ApJ, 326, 1, 1988.

THE SURFACE BRIGHTNESSES OF GALAXIES IN THE HUBBLE DEEP FIELD

BRYN JONES & MIKE DISNEY

*Department of Physics and Astronomy, University of Wales College of Cardiff,
P. O. Box 913, Cardiff, CF2 3YB, Wales, U.K.*

We present the distribution of surface brightnesses for faint disc galaxies in the
Hubble Deep Field. The observed surface brightnesses are 4 mag. too bright at a
given apparent magnitude for them to be the equivalents of nearby spirals at high
redshift. The physical sizes are $\gtrsim 2\times$ smaller than counterparts of nearby spirals.

1 Introduction

Cosmological $(1 + z)^4$ surface brightness dimming and k-correction effects
strongly influence the visibility of galaxies at intermediate and high redshifts
and will determine which galaxies can be detected in deep HST observations.
Whereas significant selection effects may be present even in conventional local
galaxy catalogues [1], samples of high-redshift galaxies will be strongly biased
towards high intrinsic surface brightnesses [2]. We use the surface brightnesses
of disc galaxies in the Hubble Deep Field as a test of their intrinsic properties.

2 The observations

A connected-pixel program was used to identify images in the Hubble Deep
Field having 8 or more contiguous pixels (equivalent radius $\theta_L = 0.064$ arcsec)
above detection thresholds μ_L of 25.91 mag. arcsec^{-2} (Vega system) for the
F450W filter and 25.22 mag. arcsec^{-2} for F814W. Visual morphological classification was performed on all images brighter than an isophotal magnitude
limit of $m_L = 28.0$ mag. in one half of the WFC frames. The galaxies classified
as spirals – based on the presence of discs and light profiles consistent with
an exponential law on visual inspection – were fitted with exponential light
profiles. Fig. 1a presents the surface brightness distribution.

3 Discussion

The results may be compared with a simple model whereby all disc galaxies
have Schechter L^* absolute magnitudes and Freeman [3] intrinsic central surface
brightnesses μ_0. Both the F450W and F814W data occupy a locus in the
$\mu_0 - m_{tot}$ plane similar in shape to this model, but are $\simeq 4$ mag. too bright in

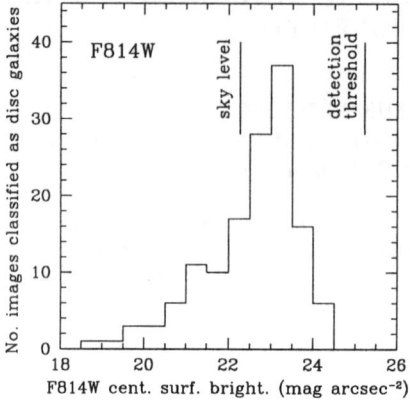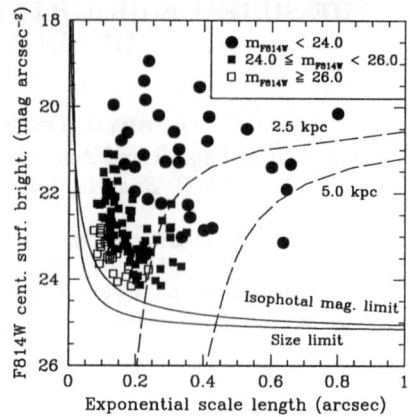

Figure 1: (a) The distribution of F814W extrapolated central surface brightness for disc galaxies to an isophotal magnitude of 28.0 mag. in one-half of the WFC area of the Hubble Deep Field. (b) The F814W central surface brightness plotted against angular exponential scale length for the same sample. Solid curves show selection boundaries imposed by the isophotal magnitude limit and minimum image area. Dashed lines represent the loci followed by Freeman surface brightness discs having physical scale lengths of 2.5 kpc and 5.0 kpc for

$$H_0 = 50\,\mathrm{kms}^{-1}(\mathrm{Mpc})^{-1},\ q_0 = 0.0,\ \Lambda = 0.0\,.$$

μ_0. An alternative model assumes that the galaxies have a broad distribution over central surface brightness and that the observed distribution is determined primarily by the selection limits used to define the image catalogue. Such a model predicts [2,1] a broader distribution for L^* galaxies than in Fig. 1a and a peak 2–3 mag. too bright. The observed angular scale lengths are small [4]; indeed a majority of images occupy a region in the scale length – μ_0 plane (Fig. 1b) which cannot be populated by conventional local spiral galaxies under any plausible cosmological model, indicating that the galaxies are physically small. Whilst they might conceivably be genuinely small spirals [5], Im *et al.* [6] propose that many brighter (I $\simeq$ 20 mag.) exponential-profile galaxies are in reality dwarf ellipticals; the galaxies in Fig. 1 may be their fainter counterparts.

References

1. M. J. Disney & S. Phillipps, MNRAS, 205, 1253, 1983.
2. S. Phillipps, J. I. Davies & M. J. Disney, MNRAS, 242, 235, 1991.
3. K. C. Freeman. ApJ, 160, 811, 1970.
4. I. Smail *et al.*, ApJ, 449, L105, 1995.
5. L. Cayón, J. Silk & S. Charlot, ApJ Lett., in press, 1996.
6. M. Im *et al.*, ApJ, 445, L15, 1995.

THE DELAYED FORMATION OF DWARF GALAXIES

J. KEPNER

Princeton University Observatory, jvkepner@astro.princeton.edu

A. BABUL

Department of Physics, NYU, babul@almuhit.physics.nyu.edu

D. SPERGEL

Princeton University Observatory, dns@astro.princeton.edu

One of the largest uncertainties in trying to understand the effect of a background UV field on galaxy formation is the intensity and evolution of the radiation field with redshift. This work attempts to address this problem by computing the quasi-hydrostatic equilibrium states of spherically symmetric objects as a function of the background UV field by integrating the full equations of radiative transfer and the corresponding reactions for nine species: H, H^+, H^-, H_2, H_2^+, H_2^-, He, He^+, He^{++}, and e^-. By characterizing the state of an object by the radiation field we can then compute its behavior for any model of the background UV flux. Our results indicate that a typical radiation field can easily prevent gas in halos with $V_c < 25$ km s^{-1} from collapsing, indicating that the less massive halos, which are typically the first virialize, are the last to form galaxies.

1 Results

Several independent lines of reasoning, both theoretical and observational[1,2], suggest that the very blue faint ($B < 24$) galaxies seen in deep images are small low-mass galaxies that are forming via a burst of star-formation at $0.5 < z < 1$, well after the larger galaxies have formed, and have since faded into low luminosity, low surface brightness objects. This is at odds with hierarchical models of galaxy formation which predict that the smallest objects formed first and subsequently merged to form larger galaxies. However, the presence of a large background UV radiation field at $z > 3$ could significantly delay the collapse of smaller objects[3]. Simulations of structure formation[4,5] indicate that a typical UV field ($J \sim J_{21} = 10^{-21} ergs^{-1} cm^{-2} ster^{-1} Hz^{-1}$) can delay the collapse of objects with circular velocities $V_c < 40$ km s^{-1}.

One of the largest uncertainties in these calculations is the intensity and the evolution of the radiation field with redshift. In addition, previous calculations do not take into account the full effects of radiative transfer which can play a crucial role in the formation of the main coolant (H_2) in the cores of these objects.

This work attempts to address this problem by computing the quasi-equilibrium states of spherically symmetric objects as a function of the background UV field; including radiative transfer and the corresponding reactions [6] for nine species: H, H^+, H^-, H_2, H_2^+, H_2^-, He, He^+, He^{++}, and e^-. By characterizing the state of an object by the radiation field we can then compute its behavior for any model of the background UV flux.

2 Results

For gas in dark matter halos specified by a circular velocity V_c and a virialization redshift z_v the state of the gas core of the object exhibits the same qualitative behavior over the range 10 km s^{-1} $< V_c <$ 50 km s^{-1} and $2 < z_v < 12$. As the flux decreases the core goes through three phases HII ($T_c \sim 20,000°$K), HI ($T_c \sim 10,000°$K) and H_2 ($T_c \sim 100°$K). In the HII state, the object is completely ionized and resembles the IGM. In the HI state the core is neutral and the object is like an inverse Stromgren sphere. In the H_2 state, conditions allow for the formation of molecular hydrogen in the core; consequently, the cooling time becomes much less than the dynamical time and the core collapses. Our results indicate that a typical radiation field can easily prevent gas in halos with $V_c <$ 25 km s^{-1} from collapsing, indicating that the less massive halos, which are typically the first virialize, are the last to form galaxies.

Acknowledgments

We gratefully acknowledge Yu Zhang and Mike Norman for the use of their non-equilibrium species solver. J. Kepner was supported by the Dept. of Energy Computational Science Fellowship Program.

1. S.P. Driver *et al*, ApJ, 453, 48, 1995.
2. A. Babul & H.C. Ferguson, ApJ, 458, 100, 1996.
3. A. Babul & M.J. Rees, MNRAS, 255, 346, 1992.
4. N. Katz, D.H. Weinberg& L. Hernquist, ApJS, 105, 19, 1996.
5. A. Thoule & D.H. Weinberg, ApJ, 465, 608, 1996.
6. T. Abel, P. Anninos, Y. Zhang, & M.L. Norman, astro-ph/9608040

KECK OBSERVATIONS OF COMPACT GALAXIES
IN THE HDF FLANKING FIELDS

D. C. KOO, A. C. PHILLIPS, R. GUZMÁN, & J. GALLEGO

UCO / Lick Observatory, University of California, Santa Cruz, CA, 95064, USA

Keck spectra of faint, compact galaxies in the HDF flanking fields reveal a population dominated by moderately luminous emission line galaxies similar to local H-II galaxies, rather than, e.g., low-luminosity galaxies at low redshifts or AGN's.

The DEEP[a] team has observed a sample of faint, compact galaxies in the HDF flanking fields. Such compact galaxies are particularly relevant for probing the mysterious population of faint blue galaxies, since several popular explanations include significant fractions of low-luminosity dwarfs at low redshifts (Driver & Phillips 1995; Im *et al.* 1995), of bursting dwarfs at high redshifts (Babul & Ferguson 1996), or even of galaxies with AGN (Rola *et al.* 1996). Spectra are critical to reveal the physical properties of such small, high surface brightness objects by yielding luminosities, look-back times, physical sizes, star formation rates, and masses.

The sample of 61 objects was chosen to be relatively faint ($20.6 < I_{814} < 23.7$ mag), to be compact with half-light radii, r_e, less than $0\farcs5$, and to have I band surface brightnesses brighter than 22.25 mag-arcsec^{-2} within r_e. Except for one AGN and ten galaxies with red colors and with weak if any emission lines (and thus consistent with high redshift early type [E/SO/Sa] galaxies), the remaining vast majority had moderate to strong emission lines.

As can be inferred from Figure 1 (R_e vs M_B) and as will be described in more detail in forthcoming papers by Phillips *et al.* and Guzmán *et al.*, we find that a significant fraction of the emission line systems have luminosities, sizes, masses, star formation rates, and excitation ratios similar to that found in local H-II galaxies. At redshifts $z \sim 0.5$, most have physical properties consistent with the 1-2 kpc sizes, L* /4 luminosities, and 10^{10} $M_\odot$ masses of bursting dwarfs postulated to explain the faint blue galaxies (Babul & Ferguson 1996). Unfortunately, we are unable to sample the same types of objects to much higher redshifts, but near $z \sim 1$, we do find larger (3 kpc) and more luminous (near L*) emission line galaxies. Limited kinematic data suggest that $\sim 60\%$ of these are indeed of low masses (i.e., less than 10^{10} $M_\odot$).

Whether the compact galaxies are some combination of bursting dwarfs and H-II galaxies, we find no evidence for 1) a large fraction of AGN activity

---

[a]Information about the Deep Extragalactic Evolution Probe project is available on the WWW at URL **http://www.ucolick.org/~deep/home.html**

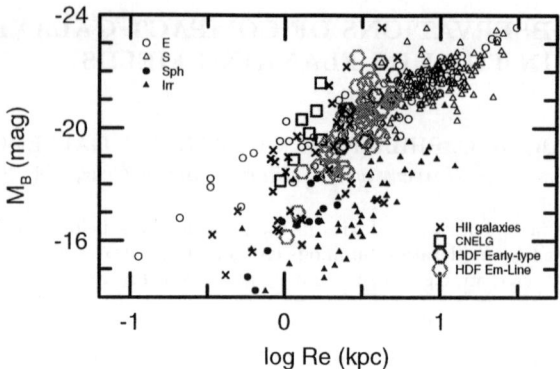

Figure 1: The half-light radius (R_e) vs B absolute magnitude (M_B) for various classes of galaxies (L^* is at $M_B \sim$ -21; we use $H_0 = 50$ km s^{-1} Mpc^{-1} and $q_0 = 0.05$).

among our faint blue compact galaxies, 2) the domination of the sample by low-luminosity galaxies at low redshifts ($z < 0.3$), or 3) significant numbers of galaxies comparable in surface brightness to the low-mass, compact narrow-emission line galaxies (CNELG) that have been suggested to be progenitors of today's spheroidals (Koo et al. 1995; Guzmán et al. 1996).

Funding was provided by the NSF Center for Particle Astrophysics grant AST-9120005 and NSF grant AST-9529098; NASA grants AR-5801.0194A, AR-06337.08-94A, and GO-05994.01-94A; and an UCM del Amo Fellowship from Spain for J.G.

References

1. A. Babul, & H. C. Ferguson, ApJ, 458, 100, 1996.
2. S. P. Driver, & S. Phillips, MNRAS, 274, 832, 1995.
3. R. Guzmán, et al.,ApJ, 460, L5, 1996.
4. M. Im, et al.,ApJ, 441, 494, 1995.
5. D. C. Koo, et al.,ApJ, 440, L49, 1995.
6. C. S. Rola, et al., preprint 1996

KECK SPECTRA OF GALAXIES AT REDSHIFT $z \sim 3$ IN THE HUBBLE DEEP FIELD

J. D. LOWENTHAL, D. C. KOO, R. GUZMÁN, J. GALLEGO, and
A. C. PHILLIPS

UCO/Lick Observatory and Board of Astronomy and Astrophysics, Kerr Hall, University of California, Santa Cruz, CA 95064

We have obtained spectra with the 10-m Keck telescope of a sample of 24 galaxies having colors consistent with star-forming galaxies at redshifts $2 \lesssim z \lesssim 4.5$ in the Hubble Deep Field (HDF). Eleven of these galaxies are confirmed to be at high redshift ($z_{med} = 3.0$). The galaxies are small but luminous, with a wide range of morphologies. The observed number density of high-redshift galaxies in the HDF is comparable to estimates of the volume density of local galaxies brighter than L*. We suggest that many may be low-mass objects in an early burst phase of star formation and that they will soon merge with other similar objects to build massive galaxies, or else consume or expel their gas and thus quench their star formation.

1 Sample and Observations

Following the pioneering work of Steidel et al. [1], the DEEP team [a] has taken advantage of the unprecedented depth of the HDF [2] and the light-gathering power of the 10-m Keck telescope to identify and study star-forming galaxies at $z \sim 3$. We used all four HDF bands (F300W, F450W, F606W, F814W) to identify 43 objects with colors implying star-forming galaxies with Lyman-continuum and Lyα breaks at $z > 2.5$, i.e. blue but with little or no flux in U (or B; "U and B-band dropouts"). 25 objects with $I_{AB} < 25.5$ ($I_{AB,median} = 25.0$) were observed in 3 nights on 10-m Keck telescope using the LRIS multi-object spectrograph.

2 Properties

Eleven objects were confirmed to be at high redshift, with $z_{med} = 3.0$ and $2.2 < z < 3.6$. We find 6 Lyα emission sources, with EW(Lyα)$_{rest} \sim 40$ Å, and several **absorption-line-only redshifts**, based on rest-UV stellar and interstellar lines (SiII, OI, CII, SiIV, CIV) typical of starburst galaxies and hot stars. Rest equivalent widths are as high as 4 Å; however, they are probably **not directly useable** to measure galaxy masses [3]. The high-redshift galaxies

---

[a] DEEP = Deep Extragalactic Evolutionary Probe, a collaboration of the University of California, The Johns Hopkins University, CalTech, and the University of Chicago. See http://www.ucolick.org/~deep/home.html for more information.

157

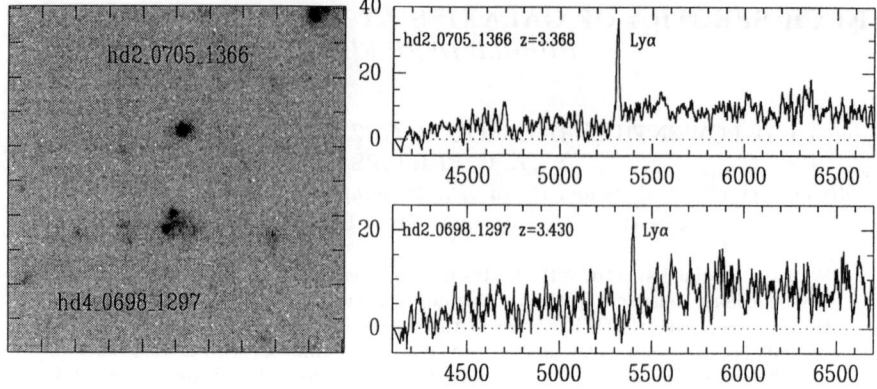

Figure 1: Image and Keck LRIS spectra of two confirmed high-redshift galaxies in the HDF. The image is extracted from the F814W HDF frame and is $10''$ on a side. Both objects have $I_{814,AB} = 25.0$. The spectra clearly show Lyα emission, a strong Lyα break, and absorption lines characteristic of hot stars and local starburst galaxies.

are small but luminous, with half-light radii $1.8 < r_{1/2} < 6.5$ h_{50}^{-1} kpc and absolute magnitudes $-21.5 < M_B < -23$. The median absolute magnitude is $M_B \sim -22.5$ ($H_0 = 50$), or 1.5 mag brighter than L*. Combined with the 5 high-redshift galaxies in the HDF previously confirmed with Keck spectra by Steidel $et\ al.$, the 16 confirmed sources yield a comoving volume density of $n \geq 2.5 \times 10^{-4}$ h_{50}^3 Mpc^{-3} ($q_0 = 0.05$), comparable to estimates of the volume density of local galaxies brighter than L*.

3 Speculations on the Nature of $z = 3$ Galaxies

The small sizes, high luminosities and star formation rates ($SFR \sim 10$ M$_\odot$ yr^{-1} – consistent with the UV absorption lines and Lyα emission), and the wide variety of morphologies (bright knots, tails, clumps, etc.) all favor a **premerger** scenario of multiple, low-mass bursts, rather than, e.g., a monolithic gE precursor. Subsequent significant merging is likely.

References

1. Steidel, C. C., Giavalisco, M., Dickinson, M., & Adelberger, K. L. 1996, ApJ, in press
2. Williams, R. $et\ al.$, 1996, AJ, in press
3. Conti, P. S., Leitherer, C., & Vacca, W. D. 1996, ApJ, 461, L87

ON THE APPLICATION OF EVOLUTIONARY SYNTHESIS OF STELLAR POPULATION

CLAUDIA MARASTON

Department of Astronomy, University of Bologna, Via Zamboni, 33,
40126 Bologna, Italy

1 Abstract

Studies concerning the nature of Faint Blue galaxies or the research of the High-Redshift galaxies, make an unavoidable use of Evolutionary synthesis models of stellar populations. A correct application of this technique implies that, first of all, these models have been tested and calibrated on the natural counterparts of a simple (i.e. single istantaneous burst and a unique chemical composition) stellar population, like a globular cluster is. An innovative tool for the construction of Evolutionary Synthesis models of Stellar Populations is presented, based on the so-called **Fuel Consumption Theorem** (Renzini & Buzzoni, 1986). The modular structure of the code allows to easily assess the impact on the synthetic spectral energy distribution of various assumptions and model ingredients, such as, for example, uncertainties in the stellar evolutionary models, color-temperature transformations, etc. As an illustrative example, we show a solar model (Y=0.27, Z=0.02) for a Simple Stellar Population (SSP) with a Scalo slope IMF and an age ranging between 30 Myr and 15 Gyr: the evolution of $(B-V)_0$, $(U-B)_0$ and $(V-K)_0$ broad band colours is compared to the Magellanic Clouds Clusters data and a very good agreement is evident. The disagreement between model and observational data, clearly visible in fig. 1 for $(B-V)_0 > 0.5$, is not real as it is mainly due to the increasing difference in metallicity between the old Magellanic clusters and the solar model. The same happens in fig. 2 for the latest SWB *type* (Searle, Wilkinson & Bagnuolo, 1980) clusters (Log t/yr > 9.4). A detailed description of the computational procedure and the other results, concerning, for example, the evolution of the mass to light ratios, will be given in a forthcoming paper, in which we'll present models for other chemical compositions.

160

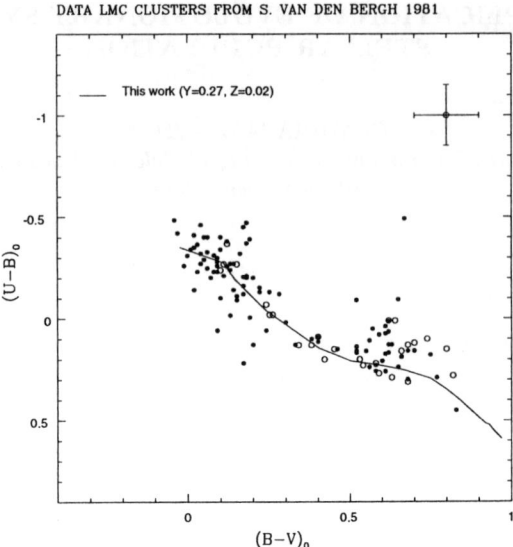

Figure 1: $(U - B)_0$ **versus** $(B - V)_0$

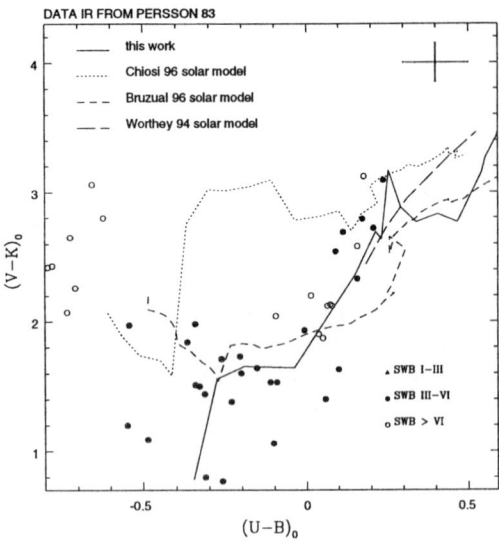

Figure 2: $(V - K)_0$ **versus** $(U - B)_0$

PHOTOMETRIC ANALYSIS OF THE HUBBLE DEEP FIELD

J.M. MIRALLES, R. PELLÓ, J.-F. LE BORGNE

UMR 5572, Observatoire Midi-Pyrénées,
14 Av. Edouard Belin, 31400 Toulouse, France

The HDF images offer a new perspective in the study of the faint galaxies and high-z populations through photometric analysis. This poster describes our method to compute the photometric redshift distribution in the HDF and the first results compared to the spectroscopic data already available. The poster with the figures can be found at http://www.obs-mip.fr/omp/unites_de_recherche /umr5572/people/miralles/poster_cambridge.ps

1 Photometric Redshift Method

The main goal is to determine the redshift distribution of faint field galaxies, beyond the reach of spectroscopic surveys.

The idea is to fit a photometric redshift, using the extremely low-resolution spectral energy distribution (SED) obtained from broad-band filters and the codes for spectrophotometric evolution of galaxies[1]. For each color, a maximun (old single burst of star formation) and a minimum (continuous star forming system) curves are computed in the color-redshift plane, limiting the permitted color-z area for any galaxy. Each observed color is compared to the predicted color-z diagram, taking into account the photometric errors, and a permitted redshift interval is obtained. The intersection of all the color-z intervals gives the final redshift interval (hereafter Δz). We do not require the SED to be coherent with a particular spectromorphological type of galaxy. The SED is mainly dominated by the continuum and absorption features (the 4000 Å and Balmer breaks in the visible), thus any "normal" galaxy is expected to fit between the two extreme models. No photometric redshift is found when we are dealing with an object which SED is different from a normal galaxy.

2 Comparison with the spectroscopic sample. Redshift distribution up to B=30

HDF images[2] were taken in 4 filters from the UV to the near-IR. We computed the Δz-phot for all the galaxies detected and the results were compared to the sample of spectro-z obtained by the Caltech, Hawaii and Berkeley groups[3][4] (53 objects at z≤1.355). The computed Δz remains close to the observed spectroscopic redshifts in most cases. For 41 galaxies the z-spec lies inside the Δz or at less than 0.05. No major systematic biases are observed, despite

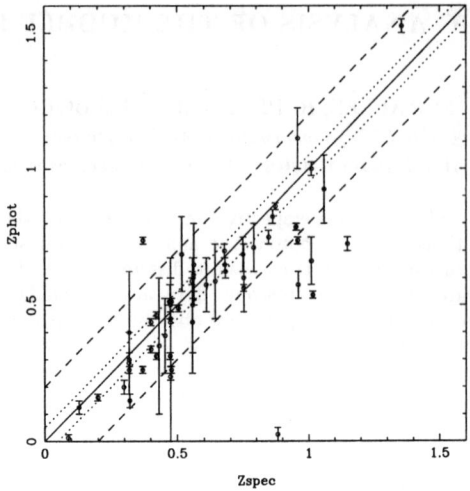

Figure 1: z-spectro versus Δz-phot for the 53 galaxies of the sample.

staigthforward simulations show that errors increase strongly at z∼1 (Miralles et al., in prep.).

The Δz of higher-redshift galaxies from Steidel et al. [5](2.5<z<3.5) were also compared. Among the 5 common galaxies with a z-spec only 2 have been detected in the F300 filter. Compared to lower redshift galaxies, the Δz are very unacurate, but they are correctly identified as z>2 galaxies. Near-IR data are necessary to better constraint the Δz for these galaxies.

When the method is applied to fainter galaxies, the distribution remains roughly the same than for the spectroscopic sample up to B≤27, and ∼70% of the objects are at a z≤1.0. For 27<B≤29, there is a significant increase of the number of objects at redshifts of 1.0≤z≤1.5. For B>29 the increase is significant for the objects at 1.5≤z≤2.5.

References

1. Bruzual G., Charlot S.,ApJ, 405, 538, 1993.
2. Williams et al., AJ in press, 1996
3. Cohen J.G. et al., ApJL submitted, 1996
4. Moustakas L., Zepf S., Davis M., in preparation, 1996
5. Steidel C.C., Giavalisco M., Dickinson M., Adelberger K.L., AJ in press, 1996

PHOTOMETRIC REDSHIFTS AND THE LUMINOSITY FUNCTIONS OF GALAXIES IN THE HDF

B. MOBASHER, M. ROWAN-ROBINSON and A. GEORGAKAKIS

Astrophysics Group, Blackett Laboratory, Imperial College, Prince Consort Road, London SW7 2BZ.

Using the rest frame Spectral Energy Distributions of Ellipticals to HII galaxies, we predict the photometric redshifts of galaxies detected in the HDF. A median redshift of 2.1 is estimated with 72% having $z > 1$ and 5% with $z > 3$. The I-band luminosity function, calculated from the HDF galaxies agrees well with that from the Canada-France survey (CFRS) for $z < 1$. This shows evidence for luminosity evolution (at $M_I < -21$) in the range $2 < z < 3$, smaller than the rate seen in quasars and starburst galaxies.

1 Introduction

One of the aims of the HDF is to examine the evolution of galaxies at deep levels. The HDF and its follow up spectroscopic observations are expected to constrain the evolutionary models for formation of different populations of galaxies and provide a natural extension to ground-based optical surveys. In this study, using the available multi-waveband information, we predict photometric redshifts of galaxies detected in the HDF. These will then be used to make a preliminary study of the luminosity function of galaxies and their evolution with redshift. We assume $H_0 = 50$ km/sec/Mpc and $q_0 = 0.5$.

2 Photometric Redshifts

The observed, rest frame spectral energy distributions (SEDs) for 6 different types of galaxies (E/S0, Sab,Sbc,Scd,Sdm from Yoshii and Takahara 1986 and HII from Calzetti and Kinney 1993) are used to produce a grid of SEDs in the $\log(1+z)$ range 0.01 to 0.7 in 0.01 intervals, ie $0 < z < 4$. These are then compared with the observed SEDs for the HDF galaxies and the best fit selected by least squares. Observations of high redshift quasars suggest a gradual extinction of the Lyman continuum starting shortward of Lyman α. Therefore, we modify the above SEDs by assuming $f_\lambda \sim \lambda^4$ for $\lambda < 912 \mathring{A}$ for all galaxy types. A median redshift of 2.1 is found with 72% of galaxies having $z > 1$ and only 5% with $z > 3$.

For an assumed magnitude error of ± 0.06 mag. for brighter galaxies (I < 25 mag.) in each band, the typical formal uncertainty in $(1 + z)$ ranges from $5 - 15\%$ for a galaxy detected in all 4 bands. The photometric redshifts

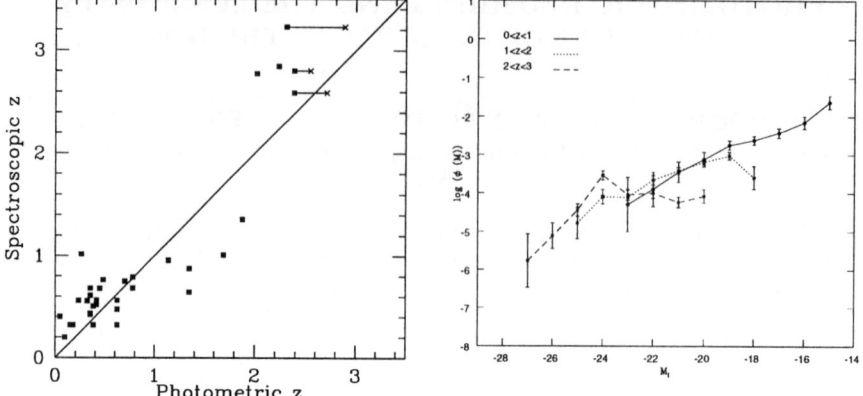

Figure 1: (left) Spectroscopic vs. photometric redshifts. Lines connected to crosses indicate cases where a more accurate redshift resulted from using U-band detections fainter than our magnitude limit of U=26.5 mag.

Figure 2: (right) I-band luminosity function for 3 redshift intervals.

are compared with the spectroscopic measurements for a sample of 50 HDF galaxies with such observations (Fig 1), giving an rms scatter of 0.15.

3 Luminosity Function of the HDF Galaxies

Using an HDF magnitude limit of I=27.5 mag., the I-band luminosity functions are calculated in three different redshift intervals (Figure. 2). The K-corrections in the I-band, for each of the 6 morphological types, are estimated using their respective SEDs as discussed above. The luminosity function for $z < 1$ shows remarkably good consistency with that found in the Canada-France survey to I = 22.5 mag. (Lilly et al 1995). They agree both in the normalisation and in the steep faint end slope, which we find to be $\alpha = 1.5$ at $M_I > -22$. There is no significant evolution of the luminosity function to $z \sim 2$. However, the rate of evolution increases in the range $2 < z < 3$ but is still smaller than the rate of luminosity evolution found for quasars and starburst galaxies.

References

1. Yoshii Y., and Takahara F., 1986, Ap.J. 326, 1.
2. Calzetti, D. and Kinney, A. L. 1992, Ap.J, 399, L39.
3. Lilly, S.J., Tresse, L., Hammer, F., Crampton, D., and Le Fevre, O. 1995, Ap.J., 455, 108.

QUANTITATIVE MORPHOLOGY OF MODERATE REDSHIFT GALAXIES

AVI NAIM, KAVAN U. RATNATUNGA and RICHARD E. GRIFFITHS
The Johns Hopkins University, Department of Physics & Astronomy, Baltimore, MD 21218, U.S.A.

Recently interest in peculiar galaxies has arisen due to their reportedly increasing numbers at moderate redshifts. These reports rely mostly on eyeball classifications, but it appears that peculiarity is currently an ill-defined entity: the agreement between experts on what constitutes a peculiar galaxy is not very good even for local universe galaxies.

As a first step towards understanding these galaxies better we propose a set of four purely morphological parameters which are measured automatically by software we wrote. We define criteria for peculiarity and classify a sample of 978 galaxies by eye, according to these criteria. We then train an artificial neural network (ANN) to distinguish peculiar galaxies from normal ones. The mixture between the two classes is considerable. We find a significant population of bulge-dominated peculiars and conclude that peculiars do not all continue the Hubble sequence beyond the irregulars. The nature of the bulge-dominated peculiars is currently under investigation.

Having succeeded in applying our parameters to a *supervised network*, we now use the same parameter space to perform *unsupervised learning* and see how galaxies arrange themselves in that space. The technique we use is a variant of Self Organising Maps (SOMs), which allows one to cast the topology of a high-dimensional space onto a two dimensional map. Galaxies organise in four major concentrations in the SOM. Figure 1 describes how subsets of the full sample, selected by I magnitude, colour, bulge dominance and eyeball class are mapped. Morphology appears to correlate with all of these quantities to varying extents. An apparent evolutionary track of galaxies is traced on the SOM.

166

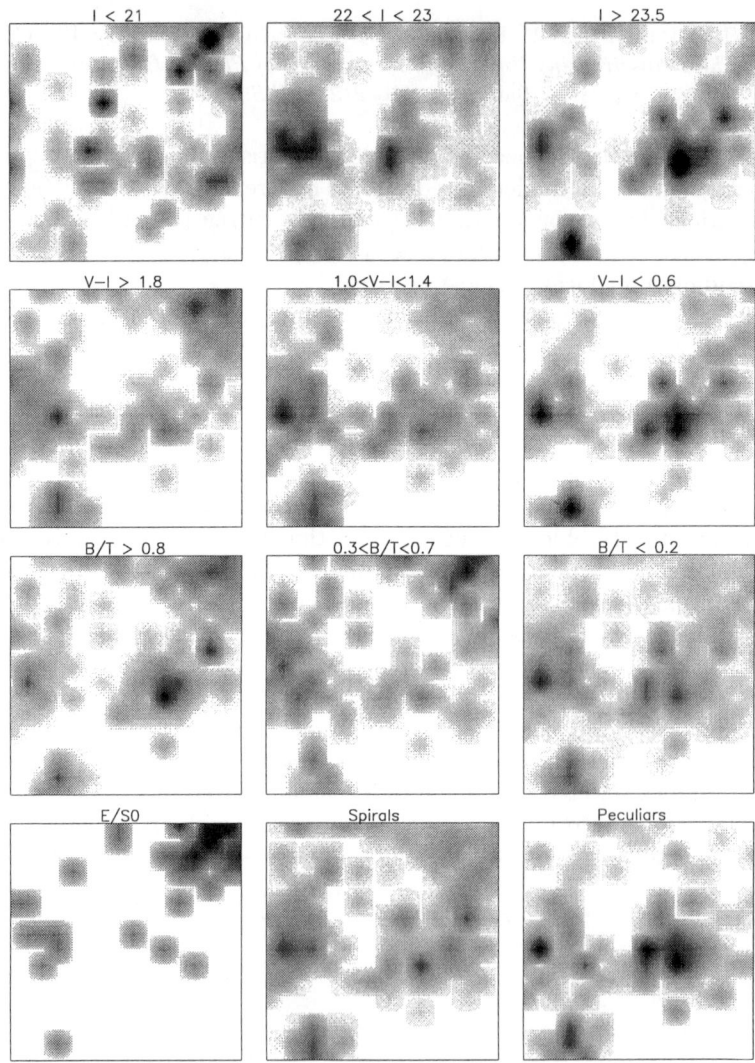

Figure 1:

AUTOMATED MORPHOLOGICAL CLASSIFICATION IN DEEP HST FIELDS: RAPIDLY AND PASSIVELY EVOLVING FAINT GALAXY POPULATIONS

S.C. ODEWAHN and ROGIER A. WINDHORST

Dept. of Physics & Astronomy, Arizona State University, Tempe, AZ85287-1504

SIMON P. DRIVER

School of Physics, University of New South Wales, Sydney, NSW 2052, Australia

WILLIAM C. KEEL

Dept. of Physics & Astronomy, Univ. of Alabama, Tuscaloosa, AL 35487-0324

We present recent results from Odewahn et al. (1996) who analyze deep HST/WFPC2 images in U, B, V, I using artificial neural network (ANN) classifiers, which are based on galaxy surface brightness (SB) and light profile (but not on color nor on scale-length, r_{hl}). The ANN distinguishes quite well between E/S0's, Sabc's, and Sd/Irr+M (M indicating merging systems) for $B_J \lesssim 27$ mag, with larger classification errors for $B_J \gtrsim 27$ mag. We discuss the effects of the cosmological SB-dimming and correct for effects from the redshifted UV-morphology on the classifications. We present eyeball classifications in $UBVI$ by four independent observers, classifications from ANN's trained with V_{606} and I_{814} images respectively, and from an ANN trained in the *rest-frame* UBV according to the expected redshift distribution as a function of B_J . These methods yield consistent galaxy counts as a function of type down to $B_J \simeq 27$. The median scale-length at $B_J \simeq 27$ mag is $r_{hl} \simeq 0\farcs25$-$0\farcs3$. Early and late-type galaxies are fairly well-separated in BVI color-magnitude diagrams for $B \lesssim 27$ mag, with E/S0's being the reddest and Sd/Irr+M's generally blue. We present the B -band galaxy counts for five WFPC2 fields as a function of morphological type for $B_J \lesssim 27$ mag. E/S0's are only marginally above the no-evolution predictions, and the Sabc's are at most 0.4 dex above these models for $B_J \gtrsim 24$ mag, suggesting that the formation of early-type galaxies was largely complete by z~1. The faint blue galaxy (FBG) counts in the B_J -band are dominated by Sd/Irr+M's, and can be explained by a moderately steep local LF undergoing strong luminosity evolution.

In Odewahn et al. (1996) we use 5 WFPC2 fields (53W002, HDF, 3 HST archival B_{450} fields) to determine B_{450} properties of faint galaxies as a function of morphological type. We developed artificial neural network (ANN) image classifiers using photometric parameter spaces to make quantitative estimates of galaxy types in the I_{814} and/or V_{606} WFPC2 images. Tests of the automated surface photometry and ANN systems were conducted using the simulation software of the MORPHO package of Odewahn 1995. We have also developed a system of ANN classifiers designed to classify galaxies in their appropriate **rest-frame** UBV filters. Although color or size information is used as input ANN

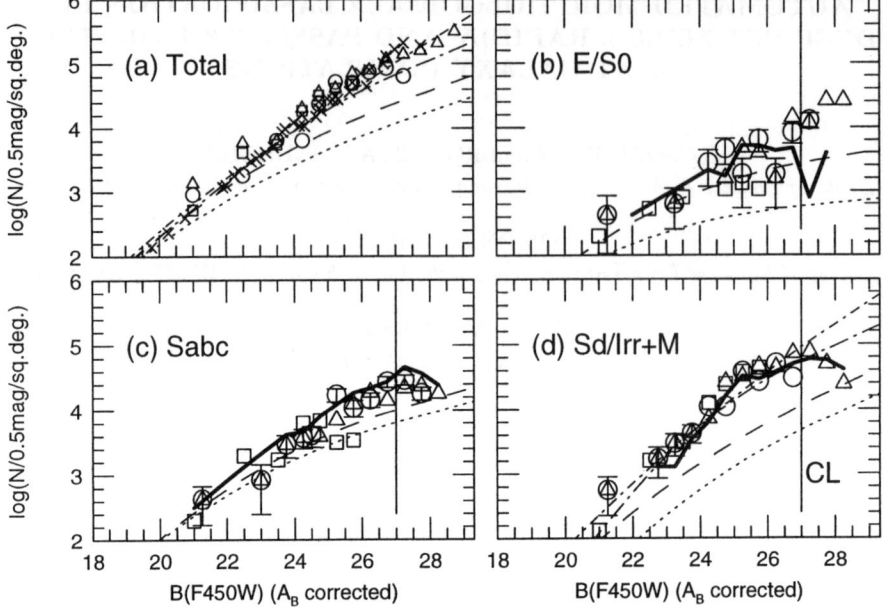

Figure 1: B_{450} number counts from five WFPC2 fields. A single filter I_{814} ANN classifier (open symbols) and a *rest-frame* ANN classifier (thick line) were used to separate the morphological samples. The thin dashed or dotted curves represent no-evolution models and the dot-dash curves represent strongly evolving and/or dwarf-dominated models for the late-types.

information, we find that E/S0 galaxies are well separated in color-magnitude spaces and that they have the smallest scale-lengths. Following Driver et al. 1995, our automated type estimates were used to construct the B_{450} number counts in Figure 1. We find evidence for modest evolution among the E/S0 and Sabc samples, but strong evolution for the late-type and merging samples.

We acknowledge support from HST grants GO.5308.0*.93A, GO.5985.0*.94A, GO.2684.03.94A, and AR.6385.01.95A.

References

1. S.P. Driver et al. ApJ, 449, L23, 1995.
2. S.C. Odewahn et al. ApJ, 472, L00, 1996.
3. S.C. Odewahn PASP, 107, 770, 1995.

CONSTRAINT ON GALAXY EVOLUTION FROM FAINT OBSERVATIONS

L. POZZETTI[1,2], MADAU P.[3], G. BRUZUAL A.[4], G. ZAMORANI[2,5]

[1] Dipartimento di Astronomia, Università di Bologna, Italy

[2] Osservatorio Astronomico di Bologna, Italy

[3] Space Telescope Science Institute, Baltimore, USA

[4] C.I.D.A., A.P. 264, Mérida 5101-A, Venezuela

[5] Istituto di Radioastronomia del CNR, Bologna, Italy

We have studied the effect of intergalactic absorption on the theoretical galaxy counts and redshift distribution at faint magnitudes, in a pure luminosity evolution (PLE) model. The results are then compared with *Hubble Deep Field* (HDF) galaxy counts. Predictions on the comoving luminosity density at different redshifts are then compared with recent observations.

1 PLE models with Intergalactic Attenuation

It has been realized only recently that the increasing opacity of the intergalactic medium (IGM) at high redshifts can be efficiently used to identify galaxies at $z \geq 2$. Indeed, the accumulated Lyman-continuum absorption from the Lyman-α forest clouds and Lyman-limit systems along the path is so severe that galaxies beyond $z \approx 3$ become effectively undetectable in the U-band (Madau 1995).

We focus here on the effects of intergalactic attenuation on galaxy number counts. We examine some simple galaxy evolution models which are able to reproduce most of the observed properties of faint field galaxies (Pozzetti et al. 1996), and show that the absorption-induced loss of sources at the depth now probed by the HDF is significant in the ultraviolet and possibly blue bands. This will produce, depending on the model, *a noticeable and characteristic flattening of the source counts at very faint magnitude limits*. We argue that such a sharp flattening may be present in the F300W HDF counts (Williams et al. 1996) at $U_{300} > 27$ mag, and that this may indicate that many of the sources at these faint levels are galaxies at $z > 2$, in agreement with the findings of Steidel et al. (1996) and Madau et al. (1996).

A PLE model with $H_0 = 50$ Km/s/Mpc, $\Omega = 0$, and formation redshift $z_f = 4.5$, provides an acceptable fits to the HDF galaxy counts derived in Williams et al.(1996).

2 Comoving luminosity density

We have compared predictions on the comoving luminosity density at different redshifts by PLE models to recent observations. At low redshifts ($0 < z < 1$) the steep and continuous increase in the blue galaxy luminosity density (L_B) derived by Lilly et al. (1996) strongly relies on the adopted estimate for the local luminosity density. The more recent determinations of the local blue luminosity function of galaxies (Ellis et al. 1996, Zucca et al. 1996), together with the Lilly et al. data, suggest a much shallower increase of L_B up to $z = 1$. This trend is well reproduced by our PLE model. At high redshifts ($2 < z < 4.5$) the same model (which assume an exponential SFR with decay time $\tau_E = 1$ Gyr for Elliptical galaxies and $z_f = 4.5$) significantly overestimates, by factors 2 (Madau et al. 1996, Giallongo et al. 1996) to 10 (Steidel et al. 1996) the observed L_{UV} for $2 < z < 4.5$. A model with $\tau_E = 0.1$ Gyr is more consistent with data at $z < 4$, but fails at $z > 4$. An acceptable fit to the same high redshift data is obtained by setting $z_f = 6$ with τ_E in the range $0.5 - 1.0$ Gyr.

3 Conclusions

A PLE model which includes the effect of intergalactic absorption provides an acceptable fits to the HDF galaxy counts. The observed counts show a flattening at faint magnitudes similar to the predictions of PLE models, but discrepancies between the model and the data are still present.

The trend of the observed comoving luminosity density at $z < 1$ is well reproduced by our PLE model, if the local LF is consistent with the determinations of Ellis et al. (1996) and Zucca et al. (1996). At high redshifts the predicted comoving luminosity density overstimates the data. Improvements may be obtained, for example, assuming a higher z_f.

References

1. Ellis R.S., et al., 1996, MNRAS, 280, 235
2. Giallongo E., et al., 1996, in "the Early Universe with VLT" (Springer)
3. Lilly S.J., Le Fèvre O., Hammer F., Crampton D., 1996, ApJ, 460, L1
4. Madau P., 1995, ApJ, 441,18
5. Madau, P., et al., 1996, MNRAS (in press)
6. Pozzetti L., Bruzual G.A., & Zamorani G. 1996, MNRAS, 281, 953
7. Steidel C.C, et al., 1996, ApJ (in press)
8. Williams et al., 1996, AJ (in press)
9. Zucca E. et al., 1996, A&A submitted

A DEEP 20 cm RADIO MOSAIC OF THE ESP GALAXY REDSHIFT SURVEY

I. PRANDONI, L. GREGORINI, P. PARMA, G. VETTOLANI

IRA - CNR, Via Gobetti 101, 40129 Bologna, Italy

H.R. de RUITER

OAB, Via Zamboni 33, 40126 Bologna, Italy

M.H. WIERINGA, R.D. EKERS

ATNF, P.O. Box 76, Epping NSW 2121, Australia

In the last two years we used the Australia Telescope Compact Array at 20 cm to image the entire region covered by the ESO Slice Project (ESP) galaxy redshift survey, with a sensitivity of $\sim$ 70 μJy (1σ). First results indicate that we will detect $\sim$ 2500 radio sources above 6σ and $\sim$ 16% of the ESP galaxies above 3σ. The aim of this radio survey is twofold. First, the sample of the radio detected ESP galaxies will give us the unique possibility of studying the existing correlations between optical (line activity, colors, morphologies, luminosities, etc.) and radio properties of galaxies. Secondly, the list of all the radio sources present in the region surveyed will represent a new homogeneous and fairly deep reference catalogue for astronomers. It will provide a unique view of the population of low power radio galaxies and it will allow better studies of the sub-mJy population.

1   Background: The Optical Sample

In two strips of $22° \times 1°$ and $5° \times 1°$ near the SGP Vettolani et al. have made a deep redshift survey as an ESO Key Project. All the galaxies down to $b_J \sim 19.4$ were observed with the OPTOPUS multi-fibre spectrograph on the 3.6 m telescope in La Silla, yielding 3348 redshifts. The survey has a typical depth of $z = 0.1$. It fully samples the optical luminosity function down to $B = -15$ and various galaxy populations (*e.g.* spirals, ellipticals, dwarfs) are present. Interestingly, emission lines (OII, Hβ, OIII) have been found in a large fraction of the galaxy spectra ($\sim$ 40%), suggesting strong evolution of the galaxy population in terms of enhanced star formation. For further information on the ESP galaxy redshift survey see *e.g.* Zucca et al., this volume.

2   The Radio Survey: Observations

We used the Australia Telescope Compact Array (ATCA) at 20 cm to image the entire area of the optical survey (27 sq. degr.).

The ATCA supports a mosaic observing mode which allows efficient coverage of large areas of sky by interleaving short observations of a grid of pointings. The observing campaign (34 blocks of 12^h) started in November '94 and has been completed in January '96. Data reduction has been completed too.

We have got 16 big mosaiced radio maps, each covering 1.7 sq. degr. with spatial resolution $16'' \times 8''$. The noise level, after cleaning, is $\sim 70\,\mu$Jy and is fairly uniform (as needed for statistical studies) within each map and from map to map.

3 The Radio Survey: First Results

On the entire region observed, we searched for radio emission associated with the redshift survey galaxies. We pushed the search down to a 3σ-threshold (which is allowed when sky positions are known).

Radio emission was found for 524 galaxies, corresponding to a detection rate of 16.4%. Spurious detections are expected to be less than 2% of the total sample of 3196 galaxies searched for, and incompleteness has been estimated to be $\sim 1\%$.

Typically radio detected ESP galaxies are associated to very faint, point-like radio sources ($\sim 86\%$ of them have $S_{peak} < 1\,$mJy).

The analysis of the correlations between optical and radio properties of ESP galaxies is now under way. As a first result, we found that a large fraction ($\sim 60\%$) of the radio detections is associated to galaxies showing one or more emission lines. This suggests that in normal galaxies radio emission is mostly induced by star formation, traced by the OII line (Kennicut 1983, Kennicut 1992).

We are also producing a new catalogue consisting of all the radio sources detected above 6σ in the region surveyed.

In a preliminary analysis of a 4 sq degr. area , we detected 360 radio sources above 0.4 mJy. A large fraction of them ($\sim 40\%$) are sub-mJy objects.

This leads us to expect a total number of ~ 2500 radio sources in the entire area observed (27 sq. degr.) and ~ 1000 sub-mJy sources. This catalogue will therefore be especially useful in studying the sub-mJy population which is still poorly understood (e.g. Condon 1984).

References

1. J.J. Condon, ApJ, 287, 461, 1984.
2. R.C. Kennicut, A&A, 120, 219, 1983.
3. R.C. Kennicut, ApJ, 388, 310, 1992.

DIAGNOSTIC METHODS FOR EMISSION-LINE GALAXIES IN REDSHIFT SURVEYS

C.S. ROLA[1,2], E. TERLEVICH[1] & R. TERLEVICH[2]

Institute of Astronomy[1], Madingley Road, Cambridge CB3 0HA, UK
Royal Greenwich Observatory[2], Madingley Road, Cambridge CB3 0EZ, UK

We present new diagnostic diagrams which permit to identify the nature of emission-line galaxies. These are specially useful for deep surveys as the classification is based in a minimum of observational data and reddening corrections and flux calibration are not essential. The new diagrams require only the [O II]λ3727,3729 and Hβ emission-lines.

Several redshift surveys have produced an extensive data-base of spectroscopy of galaxies at moderate redshifts ($z \approx 0.1 - 0.7$); still, very little is known about the nature of their observed emission-line galaxies (ELGs). Nearby ELGs are generally classified by diagnostic diagrams, using emission-line intensity ratios (see Baldwin, Philips & Terlevich 1985, Veilleux & Osterbrock 1987). However, at z beyond ≈ 0.3 most of the strong emission-lines move out of the optical spectral range. This renders the existent diagnostic methods useless for distant objects found in redshift surveys where most of the ELGs are hence classified using the [O II]λ3727 Å emission-line, namely its equivalent width, to avoid flux calibration.

Therefore, in this work we investigate new diagnostic methods that are appropriate to deep surveys. These should require a minimum of spectral coverage, preferably in the blue. For this purpose, we considered a statistically significant local sample of ELGs: H II galaxies, Seyfert 2s and LINERs, mainly from the Terlevich *et al.* (1991) catalogue. The sample was then classified in active and H II galaxies based on four standard diagnostic diagrams, combined with a grid of photoionisation models (Rola 1995). Then this classification was transferred to new diagrams using only the few emission-lines that are usually available to redshift surveys. We based our new diagnostic diagrams on the equivalent widths of [O II]λ3727 and Hβ, both lines being generally present in the optical range up to $z \approx 0.75$. In the diagrams of Figure 1, *dot* symbols represent *H II galaxies*, *triangles Seyfert 2s* and *asterisks LINERs*. The horizontal lines drawn in each diagram were defined by inspection of the distribution of the EW([O II]λ3727)/EW(Hβ) ratio. The curve and vertical lines are based in the analysis of the data, where we have found that an EW(Hβ) of 10 Å is a good upper limit to separate active from H II galaxies (given that only a few H II galaxies lie below this limit). Thus, the combination of these two limits leads to two separation zones in both diagrams of Figure 1. The first one corres-

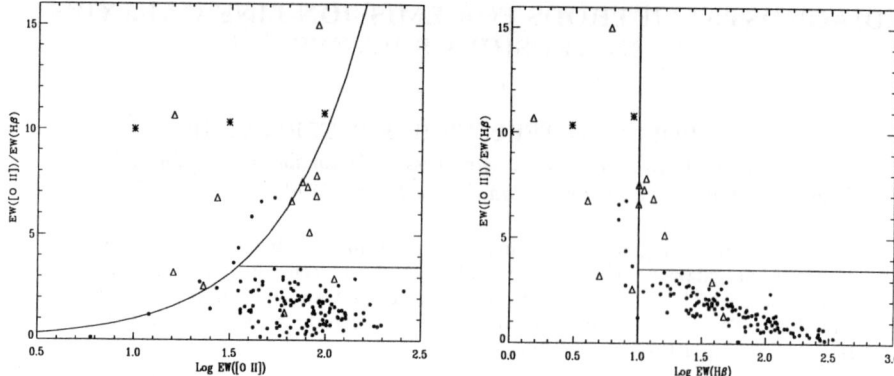

Figure 1: EW([O II])/EW(Hβ) vs. EW([O II]) (left) and EW(Hβ) (right). The left and right upper sections define the region where active galaxies lie while the H II galaxies region is the bottom right one.

ponds to the lower right region (henceforth called *H II region*), where most of the H II galaxies lie and only a few active ones fall. The second one (hereafter called *AGN region*) corresponds to the opposite region. The probability that an ELG which falls in the AGN region is an active galaxy is $\mathrm{Prob_{AGN}(AGN)} \approx 95\ \%$ and the probability of being an H II galaxy is $\mathrm{Prob_{AGN}(H\ II)} \approx 5\ \%$, while in the H II region, $\mathrm{Prob_{H\ II}(AGN)} \approx 10\ \%$ and $\mathrm{Prob_{H\ II}(H\ II)} \approx 90\ \%$. This implies that Figure 1 diagrams are very efficient in separating Seyfert 2s and LINERs from H II galaxies. This method provides a strong and valuable diagnostic of the nature of ELG's which seems not to be very much affected by galaxy evolution, at least up to intermediate redshifts as fully discussed in Rola, Terlevich & Terlevich (to be submitted to MNRAS). Further diagnostic methods are also discussed in this paper.

Acknowledgments

CR acknowledges the Conference organisation for financial support and the JNICT (Portugal) for the grant BPD/6064/95 from the PRAXIS XXI program.

References

1. Baldwin, J., Philips, M. , & Terlevich, R., 1981, PASP, 93, 5
2. Rola, C. S., 1995, Ph.D. thesis, Université de Paris VII, France
3. Terlevich, R., Melnick, J., Masegosa, J., Moles, M., Copetti, M., 1991, AAS, 91, 285
4. Veilleux, S. & Osterbrock, D., 1987, ApJS, 63, 295

PHOTOMETRIC REDSHIFTS IN THE HUBBLE DEEP FIELD

M.J. SAWICKI, H. LIN, H.K.C. YEE

Department of Astronomy, University of Toronto, Toronto, M5S 3H8, Canada

Photometric redshifts for galaxies in the Hubble Deep Field are measured. Luminosity functions show steepening of the faint-end slope and mild brightening of M^* out to $z \approx 3$, followed by a decline at higher z; an excess of faint, star-forming galaxies is seen at low z. Our results are consistent with the formation of large galaxies at $z = 2$–3, followed by that of dwarfs at $z < 1$.

1 Colour Redshifts in the HDF

Because of the extreme depth of the Hubble Deep Field (HDF), spectroscopic redshifts are not practical for all but the brightest objects. The redshift of a galaxy can, however, be estimated by comparing the observed broadband colours against a set of reference templates computed for a range of redshifts and spectral types. We computed templates by extending empirical spectral energy distributions[1] into the UV, applying Lyman blanketing[2] at high z, and convolving with HST filter transmission curves.

Object finding and photometry was done using the PPP faint galaxy photometry package[3]. There are 1003 objects with $F814W_{AB} \leq 27$, of which 90% are detected in all four HDF bandpasses. Each object's observed colours were compared against the templates (using a least-squares fitting technique) to obtain the most likely redshift and spectral type. Photometric redshifts agree well with spectroscopic ones, with a scatter of $\sigma_z = 0.13$ at $z < 1.5$, increasing to $\sigma_z = 0.32$ at $z > 2$. The catastrophic failure rate is small (2/57 objects).

2 Galaxy Population to $z = 4$

For a full discussion the reader is referred to our main paper[4]. The results are summarized in Fig. 1(a–c). The bright end of the luminosity function (LF) brightens moderately between the present epoch and $z \approx 2.5$. This brightening is accompanied by a steepening of the faint-end slope. Beyond $z \approx 3$ the LF fades to values similar to those seen locally. This fading could be a signature of the onset of galaxy formation which is expected to occur around that redshift[5]. In this scenario, the star-forming galaxies seen at $z > 2$ in Fig. 1b will become present-day ellipticals and spirals; those at $z \approx 1$ are star-forming dwarfs which are also seen in the Hubble diagram, and evolve to become the faint ($M_{F450W_{AB}} > -15$) galaxies in the $z = 0.2$–0.5 LF.

176

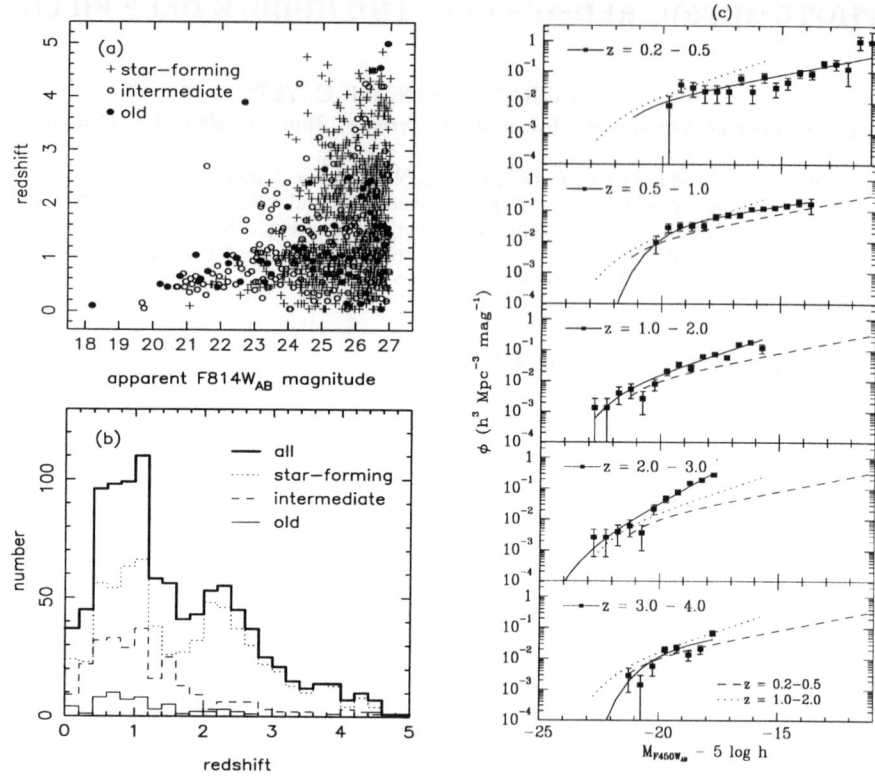

Figure 1: (a) Hubble diagram. (b) Redshift distribution for objects with F814W$_{AB}$ ≤ 27. (c) F450W$_{AB}$(≈ B$_{AB}$) luminosity functions; dashed and dotted lines are fiducials.

1. Coleman, G. D., Wu, C.-C., & Weedman, D. W. 1980, ApJS, 43, 393
2. Madau, P. 1995, ApJ, 441, 18
3. Yee, H. K. C. 1991, PASP, 103, 396
4. Sawicki, M. J., Lin, H., & Yee, H. K. C. 1996, AJ, submitted
5. Fukugita, M., Hogan, C. J., & Peebles, P. J. E. 1996, Nature, 381, 489

OPTICAL ROTATION CURVES OF DISTANT GALAXIES : KINEMATICS AND EVOLUTION OUT TO $z \sim 1$

N. P. VOGT & A. C. PHILLIPS

UCO / Lick Observatory, UCSC, Santa Cruz, CA, 95064, USA

Spatially resolved velocity profiles are presented for 12 faint field galaxies in the redshift range $0.1 \lesssim z \lesssim 1$ based on moderate–resolution spectroscopy obtained with the Keck 10m telescope. These data were augmented with high–resolution *Hubble Space Telescope* images from *WFPC2*, which provided V and I photometry, galaxy type, orientation, and inclination. The effects of seeing, slit width, and slit misalignment with respect to galaxy major axis were modeled along with inclination for each source, in order to derive a maximum circular velocity from the observed rotation curve. The rotation curves appear similar to those of local galaxies in both form and amplitude, implying that some massive disks were in place at $z \sim 1$. The key result is that the kinematics of these distant galaxies show evidence for only a modest increase in luminosity ($\Delta M_B \lesssim 0.6$) compared to velocity–luminosity (Tully–Fisher) relations for local galaxies.

The correlation between rotational velocity and luminosity of disk galaxies provides a powerful tool with which to study distant galaxies and to investigate galaxy evolution. Galaxy evolution models range from those with mild amounts of luminosity brightening in the past (Gronwall & Koo 1995) to those requiring more dramatic changes (Glazebrook *et al.* 1995). By comparing a distant sample of rotation curves to local TF relations, we can directly constrain the global brightening of disk galaxies in the past.

Twelve inclined disk galaxies with *WFPC2* images have been observed on the Keck telescope; spatially resolved major axis velocity profiles have been acquired using a 600 line mm^{-1} grating (1.26 Å and (0.''215 per pixel) for emission lines ranging from Hα through [O II] $\lambda 3727$. Gaussian profiles were fit to these rotation curves, and an exponential disk model with flat terminal velocity was used to deconvolve the effects of slit width and position angle, inclination, and seeing to derive a maximum circular velocity for each galaxy.

The shapes of the rotation curves of these high redshift galaxies are similar to those of local galaxies. The high–redshift rotation curves are relatively symmetric, show a "solid–body" rise in the inner regions, and turn over to a relatively constant circular velocity in the outer parts. The maximum velocities are comparable to those of local spirals. Rough calculations yield masses between 1 and 5×10^{11} $M_\odot$, well within the range of masses found for nearby spiral galaxies. Notably, the kinematics of these distant galaxies show evidence for only a modest increase in luminosity of $\Delta M_B \lesssim 0.6$ compared to

Tully-Fisher relations for local galaxies.

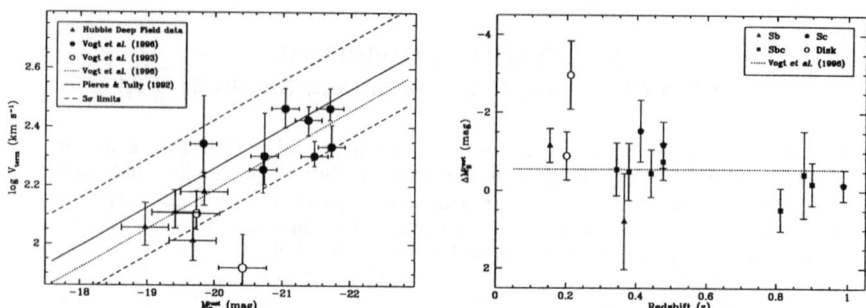

Figure 1: (A) Four recently observed galaxies from the *Hubble Deep Field* area and ten galaxies from Vogt *et al.* (1996,1993) are compared to the Pierce & Tully (1992) relationship based on HI velocity width measurements for a set of cluster spirals, shown by the best–fit B-band relation (solid line) and 3σ limits (dashed lines). The weighted fit to the eight galaxies from Vogt *et al.* (1996) (dotted line; assuming the same slope) produces an offset of 0.55 ± 0.16 mag, and the *Hubble Deep Field* area data confirms this trend, and extends it towards intrinsically lower luminosity galaxies (L^* corresponds to $M_B \sim -20.4$). Data have been corrected for internal extinction, and we assume $H_0 = 75$ km s^{-1} Mpc^{-1} and $q_0 = 0$. (B) Brightening relative to the Pierce and Tully relationship is shown, and the weighted fit to the eight galaxies from Vogt *et al.* (1996) (dotted line). There is no notable increase in brightening as a function of redshift, nor is there a clear trend with type.

Acknowledgments

Funding was provided by the Center for Particle Astrophysics and by NSF grants 9529098, AST-922540, AST-9120005, AR-06337.08-94A, and AR-06337.21-94A; NASA grant AR-5801.0194A. The authors thank the UCSC members of the DEEP project for their assistance in these endeavors.

References

1. K. Glazebrook, R. Ellis, B. Santiago, & R. Griffiths, MNRAS, 275, L19, 1995.
2. C. Gronwall & D. C. Koo, ApJ, 440, L1, 1995.
3. M. J. Pierce & R. B. Tully, ApJ, 387, 47, 1992.
4. N. P. Vogt, T. Herter, M. P. Haynes, & S. Courteau, ApJ, 415, L95, 1993.
5. N. P. Vogt, D. A. Forbes, A. C. Phillips, C. Gronwall, S. M. Faber, G. D. Illingworth, & D. C. Koo, ApJ, 465, L15, 1996.

TESTS OF MORPHOLOGICAL PECULIARITY INDICES FOR DISTANT AND LOCAL GALAXIES

K. L. WU, S. M. FABER

Astronomy & Astrophysics, UCSC, UCO/Lick Observatory,
Santa Cruz, CA 95064, USA

T.R. LAUER

KPNO/NOAO, P.O. Box 26732, Tucson, AZ 85726, USA

Two indices of morphological peculiarity are tested on images of three galaxies: two spirals and one peculiar. The indices are (1) the ratio of the perimeter of a given light contour to the circumference of a circle with the same area, and (2) a concentration index. The concentration index is robust to smoothing, while moderate amounts of smoothing are required to achieve contours that are neither dominated by noise nor completely featureless due to over-smoothing. We find that, for both indices, it *is* possible to discriminate between the spirals and the peculiar. This holds also when the galaxies are "redshifted" to $z \sim 0.6$ or 0.8.

The morphological classification of nearby normal galaxies is embodied in the Hubble system, in which galaxies range from ellipticals through spirals to irregulars. Some galaxies, however, do not fit neatly into any of the Hubble categories, *e.g.*, galaxies that have undergone a recent merger often display tidal tails. The frequency and types of these *peculiar* galaxies play a key role in our understanding of the merging history of galaxies and their progenitors.

From early deep HST images, the frequency of peculiar galaxies at high redshifts has been claimed to be much higher than it is locally (*e.g.*, Driver *et al.* 1995). Nevertheless, little has been done to quantify the types and degrees of peculiarities that are seen.

We are developing several algorithms to quantify peculiarities in galaxy morphology. These algorithms will be applied to galaxy images and are sensitive to several different features. The two indices presented here are:

- **P/C ratio:** Ratio of the perimeter of a contour to the circumference of a circle with the same area.
- **Concentration Index:** Ratio of radii enclosing 80% and 20% of the total light (measured through circular apertures).

These indices were applied to three local galaxies: two normal spirals (NGC 4535, NGC 5334 [Frei *et al.* 1996]) and one peculiar (NGC 520 [Hibbard & van Gorkom 1996]), believed to be a merger remnant from its distorted core and tidal tail. The indices were also measured for simulations of distant galaxies, made by adding noise to and rebinning the local images to match the

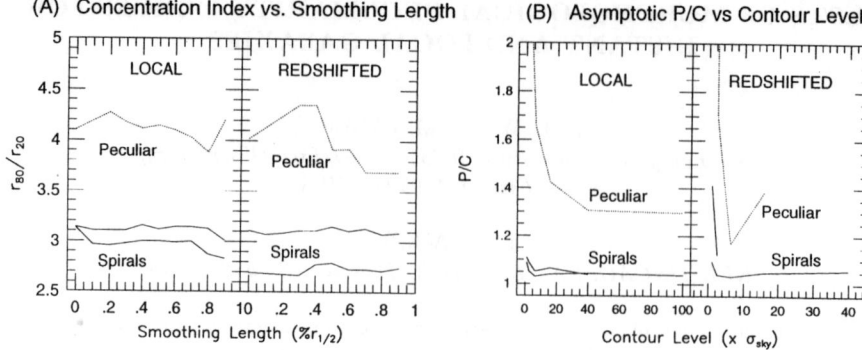

Figure 1: (A) The concentration index is roughly independent of the smoothing length chosen. Its value clearly differentiates the spirals from the peculiar, both locally and at high z. (B) After testing several smoothing lengths (ranging from 0.1 to $0.9r_{1/2}$), we find that moderate smoothing ($\sim 0.5r_{1/2}$) is required for meaningful values of the P/C ratio. As illustrated in (B), we also find that, for the local sample, meaningful P/C ratios are found only for moderate brightness contour levels ($\sim 5\sigma_{sky}$). At shorter smoothing lengths and fainter contours, noise dominates, contours are broken, and P/C $\gg 1.00$. At longer smoothing lengths, the galaxy becomes over-smoothed and P/C ≈ 1.00. At brighter contours, only the core of the galaxy is sampled, and again, P/C rapidly approaches 1.00. Note also that the P/C ratios truncate at lower contours for the redshifted images, primarily because heavy smoothing washes out bright galaxy cores. One of the spirals (NGC 5334) truncates at a particularly low level because it is a low surface brightness object.

signal-to-noise ratio and sampling of HST WFPC-2 images. Various degrees of smoothing were applied to test the robustness of the indices against smoothing.

As illustrated in Figure 1, both the concentration index and the P/C ratio can discriminate between spiral and peculiar galaxies. It will be interesting to see if this also holds true for the distant ($z \sim 0.8$ to 1) galaxies seen by HST.

Acknowledgments

We thank J Hibbard and P Guhathakurta for many very helpful discussions. KLW was supported by NASA Grant No. NAS-5-1661 to the WF/PC IDT.

References

1. S.P. Driver, R.A. Windhorst, E.J. Ostrander, W.C. Keel, R.E. Griffiths, & K.U. Ratnatunga, Astrophys. Lett., 449, L23, 1995.
2. Z. Frei, P. Guhathakurta, J.E. Gunn, & J.A. Tyson, AJ, 111, 174, 1996.
3. J.E. Hibbard & J.H. van Gorkom, AJ, 111, 655, 1996.

THE LUMINOSITY FUNCTION AND MEAN DENSITY OF GALAXIES FROM THE ESO SLICE PROJECT (ESP) REDSHIFT SURVEY

E. ZUCCA[1,2], G. VETTOLANI[2], A. CAPPI[1], R. MERIGHI[1],

M. MIGNOLI[1], G. STIRPE[1], G. ZAMORANI[1], H. MacGILLIVRAY[3],

C.COLLINS[4], C. BALKOWSKI[5], V. CAYATTE[5], S. MAUROGORDATO[5],

D. PROUST[5], G. CHINCARINI[6], L. GUZZO[6], D. MACCAGNI[7],

R. SCARAMELLA[8], A. BLANCHARD[9], M. RAMELLA[10]

[1] *Osservatorio Astronomico di Bologna, Bologna, Italy*

[2] *Istituto di Radioastronomia del CNR, Bologna, Italy*

[3] *Royal Observatory Edinburgh, Edinburgh, United Kingdom*

[4] *Liverpool John Moores University, Liverpool, United Kingdom*

[5] *DAEC, Observatoire de Paris–Meudon, Meudon, France*

[6] *Osservatorio Astronomico di Brera–Merate, Milano, Italy*

[7] *Istituto di Fisica Cosmica e Tecnologie Relative, Milano, Italy*

[8] *Osservatorio Astronomico di Roma, Monteporzio Catone, Italy*

[9] *Universitè Louis Pasteur, Strasbourg, France*

[10] *Osservatorio Astronomico di Trieste, Trieste, Italy*

We present the results about the luminosity function and mean density of galaxies from the ESO Slice Project redshift survey, which consists of more than three thousands galaxies with $b_J \leq 19.4$, over an area of about 30 square degrees.

1 The ESO Slice Project redshift survey

The ESO Slice Project (ESP) is a galaxy redshift survey we have recently completed as an ESO Key–Project over about 30 square degrees, in a region near the South Galactic Pole (Vettolani et al., submitted to A&A). The survey is nearly complete to the limiting magnitude $b_J = 19.4$ and consists of more than three thousands galaxies with reliable redshift determination.

The ESP survey is intermediate between shallow, wide angle samples and very deep, monodimensional pencil beams: spanning a volume of $\sim 10^5$ h^{-3} Mpc^3 at the sensitivity peak ($z \sim 0.1$) it can provide an accurate determination of the "local" luminosity function and the mean galaxy density (Zucca et al., submitted to A&A).

2 The luminosity function

We find that, although a Schechter function (with $\alpha = -1.22$, $M_{b_J}^* = -19.61$ $+5\log h$ and $\phi^* = 0.020\ h^3$ Mpc^{-3}) is an acceptable representation of the luminosity function over the entire range of magnitudes ($M_{b_J} \leq -12.4$ $+5\log h$), our data strongly suggest a steepening of the luminosity function for $M_{b_J} \geq -17 +5\log h$. Such a steepening, well fitted by a power law with slope $\beta \sim 1.6$, is in agreement with what has been recently found by similar analyses for both field galaxies (Marzke et al. [1]) and galaxies in clusters (f.i. Driver & Phillipps [2]).

This steepening at the faint end of the luminosity function is almost completely due to galaxies with emission lines: in fact dividing galaxies into two samples, i.e. galaxies with and without emission lines, we find significant differences in their luminosity functions. In particular, galaxies with emission lines show a steeper slope and a fainter M^*. The galaxies in the faint end are mainly compact objects, very nearby ($v < 10000$ km/s), but note that this excess is not due to a local overdensity (i.e. a cluster or a group): in fact in the surveyed area there is a significant underdensity for $v < 15000$ km/s.

The normalization and the α and M^* parameters of our luminosity function are in excellent agreement with those of the AUTOFIB redshift survey (Ellis et al. [3]). Viceversa, our normalization is a factor ~ 2 higher than that found for both the APM (Loveday et al. [4]) and the Las Campanas (Lin et al. [5]) redshift surveys. Also the faint end slope of our luminosity function is significantly steeper than that found in these two surveys.

The galaxy number density for $M_{b_J} \leq -16 +5\log h$ is well determined ($\bar{n} = 0.08\ h^3$ Mpc^{-3}). Its estimate for $M_{b_J} \leq -12.4 +5\log h$ is more uncertain, ranging from $\bar{n} = 0.28\ h^3$ Mpc^{-3} , in the case of a fit with a single Schechter function, to $\bar{n} = 0.54\ h^3$ Mpc^{-3} , in the case of Schechter function and power law fit. The corresponding luminosity densities in these three cases are $\rho_{LUM} = (2.03, 2.23, 2.31) \times 10^8\ h\ L_\odot$ Mpc^{-3}, respectively.

References

1. Marzke, R.O., Huchra, J.P., Geller, M.J., 1994, ApJ 428, 43
2. Driver, S.P., Phillipps, S., 1996, ApJ in press
3. Ellis, R.S., Colless, M., Broadhurst, T., Heyl, J., Glazebrook, K., 1996, MNRAS 280, 235
4. Loveday, J., Peterson, B.A., Efstathiou, G., Maddox, S.J., 1992, ApJ 390, 338
5. Lin, H., Kirshner, R.P., Shectman, S.A., et al. 1996, ApJ 464, 60

SECTION 2

GALAXIES IN DISTANT CLUSTERS

SECTION 2

HST OBSERVATIONS OF DISTANT CLUSTERS: IMPLICATIONS FOR GALAXY EVOLUTION

A. DRESSLER

Carnegie Observatories, 813 Santa Barbara Street, Pasadena, CA 91101, USA

I. SMAIL[a]

Department of Physics, University of Durham, South Road, Durham DH1 3LE, England

The "MORPHS" group has completed the cataloging, parameterization, and morphological classification of $\sim$2000 galaxies in 10 rich clusters from $0.36 < z < 0.56$. From a weak lensing analysis using these data, which compares the X-ray properties (L_X) of the clusters with virial temperature estimates (T_v) from the lensing shear strength, we find little evidence for evolution in the L_X–T_v relation from that observed for local clusters. We discuss how this observation constrains models for the X-ray evolution of clusters. The data have also been used to study the color dispersion of bona-fide ellipticals in high-z clusters: we find the spread to be very small, suggesting an early formation epoch for the stellar populations of cluster ellipticals. This is consistent with the evolution of the morphology-density relationship, in which we find ellipticals to be as abundant at $z = 0.5$ as in clusters today, and already well ensconced in the dense regions. In contrast, S0's are less plentiful and less well-concentrated compared to the present epoch, and spiral galaxies everywhere more abundant. Combined with other spectroscopic and morphological data, these observations suggest that most of these rapidly evolving systems are not likely to become bright ellipticals, which were more likely formed at early epochs. Cluster S0 galaxies, on the other hand, are likely to have been produced in large numbers in the recent past.

1 Introduction

Substantial progress has been made in the study of galaxy evolution through use of the lookback in cosmic time afforded by the observations of distant clusters and field galaxies. Very different histories of star formation produce spectral and integrated color characteristics that are essentially indistinguishable by the present epoch. Observations of the state of galaxies at much earlier times help break this degeneracy and offer a clearer picture of the evolutionary path taken by galaxies of various types and masses, and in different environments.

A group of us we call the "MORPHS" — Richard Ellis, Warrick Couch, Gus Oemler, Harvey Butcher, Ray Sharples, Bianca Poggianti, Amy Barger,

---

[a]Visiting Research Associate at the Carnegie Observatories.

and ourselves, has been using images from the Hubble Space Telescope Wide Field Planetary Camera 2 (WFPC-2) and extensive ground-based photometry and spectroscopy to study the properties of, and galaxy populations in, rich clusters of galaxies at $z \sim 0.5$. Here we report on various results from our group, which should be referred to by the specific papers named in each section.

2   Mass Estimates for Distant Clusters

Rich clusters of galaxies can be identified to high redshift and can thus be used as tracers of the evolution of structure in the universe. Moreover, as clusters represent the extreme tail of the mass fluctuation spectrum they provide a particularly sensitive probe of the form of the primordial power spectrum. One of the most widely used techniques to identify and study the masses of distant clusters is X-ray imaging of the hot intracluster gas bound to the cluster potential. Published X-ray surveys indicate a reduction in the volume density of luminous clusters at intermediate redshift. Unfortunately, without corroborating evidence it is difficult to determine if this arises from a real decline in the number of *massive* clusters in the past, or a change in the thermal properties of the cluster gas. The relatively new field of gravitational lensing provides an unique opportunity to tackle this issue by determining independent estimates of the cluster masses, from their effects on the shapes of background faint field galaxies. Using the homogenous, high-quality WFPC-2 imaging obtained for distant clusters by our collaboration, we have made a first attempt at combining X-ray and lensing observations of a large sample of distant clusters to determine the relative importance of cluster mass and the thermal history of the gas on the X-ray luminosities. The following discussion is abstracted from Smail et al. (1996a).

For the lensing analysis we have used deep WFPC-2 imaging of 12 distant clusters spanning the redshift range $z = 0.17$–0.56. Working from catalogs of faint galaxies ($I_{814} = 24.0$–26.0) detected in these fields we measure the mean shear strength — the average, coherent elongation of galaxy images around the cluster lens center — within a $200h^{-1}$ kpc aperture. We detect the signature of gravitational lensing in 11 of the 12 clusters; spanning nearly an order of magnitude in lensing strength. Moreover, the shear strength measured on these large scales correlates well with the presence of multiply-imaged arcs and pairs in the very central regions of the clusters, indicating that the clusters all share similar mass profiles. We have examined the correlation between the cluster X-ray luminosities and our mean gravitational shear strengths (linearly related to the central mass and the cluster virial temperature) and develop a model which allows us to predict the relationship expected from the properties

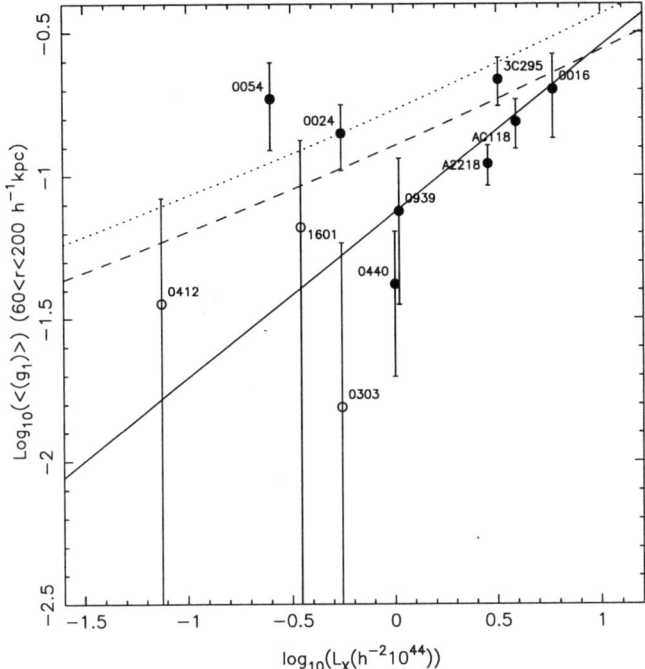

Figure 1: The correlation between the cluster X-ray luminosity and the mean shear strength, $<g_1>$, linearly related to central cluster mass. The error bars are 1σ boot-strap estimates and the solid line shows the best fit relationship for the data. The dotted line indicates the upper limit expected, assuming 100% measurement efficiency, in the case of our simple model. The dashed line represents a 75% efficiency. Filled symbols denotes those clusters which have candidate strongly-lensed features.

of local clusters. After allowing for various observational effects, we find that the predicted correlation is a reasonable match to the available data (Fig. 1), indicating that there has been little evolution in the X-ray luminosity–central mass relationship between $z \sim 0.4$ and now. Such limited evolution in the X-ray luminosity–central mass relation can be reproduced by models introducing a modest initial entropy into the gas prior to cluster formation, possibly resulting from pre-heating by AGN or galactic winds. Our results demonstrate the important role weak gravitational lensing can play in the study of the evolution of distant clusters, as the most direct and least biased probe of their growth.

3   The Ages of Elliptical Galaxies in Distant Rich Clusters

Elliptical galaxies are conventionally regarded as old galactic systems whose star formation history can be approximated as a single burst that occurred 12–16 Gyr ago. However, in recent years, this simple picture has been challenged from various viewpoints. Numerous cases have been found of ellipticals with intermediate-age stellar populations and dynamical arguments suggest that many peculiarities seen in ellipticals (shells and dust-lanes) are best explained via recent formation from the merger of gas-rich systems. Nevertheless, the small scatter observed for the $(U - V)$ colors of spheroidal galaxies in nearby clusters of galaxies still provides a basic constraint on the history of star formation in dense environments (Bower, Lucey & Ellis 1992). Using these local data and assuming the spheroidals formed stochastically, it is possible to limit their formation epoch to $z \geq 2$. However, introducing some degree of synchronicity serious weakens this limit, allowing the spheriodal population to form at more recent epochs.

In Ellis et al. (1996) we address this ambiguity using high precision rest-frame $(U-V)$ photometry of a large sample of morphologically-selected spheroidal galaxies in three $z \sim 0.54$ clusters which have been observed as part of our HST program. We use our $F555W$ and $F814W$ imaging to determine accurate rest-frame $(U - V)$ colors for spheroidal galaxies in the three clusters: Cl0016+16 ($z = 0.55$), Cl0054−27 ($z = 0.56$) and Cl0412−65 (= 0.51). Using these new data we repeat the color-scatter analysis conducted locally at a significant look-back time. Matching our aperture sizes, luminosity range and color system to those used locally we find a small scatter (≤ 0.07 mag rms, not much greater than that observed at $z \sim 0$) for galaxies classed as Es and E/S0s, both internally within each of the three clusters and externally from cluster to cluster. We do not find any trend for the scatter to increase with decreasing galaxy luminosity beyond that due to observational error. Our result thus provides a new constraint on the star formation history of cluster spheroidals prior to $z \simeq 0.5$. Although we cannot rule out the continued production of *some* ellipticals, our results do indicate that the bulk of the stars seen in luminous elliptical cluster galaxies were formed by $z \simeq 3$.

4   The Morphology-Density Relation at High Redshift

A principal goal of our group has been to study the evolution of morphological types in the rich cluster environment. To this end we have morphologically classified 1857 objects brighter than $R_{702} < 23.0$ or $I_{814} < 23.5$, in the 11 fields, as described in Smail et al. (1996b) While the addition of spectroscopically-derived parameters, such as cluster membership or stellar population, is important for understanding the evolutionary state of these populations, photometric/morphological information alone allows a simple and important comparison with the properties of present day clusters. This comparison offers clues as to how clusters of galaxies came to hold their atypical complements of galaxy types.

With a resolution approaching $0.1''$, our WFPC-2 images show detail at the level of 500 pc in the clusters, equivalent to observing galaxies in the Coma cluster with $1''$ seeing. While cruder than than the resolution usually available for morphological classification of nearby galaxies, it is sufficient for the identification of basic morphological information, and, in particular, is comparable to that presented by Dressler (1980) in the study of galaxy morphology in 55 low-redshift clusters. Our morphological samples in the distant clusters have also been selected in as similar a fashion as possible to the local data, for example, over the same area and to comparable absolute magnitude limits.

Dressler found a strong relation between the fractions of E, S0, and spiral galaxies with the local projected density where they were found, in the well known sense that ellipticals became more prevalent, and spirals less so, in regions of higher surface density. He concluded that, to first order at least, the morphology-density relation is universal, that is, representative of every cluster in the sample, regardless of its global properties.

Dressler's original data have been reanalyzed and are presented in Dressler et al. (1996) along with the morphology-density relation for the HST sample described here. That paper also reviews some of the challenges to the morphology-density relation, for example, the contention by Whitmore, Gilmore, & Jones (1993) that the principal determinant of galaxy type within rich clusters is the radial distance from the cluster center. For our purposes here we simply analyze the morphology-density relation in our $z \sim 0.5$ sample.

In Fig. 2 we show the morphology-density relation for the entire $z \sim 0.5$ sample. The density range encompassed by the more distant sample is shifted by half a dex to higher density, probably reflecting the fact that these clusters are systematically richer than the typical clusters of Dressler's local sample.

Before addressing the question of gradients in Fig. 2, we take note of differences between this and the nearby cluster sample. As is now well known, spirals are greatly overabundant at these high densities compared to present-epoch clusters, but, perhaps surprisingly, the difference seems to made up entirely by a paucity of S0 galaxies rather than an underabundance of *both* S0 and E galaxies. In fact, E galaxies appear to be in even greater abundance! At comparable densities, spirals are a factor of 2 overabundant, S0's are a factor of 2–3 underabundant, and ellipticals are a factor of 1.5 overabundant in the

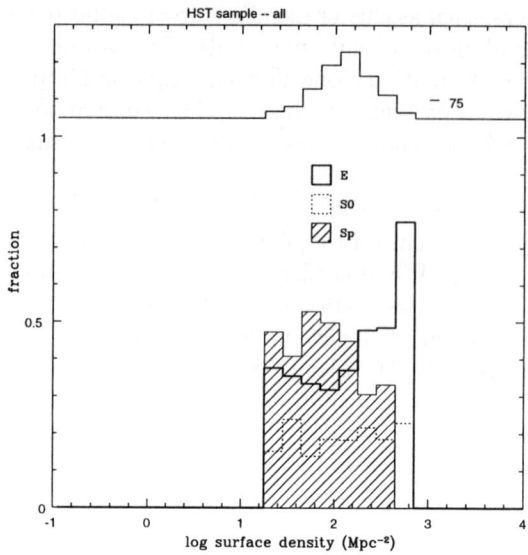

Figure 2: The morphology-density relation for 10 clusters at redshifts $0.36 < z < 0.57$. The plot shows the relative proportions of the different morphological classes (E, S0, Sp) as a function of projected density. The upper histogram shows the total number of galaxies in each bin.

$z \sim 0.5$ sample compared to Dressler's sample of nearby clusters. The paucity of S0 galaxies is particularly noteworthy. As explained in Smail et al. (1996b), we have compared the distribution in flattenings of S0 galaxies in our sample with that for the Coma cluster, to see if we have systematically misclassified S0 galaxies as ellipticals, particularly for the face-on cases. The good agreement of these distributions indicates that this is probably not the case, but at any rate it is hard to see how we could be missing more than $\sim 25\%$, which is of little consequence to the gross deficiencies in S0's found here.

We now ask whether any trend of morphology with density is apparent for the distant sample. From Fig. 2 it appears that a modest relation is present, but it is only for the bins of highest surface density — over the last factor of 5 in surface density. Over this range the spiral fraction plummets and the elliptical fraction rises sharply, but for the lower density zones, over which there is a very noticeable gradient in the nearby clusters, the relationships are basically flat. However, when the sample is divided by concentration and the degree of regularity, which to a large extent go together, a very different picture

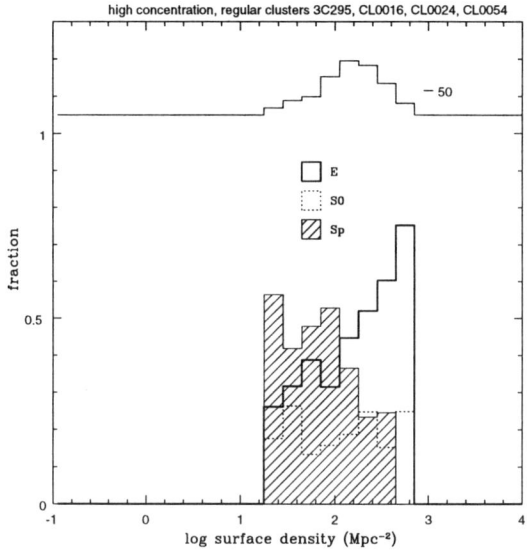

Figure 3: The morphology-density relation for 4 high-concentration, regular clusters at intermediate redshift, 3C295, Cl0016+16, Cl0024+16, and Cl0054−27.

emerges. Fig. 3 shows that, for the 4 highest concentration, regular clusters of the $z \sim 0.5$, the morphology-density relation is steep and well defined over the entire density range. In contrast, however, there are no correlations at all for the 4 lowest concentration, irregular clusters. This is a strikingly different result from the situation for present epoch clusters, for which Dressler found a strong morphology-density relationship for both irregular and regular clusters. A more detailed discussion of this difference can be found in Dressler et al. (1996).

Perhaps our most important result of this analysis, however, is simply that irrespective of whether the clusters appear dynamically "mature" or not, the incidence of elliptical galaxies is already very high, and independent of whether they are collected into dense, central regions or not. We suggest, based on this result, that elliptical galaxies predate, and are basically independent of, the virialization of a rich cluster. This is, of course, consistent with the Ellis et al. (1996) result described above of early formation of the stars in elliptical galaxies. Furthermore, we find that that the fractional representation of highly asymmetric or disturbed morphologies ($D > 1$, see Smail et al. 1996b) with local surface density mirrors the S0 or spiral trend rather than the trend for ellipticals. Together, these three observations suggest that, for the environments of rich clusters at least, the stars seen in elliptical galaxies are not the result of mergers of starforming, gas-rich systems after a redshift $z = 3$. This does not preclude the possibility of dissipationless mergers at $z = 1$, say, when these clusters might have been in the process of amalgamating small groups with lower velocity dispersion, but both the distribution and numbers of ellipticals we have found here, and their photometric and spectral properties, suggest that the stellar populations of ellipticals in these regions are not produced by late mergers, or in any process that depended on the dynamical evolution of a rich cluster. Instead, gaseous mergers or coherent collapse at high redshift, or growth of spheroids through dissipationless mergers until later epochs, seems to be the history indicated for ellipticals. It is remarkable, we think, that the environment of proto-clusters of this richness was able to produce such a large population of ellipticals before the identities of the clusters themselves was well established.

The situation for the S0 galaxies seems to be just the opposite. Though the ones we find are, like the ellipticals, red and with little scatter in color, their numbers are so deficient as to suggest that many need to be added since $z = 1$, in order to reach the populations of present-epoch clusters. The source of these S0's seems clear: the overabundance of spirals provides a resevoir of galaxies which may be stripped by ram pressure, tidally harassed (Moore et al. 1995) merged, or subject to strong 2-body gravitational interactions, with the result of producing today's dormant disk galaxies in clusters. Our $z \sim 0.5$ cluster sample includes a significant number of disturbed, distorted morphologies, often with spectroscopic evidence of strong episodes of star formation. These may be the result of mergers, strong interactions, accretions, harassment, or stripping — we are still unable to tell which of these processes are responsible. But, we do know from our morphological classifications that most of these are *disk* systems — they do not seem destined to settle into ellipticals galaxies when their jostling and bursts of star formation have ceased. Though the exact mechanism(s) may be yet unspecified, it seems that at least half of the S0 galaxies in today's clusters have been made by such processes since $z = 0.5$

5 Conclusions

Our HST images of distant clusters exhibit robust shear fields due to gravitational lensing. These allow us to estimate the cluster mass (or virial temperature) and compare these to the cluster X-ray luminosity. The relation between L_X and T_v in these distant clusters is similar to that observed locally and we thus claim that there is no strong evolution in the L_X–T_v relationship out to

$z \sim 0.4$.

We find a remarkably small scatter of restframe $(U - V)$ color for galaxies that we have classified as ellipticals. This suggests an early epoch of formation for the stars in these galaxies, and perhaps of the galaxies themselves. Furthermore, the large number of elliptical galaxies in these clusters, $\sim 40\%$, suggests that the formation of ellipticals predates cluster virialization. If mergers are responsible for making the ellipticals that now inhabit these rich clusters, they must have been dissipationless, in the "group phase" at $z \sim 1$, or much earlier, $z > 3$, if significant dissipation and star formation were involved. In contrast, the relative paucity of S0's in the intermediate redshift clusters suggests that many of them have indeed been added since $z \sim 0.5$, by mechanisms that acted on the excessive numbers, compared to today's clusters, of spirals and irregulars.

1. Bower, R.G., Lucey, J.R. & Ellis, R.S., 1992, MNRAS, 254, 601.
2. Dressler, A. 1980, ApJ, 236, 351.
3. Dressler, A., Oemler, A. Jr., Smail, I., Couch, W.J., Ellis, R.S., Barger, A., Butcher, H., Poggianti, B.M., & Sharples, R.M. 1996, Ap.J., submitted.
4. Ellis, R.S., Smail, I., Dressler, A., Couch, W.J., Oemler, A., Butcher, H. & Sharples, R.M., 1996, Ap.J., submitted. (astro-ph/9607154).
5. Moore, B., Katz, N., Lake, G., Dressler, A., & Oemler, A. Jr. 1996, Nature, 379, 613.
6. Smail, I., Dressler, A., Couch, W.J., Ellis, R.S., Oemler, A. Jr., Butcher, H., & Sharples, R.M., 1996a, Ap.J., in press. (astro-ph/9610270).
7. Smail, I., Ellis, R.S., Dressler, A., Couch, W.J., Oemler, A. Jr., Sharples, R.M., & Butcher, H., 1996b, Ap.J., submitted.
8. Whitmore, B.C., Gilmore, D.M., & Jones, C. 1993, ApJ, 407, 489.

THE EVOLUTION OF ELLIPTICAL GALAXIES IN CLUSTERS TO $z \sim 0.55$ FROM THE *HUBBLE SPACE TELESCOPE*

A.J. BARGER

Institute of Astronomy, Madingley Road,
Cambridge, CB3 0HA, England

We analyse surface photometry measurements made from *Hubble Space Telescope* images of morphologically classified elliptical galaxies in the cores of 12 rich clusters with redshifts between $z = 0.17$ and $z = 0.56$. We also present new near-infrared images for 6 of these clusters over the same fields. We use the mean surface brightness, $\langle\mu\rangle_e$, versus effective metric radius, R_e, projection of the fundamental plane to determine mean surface brightnesses at the standard condition $R_e = 3\,\mathrm{kpc}$ for the cluster galaxies grouped by redshift. The variation of these $\langle\mu\rangle_e^{SC}$ values with redshift shows luminosity evolution in both the B and K bands comparable to that expected from passive evolutionary models in which the bulk of the stellar population has formed by redshifts $z > 1$.

1 Introduction

A remarkable empirical correlation of size, surface brightness, and central velocity dispersion known as the *fundamental plane*[4,5] exists with very low scatter for ellipticals. At high redshifts it is difficult to obtain velocity dispersions because such measurements require long telescope exposures. However, relatively tight correlations also exist in the Kormendy relation[7] projection of the fundamental plane, $\langle\mu\rangle_e$ versus R_e. The low scatter allows a reasonably accurate determination of $\langle\mu\rangle_e$ at a standard condition (SC) and the relation can be explored with large data sets at high redshifts. The dependence of $\langle\mu\rangle_e^{SC}$ with redshift is determined by $(1 + z)^{-4}$ general relativistic surface brightness dimming, k-corrections, passive luminosity evolution, and dynamical evolution. Since the surface brightness dimming is well established and the k-corrections can be accurately deduced, we can infer the luminosity evolution in the elliptical population and compare with predictions from evolutionary synthesis models.

2 Observations

The data set used in our analysis comprises 12 rich clusters with redshifts between $z = 0.17$ and $z = 0.56$ imaged with WFPC-2. Ten clusters were observed in two studies[3,9] of the morphologies of galaxies in distant clusters, and the remaining two were kindly provided to us by Profs. Turner and Fort.

All 12 clusters were observed in either the F702W or F814W filters.

In addition to our high-resolution imaging with the *HST*, we obtained new near-infrared (K) observations over the same fields for 6 of these clusters. Cl0016+16, Cl1601+42, 3C295, and Cl0024+16 were imaged in K_s with a near-infrared camera on the Palomar Observatory 1.52-m telescope using a 256×256 NICMOS-3 array; Cl0054-27 and Cl0412-65 were imaged in K' with IRIS on the Anglo-Australian Telescope using a 128×128 HgCdTe detector array. We have previous K-band imaging for a further two clusters in our *HST* sample (AC103, AC114)[1]. All near-infrared observations were converted to standard K for an homogeneous data set.

3 Profile Fitting

We select ellipticals for this study from visual morphological classification lists limited in magnitude to $R_{702} \leq 23.5$ and $I_{814} \leq 23.0$[3,9]. The elliptical classifications are robust to $I_{814} \leq 21.0$, but fainter than this the distinction between the early-type classes becomes more uncertain[6]. We therefore impose a lower absolute magnitude limit of $M_B < -19.5$ ($H_o = 50 \, \mathrm{km\,sec^{-1}\,Mpc^{-1}}$, $q_o = 0.5$) to our selection, which corresponds to $I_{814} \sim 21.6$ at $z \sim 0.55$. Since galaxies brighter than -22.5 are likely to be brightest cluster galaxies, which follow a Kormendy relation with a steeper slope, we restrict our samples to ellipticals with $M_B > -22.5$. To reduce field contamination we use all available colour and spectroscopic information. The colour criteria used to isolate ellipticals only remove galaxies significantly foreground or background (± 0.3 mag around the colour-magnitude relation).

We obtain surface brightness profiles using the ELLIPSE task in IRAF. ELLIPSE finds the best-fitting elliptical isophotes at a given semi-major axis, which we increment from 1.5 to 10 pixels in half-integer steps. Circular radii are then calculated and linear de Vaucouleurs laws fit to determine half-light radii, r_e, and mean surface brightnesses interior to these radii, $\langle \mu \rangle_e$. All $\langle \mu \rangle_e$ are corrected for $(1 + z)^{-4}$ surface brightness dimming. For the clusters with near-infrared imaging, the *HST* $\langle \mu \rangle_e$ measurements are converted into $\langle \mu_K \rangle_e$ using $\langle \mu_{R702} \rangle_e - \langle \mu_K \rangle_e = (R_{702} - K)$ and $\langle \mu_{I814} \rangle_e - \langle \mu_K \rangle_e = (I_{814} - K)$. We measure the $(R_{702} - K)$ and $(I_{814} - K)$ colours in 5 arcsec diameter circular apertures and correct them for mean aperture colour effects. All $\langle \mu \rangle_e$ values are corrected for foreground Galactic extinction. The k-corrections in K are well determined and range from -0.36 mag at $z \simeq 0.17$ to -0.56 mag at $z \simeq 0.56$. The uncertainties associated with optical k-corrections are minimized because the F814W and F702W passbands at high redshift nearly overlap the rest-frame B band.

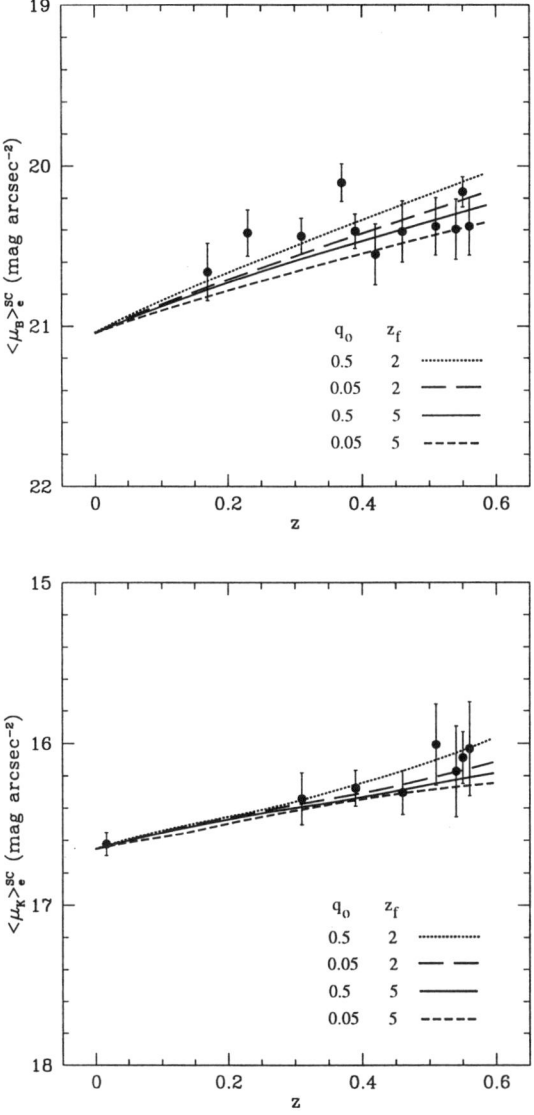

Figure 1: Mean effective surface brightnesses in rest-frame B and K versus redshift at standard condition $R_e = 3\,$kpc. The local cluster data point in the K-band figure was deduced from Pahre, Djorgovski, & de Carvalho (1996) data using our slope parameter and absolute magnitude cuts. The curves represent passive evolutionary model predictions (Bruzual & Charlot 1996) for $q_o = 0.5, 0.05$ and $z_f = 2, 5$, assuming $H_o = 50\,\mathrm{km\,sec^{-1}\,Mpc^{-1}}$.

4 Evolution in Surface Brightness

We first compare our data with the Kormendy relation, $\langle\mu\rangle_e = a + b\log R_e$, where a and b are constants for a specific passband and redshift. In both passbands the data are well described by a linear relationship. Additionally, within a given passband the slope does not appear to be a function of redshift; therefore, we assume an universal z-independent slope and determine its value from the full data set in a given passband using ordinary least squares linear regression. We condense the information contained in the rest-frame Kormendy relation plots in each passband and for each cluster to a single number, $\langle\mu\rangle_e^{SC}$, by employing the SC $R_e = 3\,\mathrm{kpc}$. Since this value falls at the midpoint of the R_e range of our data, the deduced value of $\langle\mu\rangle_e^{SC}$ is insensitive to the precise value of the slope, $d\langle\mu\rangle_e/d\log R_e$. Figure 1 shows the relation between $\langle\mu\rangle_e^{SC}$ and the redshift, z. In both the B and K bands there is a slow but steady brightening in $\langle\mu\rangle_e^{SC}$. A comparison with an evolutionary synthesis code [2] 1 Gyr initial burst c-model shows that this observed brightening is in approximate accord with passive evolution expectations.

Acknowledgments

I would like to thank my colleagues A. Aragón-Salamanca, I. Smail, H. Butcher, W. J. Couch, A. Dressler, R. S. Ellis, A. Oemler, B. Poggianti, and R. M. Sharples for permission to present results in advance of publication.

References

1. A. J. Barger, Aragón-Salamanca, A., R. S. Ellis, W. J. Couch, I. Smail and R. M. Sharples, MNRAS, 279, 1, 1996.
2. A. G. Bruzual and S. Charlot, 1996, in preparation
3. W. J. Couch et al., 1996, in preparation
4. S. G. Djorgovski and M. Davis, ApJ, 313, 49, 1987.
5. A. Dressler, D. Lynden-Bell, D. Burstein, R. L. Davies, S. M. Faber, R. J. Terlevich, G. Wegner, ApJ, 313, 42, 1987.
6. R. S. Ellis, I. Smail, A. Dressler, A. Oemler Jr., W. J. Couch, R. M. Sharples, 1996, submitted, astro-ph/9607154
7. J. Kormendy, ApJ, 218, 333, 1977.
8. M. A. Pahre, S. G. Djorgovski, R. de Carvalho, ApJ, 456, L79, 1996.
9. I. Smail, A. Dressler, W. J. Couch, R. S. Ellis, A. Oemler Jr., H. Butcher, R. M. Sharples, 1996, submitted

SURFACE-BRIGHTNESS EVOLUTION OF CLUSTER GALAXIES

DAVID SCHADE

Dept. of Astronomy, University of Toronto,
60 St. George St., Toronto, M5S 3H8

Surface brightness evolution has been detected in elliptical galaxies (consistent with passive evolution models of old stellar populations) and in disk galaxies (presumably due to enhanced star-formation rates). The rates of evolution in clusters and the field are not measurably different. In addition to this similarity, the high-redshift populations in both environments exhibit a "blue-excess" population, increased rates of star formation, and high frequency of peculiar structure. Thus, there are several parallels between evolving cluster and field galaxies and the high-redshift cluster environment will be understood only by comparison with the field population *at the same epoch*.

1 Introduction

In the course of programs to understand the evolution of galaxies by focusing on morphological properties, two-dimensional surface photometry has been done for samples of field galaxies[22,23] and cluster galaxies[23,24] using HST, ground-based imaging, and HST archival data. The cluster and field galaxy populations were much more similar at $z \sim 0.5$ than they are at the present time.

2 Recent results on high-redshift clusters

2.1 Elliptical galaxies

Ellipticals are present in clusters at high redshift[8] and their colours are consistent with passive evolution models.[10,1,20,18] The small dispersion in the colour-luminosity relation at $z \sim 0.5$[11] suggests they formed at substantially earlier epochs. Bender, Ziegler & Bruzual[3] find evidence from the Mgb-σ and Faber-Jackson relations for passive evolution in the B-band of 0.5 ± 0.1 in a cluster at $z = 0.37$, a result consistent with fundamental plane work at $z = 0.39$.[27]

Imaging has been used to search for evolution in the size-luminosity relation—one projection of the fundamental plane—of elliptical galaxies. Schade et al.[25,26] analysed ground-based and HST imaging and find an increase with redshift in surface brightness or luminosity (at a given size) of $\Delta M_B \sim -z$, consistent with both the spectroscopic studies cited above and with other imaging work.[19,2] Thus, the luminosity evolution expected from passive evolution models of elliptical galaxies has been detected by several groups.

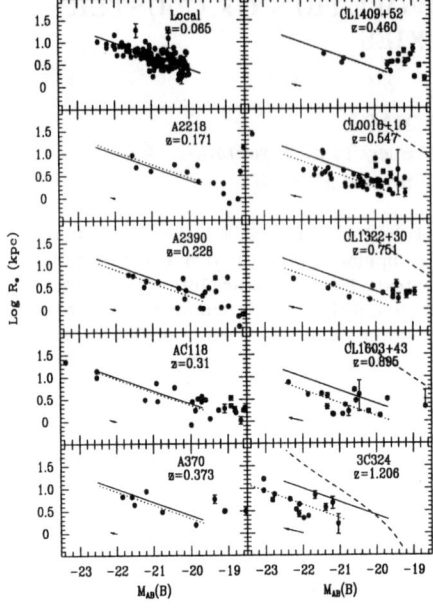

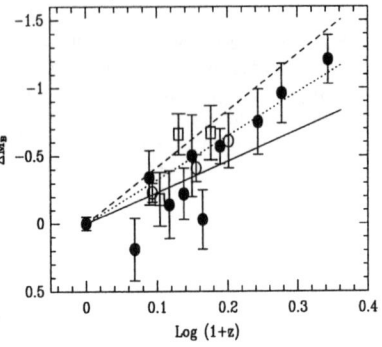

Figure 1: The evolving size-luminosity relation for elliptical galaxies derived from *HST* archival imaging of 9 clusters with $0 < z < 1.2$[26]. Galaxies of a given size are more luminous by ~ 1 mag at $z = 1$. Solid lines show the local relation and dotted lines indicate the evolved relation measured in each cluster. (Long-dashed lines indicate the surface-brightness selection.)

Figure 2: ΔM_B is the shift in luminosity at a given size as measured, e.g., from figure 1. Solid symbols are for cluster elliptical galaxies using *HST* imaging (but no membership information) and open symbols are from ground-based imaging of CNOC fields[25] where all of the galaxies have redshift information (open circles=cluster E's, open squares=field E's.)

2.2 Blue cluster galaxies

The blue fraction of the cluster galaxy population increases from a few percent locally to 25% at $z \sim 0.5$[4] and, by $z \sim 0.9$, perhaps 80% of cluster galaxies were blue.[20] This blue population shows spectroscopic signs of enhanced levels of star-formation[6] and is made up largely of disk-like galaxies with a high frequency of peculiar/irregular structure.[21,16,9,7] Surface photometry of galaxies (with redshifts) from the Canadian Network for Observational Cosmology (CNOC) cluster survey[5] shows that galactic disks in 3 clusters have surface brightness higher than the Freeman[12] value by ~ 1 mag at $z = 0.55$[24]. Furthermore, this disk brightening is consistent with observations of field galaxies.[23,22,24]

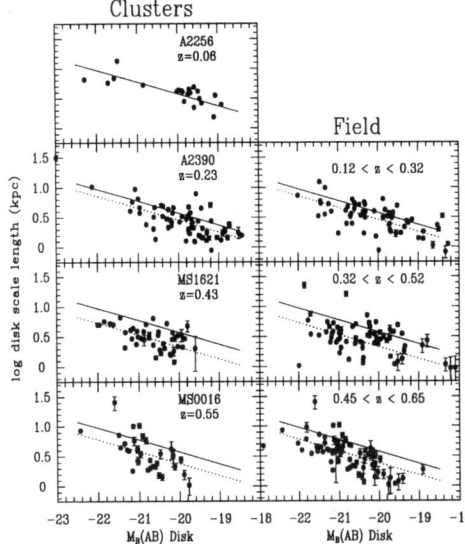

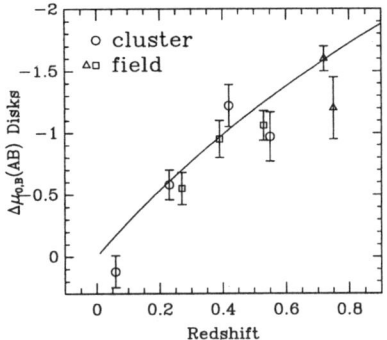

Figure 4: ΔM_B is the shift in surface brightness relative to the Freeman law measured, e.g. from plots like Figure 3. Open circles are CNOC cluster galaxies and squares are CNOC field galaxies. Triangles are field galaxies from Schade et al.[22,24]. The line is the evolution of the luminosity density $((1 + z)^{2.7})$ in the B-band from Lilly et al.[15].

Figure 3: The evolving size-luminosity relation for galactic disks in clusters and the field[24]. Solid lines show the Freeman[12] law and dotted lines indicate the mean evolved relation.

3 Parallels of cluster and field population at high-z

Cluster and field populations are dominated at high redshift by a "blue-excess" population that is mysteriously "absent" from the local population[14,9]. Both populations show a high (and similar) frequency of peculiar/irregular structure,[13,7] and both populations show elevated rates of star formation relative to local populations[17,6]. It has been argued here that field and cluster populations also show evolution in surface brightness (both disks and ellipticals) and that the rates of evolution are not measurably different (although the cluster population we have studied consists largely of galaxies far from the dense cluster core).

These similarities suggest that much of the physics of galaxy evolution is common to cluster and field populations and that the development of the high-z cluster population, largely through infall of field galaxies, needs to be understood in the context of the field population *at that redshift*, whose state differs markedly from the population we see in the present-day universe.

Acknowledgments

It is a pleasure to thank the many collaborators in the Canada-France Redshift Survey and the Canadian Network for Observational Cosmology groups, in particular Simon Lilly, Ray Carlberg, and Felipe Barrientos.

References

1. Aragon-Salamanca,A., Ellis,R., Couch,W., & Carter,D., MNRAS, 262, 764, 1993.
2. Barrientos,F., Schade, D., & López-Cruz, O. ApJ, 460, 89, 1996.
3. Bender,R. Ziegler, B., & Bruzual, G. ApJ, 463, 51, 1996.
4. Butcher, H., & Oemler, A. ApJ, 285, 426, 1984.
5. Carlberg, R. G., et al. JRASC, 88, 39, 1994.
6. Couch, W., & Sharples, R. MNRAS, 229, 423, 1987.
7. Couch, W., Ellis, R., Sharples, R., and Smail, I., ApJ, 430, 121, 1994.
8. Dickinson, M. 1995 in *Fresh Views of Elliptical Galaxies*, A.S.P. Conference Series, ed. Buzzoni,A., Renzini, A., & Serrano, A. p. 283
9. Dressler, A., Oemler, A., Butcher, H., and Gunn, J. ApJ, 430, 1, 1994.
10. Dressler, A., & Gunn, J. 1990, in *Evolution of the Universe of Galaxies*, San Francisco,Astronomical Society of the Pacific
11. Ellis,R., Smail,R., Dressler,A., Couch,W., Oemler,A., Butcher, H., & Sharples,R. 1996, preprint, astro-ph/9607154
12. Freeman, K. ApJ, 160, 811, 1970.
13. Glazebrook, K., Ellis, R., Santiago, B., & Griffiths, R. mnras, 275, L19, 1985.
14. Lilly, S. ApJ, 411, 501, 1993.
15. Lilly, S., Le Fevre, O., Hammer, F., Crampton, D. ApJ, 460, L1, 1996.
16. Lavery, R., Henry, P. ApJ, 426, 524, 1994.
17. Hammer, F., Flores, H., Lilly, S., Crampton, D., LeFevre, O., Rola, C., Mallen-Ornelas, G., Schade, D., Tresse, L. 1996 A & A, submitted
18. Oke, J., Gunn, J., & Hoessel, J. AJ, 111, 29, 1996.
19. Pahre, M., Djorgovski, S., & de Carvalho, R. ApJ, 456, 79, 1996.
20. Rakos,K., & Schombert,J. ApJ, 439, 47, 1995.
21. Thompson, L. ApJ, 324, 112, 1988.
22. Schade, D., Lilly, S., Crampton, D., Le Fèvre, O., Hammer, F., & Tresse, L. ApJ, 451, 1, 1995.
23. Schade, D., Lilly, S., Le Fèvre, O., Hammer, F., & Crampton, D. ApJ, 464, 79, 1996.
24. Schade, D., Carlberg, R., Yee, H., López-Cruz, O., & Ellingson, E. ApJ, 465, L103, 1996.
25. Schade, D., Carlberg, R., Yee, H., López-Cruz, O., & Ellingson, E. ApJ, 464, L63, 1996.
26. Schade, D., Barrientos,F., & López-Cruz, O. submitted to ApJ Letters
27. van Dokkum, P. & Franx, M. MNRAS, 281, 985, 1996.

THE FUNDAMENTAL PLANE IN INTERMEDIATE REDSHIFT CLUSTERS: THE EVOLUTION OF M/L RATIO

DANIEL D. KELSON

University of California Observatories / Lick Observatory, Board of Studies in Astronomy and Astrophysics, University of California, Santa Cruz, CA 95064, USA

PIETER G. VAN DOKKUM, MARIJN FRANX

Kapteyn Astronomical Institute, P.O. Box 800, NL-9700 AV Groningen, The Netherlands

GARTH D. ILLINGWORTH

University of California Observatories / Lick Observatory, Board of Studies in Astronomy and Astrophysics, University of California, Santa Cruz, CA 95064, USA

We have started a program to make a detailed study of the structural, kinematic, and spectroscopic properties of early-type galaxies in intermediate redshift clusters. We present initial results using the Fundamental Plane of early-type galaxies, derived from Keck spectroscopy and HST imaging. Early-type galaxies in two distant, rich EMSS clusters, Cl1358+62 at $z = 0.33$ and Cl2053–04 at $z = 0.58$ have been observed using Keck LRIS multi-slit spectroscopy to derive velocity dispersions. WFPC2 images were used to derive length scales, r_e, and surface brightnesses, I_e. Fundamental plane relations were derived for these clusters and compared to that recently derived for Cl0024+16 (z=0.39), and that for a rich nearby cluster, Coma. The Fundamental Plane in Cl1358+62 and Cl2053–04 is similar to that in Coma. Preliminary indications suggest that cluster early-type galaxies formed at very early epochs, with $z_f \gg 1$.

1 Introduction

The Fundamental Plane (FP) is an empirical relation between effective radius, velocity dispersion, and surface brightness, for early-type galaxies. (*e.g.* Djorgovski and Davis 1987). In a study of 10 nearby clusters Jørgensen *et al.* (1995, JFK95) found

$$r_e \propto \sigma^{1.24} I_e^{-0.82}, \tag{1}$$

with an rms scatter of ± 0.084. The implication is that the Mass-to-Light ratio is a function of the structural and kinematical observables (Faber *et al.* 1987), such that

$$M/L_V \propto r_e^{0.22} \sigma^{0.49} \tag{2}$$

However, the luminosity evolution of single burst stellar populations is mostly a function of IMF (Tinsley & Gunn 1976), such that

$$\log \frac{M}{L_V} \propto (1.3 - 0.3x) \times \log t, \tag{3}$$

where x is the slope of the IMF, and t is the time elapsed since the epoch of star formation. Therefore, the small scatter in M/L_V (23% in Coma; JFK95) suggests a small scatter in *age*, for a given σ and r_e.

As a result, the Fundamental Plane can be an extremely effective tool for studying early-type galaxy evolution. Franx (1995) and van Dokkum & Franx (1996, vDF) have successfully measured the evolution of early-type galaxies from comparisons of A665 and Cl0024+16 with Coma. In the latter work, they found only modest brightening (30% for $q_0 = 0.5$) between $z = 0$ and $z = 0.39$. We are extending this work to more clusters, and to higher redshift.

2 Data in Cl1358+62 and Cl2053–04

The spectroscopic observations were made through multi-slit masks using the Low Resolution Imaging Spectrograph (LRIS) of the Keck telescope. The Cl1358+62 spectra had typical resolutions (instrumental σ_i) of about ~60 km s^{-1}. The Cl2053–04 spectral resolution was about $\sigma_i \sim 85$ km s^{-1}. The procedures outlined by vDF were followed to give suitable templates. Velocity dispersions were determined using real and Fourier fitting techniques, with final uncertainties of about 5%. The dispersions were corrected for aperture to an effective aperture of $3\overset{''}{.}4$ at the distance of Coma.

The imaging observations are from the WFPC2 of HST. For Cl1358+62, multiple pointings consisting of 3600 s in F814W and 3600 s in F606W (in three exposures in each band) were obtained across a field of $\sim 8' \times 8'$ that matches the $6' \times 8'$ field of LRIS. For Cl2053–04, only a single pointing with 2400 s in F814W and 2400 s in F702W (with 2 exposures in each band) is currently available.

Structural parameters, r_e and I_e, were derived while explicitly accounting for the effects of the WFPC2 Point Spread Function (PSF). The photometry was transformed to Johnson V band directly from the observed WFPC2 filters to allow direct comparisons with Coma.

3 Evolution of the Mass-to-Light Ratio

Fig. 1 shows the Fundamental Plane relations for four clusters: Coma (JFK95, open symbols), Cl1358+62, Cl0024+16 (vDF, open symbols), and Cl2053–04. Despite the offset arising from cosmological surface brightness dimming, the

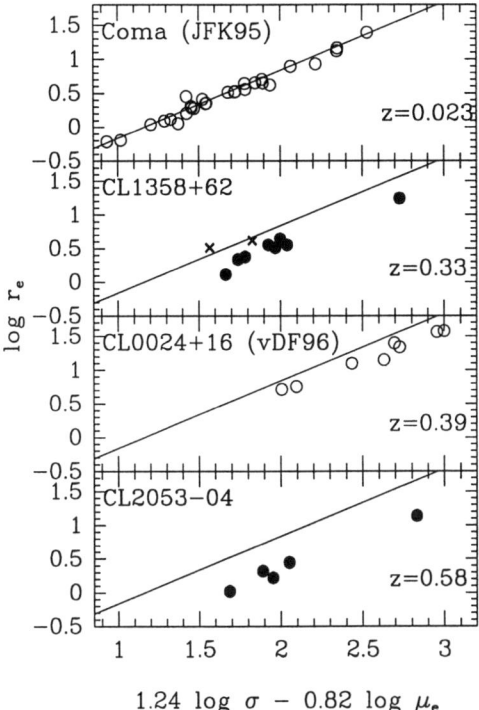

Figure 1: The Fundamental Plane of early-type galaxies as a function of redshift. The data of JFK95 and vDF are shown as open symbols; the new cluster data are consistent with those found by vDF. Two E+As in Cl1358+62 are shown as "x"s.

Fundamental Plane is very similar at high redshift to that at low redshift. The implication is that these early-type galaxies were in place by $z = 0.58$.

Relative zero-point offsets can be derived from the M/L_V relations (Eq. 3) between the high-z cluster galaxies and the Coma sample. These M/L_V offsets are shown in Fig. 2

The simple analytical model for luminosity evolution is quite instructive in this context. Overlayed in Fig. 2 are regions of likely values of x (the slope of the IMF), for burst (formation) epochs of $z_f = 1$ and $z_f = \infty$. Note that the early-type galaxies in these clusters are more consistent with an early epoch of formation than with a recent ($z_f \approx 1$) formation scenario.

The next phase of this program utilizes larger samples in Cl1358+62 and Cl2053–04. High S/N, high resolution ($\sim$ 80 km/s) spectra have recently

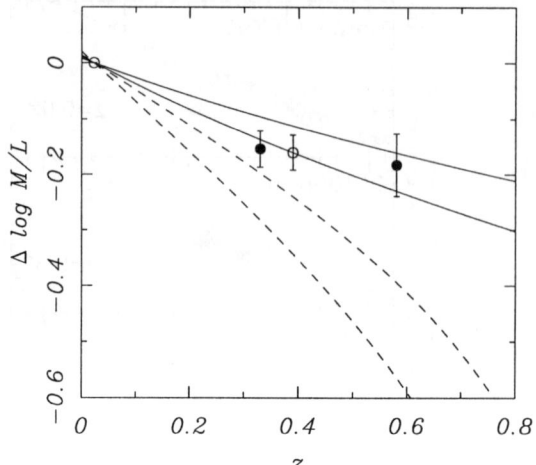

Figure 2: The evolution of M/L_V ratio with redshift. Symbols are as in Fig. 1. The solid lines enclose single burst models with $z_f = \infty$, using a range of IMF slopes and $q_0 = 0.05$. The dashed lines enclose $z_f = 1$ models. For $q_0 = 0.1$, the $z_f = \infty$ models pass directly through the data and, for $q_0 = 0.5$, the data lie above the models.

been obtained for more than 40 early-type members of Cl1358+62, as have spectra on nearly two dozen members of Cl2053–04. An HST WFPC2 mosaic of 6 pointings will also be taken of Cl2053–04 in 1997. Together, these data will allow us to determine how the shape and scatter of the Fundamental Plane evolve with redshift. Such descriptions of the evolution of early-type galaxies will place tight constraints on the formation epochs of these galaxies. In addition, line-strengths and spectral indices will be used to supplement the Fundamental Plane results.

References

1. Djorgovski, S., & Davis, M., ApJ, 313, 59, 1987.
2. Faber, S. M., *et al.*, 1987, in *Nearly Normal Galaxies*, ed. S. M. Faber (New York: Springer), 175
3. Franx, M. 1995, in IAU Symposium 164, "Stellar Populations" ed. P.C. van der Kruit & G. Gilmore (Dordrecht: Kluwer), 269
4. Jørgensen, I., Franx, M. & Kjærgaard, P., MNRAS, 280, 167, 1995.
5. Tinsley, B. M., & Gunn, J. E., ApJ, 203, 52, 1976.
6. van Dokkum, P. G., & Franx, M., MNRAS, 281, 985, 1996.

CLUSTERS OF GALAXIES AT $z > 1$

MARK DICKINSON
STScI, 3700 San Martin Dr., Baltimore MD 21218 USA

Although field galaxy studies have begun to probe the universe at $z > 1$, evidence for galaxy clusters at such redshifts has been sparse. New observations are accumulating rapidly, however, providing new data on the early evolution of elliptical galaxies, the blue "Butcher–Oemler" population, and of large scale structure at unprecedentedly large lookback times. I briefly review some of these new observations, discussing morphological and spectral characteristics of cluster galaxies at $z > 1$, x–ray evidence for massive, collapsed clusters out to $z = 1.8$, and tantalizing indications from the literature for clusters at $2 < z < 5$.

1 Introduction

Until quite recently, a talk with the title given above would have been either exceedingly short or entirely theoretical. Although rich clusters of galaxies have been popular and productive laboratories for studying galaxy evolution for many years (cf. the contributions of Dressler and others to this volume), published surveys for distant clusters largely run out of steam by $z = 1$. The most distant clusters with measured redshifts in optical or x–ray surveys have $z \approx 0.9$. While this might reflect an evolution in the cluster space density (perhaps there simply are no rich clusters at $z > 1$?), it is more probably an effect of observational selection. At $z > 1$, strong k–corrections, particularly for the red, early–type galaxies which dominate nearby clusters, may greatly reduce the visibility of distant cluster galaxies when observed at optical wavelengths, and the contrast of even a rich cluster against the tremendously numerous population of faint field galaxies may be diminished to the point of near invisibility.

From my perspective, the search for clusters beyond the limits of present surveys is primarily motivated by three interests: (1) the hunt for the formation epoch of early–type cluster galaxies, (2) the early evolution of the "Butcher–Oemler" blue cluster population, and (3) evidence for massive, collapsed clusters at high redshift as a constraint on cosmology and theories of large scale structure formation. I will touch upon all of these points briefly below, and provide a very brief summary of the most enticing evidence from the literature for galaxy clusters at $z > 2$. Because of space constraints, I primarily limit my discussion to clusters identified as collections of *galaxies;* further evidence for cluster*ing* at high redshift (e.g. from QSO absorption lines, quasar angular correlations, etc.) will unfortunately be neglected here.

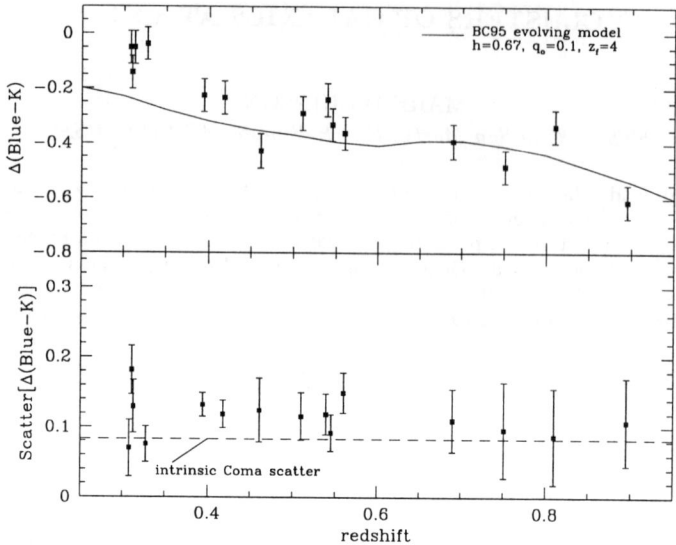

Figure 1: Evolution of cluster ellipticals from $0.3 < z < 0.9$, from Stanford, Eisenhardt and Dickinson 1997. The galaxy sample consists of morphologically selected E/S0 galaxies from 16 clusters with 5–band optical–IR ground based and HST WFPC2 imaging. The "blue" band shifts with cluster redshift so as always to measure approximately rest–frame U–band light, while the K–band is fixed in the observed frame. *Top panel:* color evolution of the E/S0 population. The vertical axis represents the color *difference* relative to the same rest–frame wavelengths in the Coma cluster, i.e. "no evolution" would be a horizontal line at $\Delta(\text{blue-}K)$ $= 0$. *Bottom panel:* Scatter in E/S0 colors around the mean color–magnitude locus. The scatter in high–z clusters is somewhat higher than for Coma, but changes negligibly with redshift from $0.3 < z < 0.9$.

2   Clusters at $z < 2$

Although the evolutionary history of elliptical galaxies has been the subject of debate, the evidence from rich clusters at $z < 1$ seems relatively unambiguous. Bower, Lucey & Ellis (1992) have cited the small color scatter in Coma and Virgo cluster ellipticals to argue for either a high redshift of formation, strongly synchronized coevality, or both. At higher redshifts, early spectroscopic observations (Dressler & Gunn 1990), optical–IR colors (Aragón–Salamanca *et al.* 1993) and analysis of the fundamental plane and its projections (Van Dokkum & Franx 1996; Dickinson 1995; Pahre *et al.* 1996; Schade *et al.* 1996) have all shown that the "red envelope" population of cluster ellipticals evolves slowly and in a fashion consistent with simple passive evolution

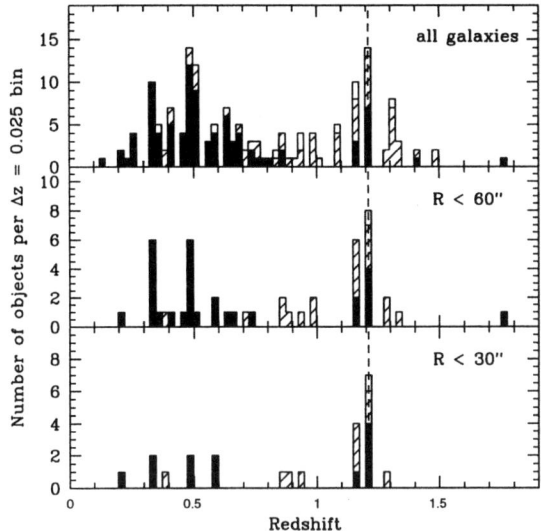

Figure 2: Redshift distribution in the field of the $z = 1.206$ radio galaxy 3C 324. The panels show subsamples restricted at several radii from the radio galaxy. The shadings indicate redshift "quality classes;" the filled area show redshifts from multiple spectral features, while the hatched redshifts are mostly based on single emission lines, generally assumed to be [OII]. The vertical dashed line marks the radio galaxy redshift.

and a high redshift of formation. Recently, Stanford, Eisenhardt & Dickinson (1997; see also contribution by Stanford to this volume) have gathered deep, wide–field 5–band IR/optical imaging on 45 clusters from Coma out to $z = 0.9$. For 16 of these, HST WFPC2 imaging allows us to morphologically select early–type galaxies from the cluster cores. As seen in figure 1, the elliptical galaxy color–magnitude (c–m) relation shows only very mild color evolution out to $z = 0.9$, and there is virtually no redshift evolution in the *scatter* of the galaxy colors around the mean c–m relation, extending the Bower *et al.* result to large lookback times where the *fractional* age differences between non–coeval galaxies would be larger. The c–m slope (not shown) also shows no significant change with redshift out to $z = 0.9$. These facts all point to a strongly coeval population and a high redshift of formation for cluster ellipticals, well beyond $z = 1$. This is one strong motivation for our search for $z > 1$ galaxy clusters: evidently by $z = 0.9$ we have still not yet approached the epoch of star forming activity for most cluster ellipticals.

Infrared imaging surveys offer the best promise for finding clusters (particularly those rich in early–type galaxies) at $z > 1$. Because of the current limitations imposed by the small size of infrared arrays, it is difficult to survey the large solid angles needed to find rare objects like rich clusters. We have chosen to narrow the search to *a priori* likely sites, namely the environments of radio loud AGN. Peter Eisenhardt and I have obtained deep R, J and K imaging of 25 powerful radio galaxies at $0.8 < z < 1.4$ and have identified several good candidates for rich clusters, primarily visible in the infrared data thanks to large overdensities of very red $(R - K \approx 6)$ galaxies with the colors expected for weakly evolved ellipticals at $z \gtrsim 1$. Even here, however, spectroscopic confirmation is essential in order to sort out projection effects and study the galaxy population.

For one field from our sample, that around 3C 324 at $z = 1.206$, Spinrad, Dey, Stern, LeFèvre and I have measured ~ 150 galaxy redshifts down to $R = 25$ and $K = 20$ using Keck and the NTT. 3C 324 shows one of the most prominent excesses of faint red galaxies in our infrared imaging data. The redshift distribution is shown in figure 2. Considering the entire $3' \times 8'$ survey field, >80% of the galaxies are foreground or background to the radio galaxy, emphasizing the difficulty of identifying rich clusters at $z > 1$ even with extensive redshift data. However, two prominent spikes are visible at $z = 1.15$ and 1.21, particularly when attention is restricted to radii within 1 arcmin of the radio galaxy. Evidently, the "3C 324 cluster" seen as an excess of galaxies in the near–IR divides into two distinct structures separated by ~ 7500 km s^{-1} in their rest frame. The red ellipticals appear to be present in both structures.

With few exceptions, the red galaxies in the 3C 324 cluster(s) have simple elliptical morphologies. Their $R - K$ colors are ~ 0.6 magnitudes bluer than similar rest–frame colors for giant ellipticals today, consistent with an extrapolation of the color vs. z relation for $z < 1$ cluster ellipticals found by Stanford *et al.*, and the scatter in their color–magnitude relation remains remarkably small, ~ 0.07 magnitudes. The rest–frame ultraviolet spectra of the 3C 324 ellipticals are qualitatively very similar to that of the red $z \approx 1.5$ radio galaxies 53W091 and 53W069 (see Dunlop *et al.* 1996, and Dey, this volume). Their surface brightnesses suggest ~ 1 magnitude of rest–frame B–band luminosity evolution (cf. Dickinson 1995), fully consistent with passive evolutionary models. The other spectroscopically confirmed cluster members (figure 3) exhibit a bewildering range of morphologies. Very few, if any, "normal" disk galaxies have been identified, although the degree to which this is a consequence of the ultraviolet rest–frame wavelengths imaged by WFPC2 at $z = 1.2$ (3200Å for the F702W filter) is unclear: NICMOS imaging planned as a HST Cycle 7 GTO program should help clarify this.

z = 1.15   z = 1.21

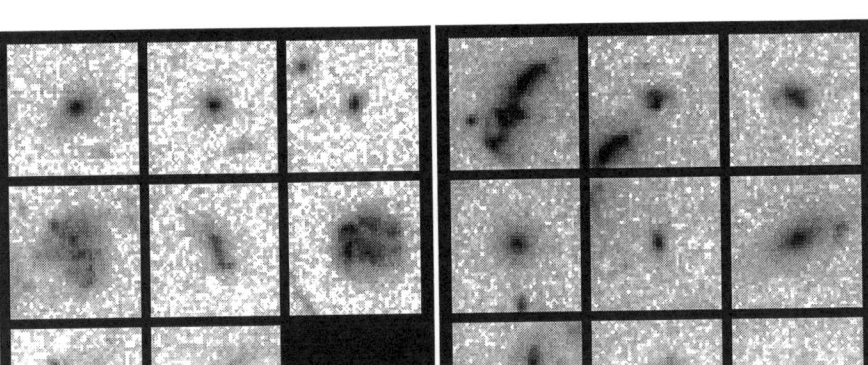

Figure 3: WFPC2 F702W images of spectroscopically confirmed galaxies in the 3C 324 cluster(s). The galaxies at left are members of the $z = 1.15$ structure, including ellipticals, some irregular disks, and two large, diffuse galaxies (bottom). Unusually, the bottom central object is extremely red and bright in the IR, suggesting perhaps a "buried" starburst. The galaxies at right are at $z = 1.21$, including the radio galaxy (top left), several ellipticals, and peculiar objects (bottom).

We have also found evidence that the 3C 324 cluster is a *massive, collapsed system*. In the course of a Rosat PSPC survey of $z > 1$ radio galaxies, we detected a faint x–ray source coincident (within $10''$) of 3C 324. The poor angular resolution of the PSPC did not allow us to resolve the x–ray emission at the low signal–to–noise of the source, however, leaving ambiguous whether the x–rays arose from the radio galaxy AGN or from surrounding cluster gas. A subsequent 72.1 ksec Rosat HRI exposure, however, clearly shows that the x–ray emission is resolved over a detectable diameter of $\sim 60''$ (~ 0.5 Mpc) (figure 4). Our spectroscopy shows no other substantial galaxy groups along the line of sight so closely aligned with the radio/x–ray source. This strongly suggests that the x–ray emission comes from resolved cluster gas at $z \approx 1.2$ with a bolometric $L_X = (8.1 \pm 1.6) \times 10^{44}$ erg s^{-1} (for $H_0 = 50$, $q_0 = 0.5$), comparable to that of the Coma cluster. We have subsequently resolved two additional radio galaxy x–ray sources at $z > 1$ with the Rosat HRI, and Crawford & Fabian (1996) have resolved a fourth. It seems quite unlikely that these multiple examples of extended x–ray emission around radio galaxies could all arise from foreground group/cluster sources seen in projection. The most dis-

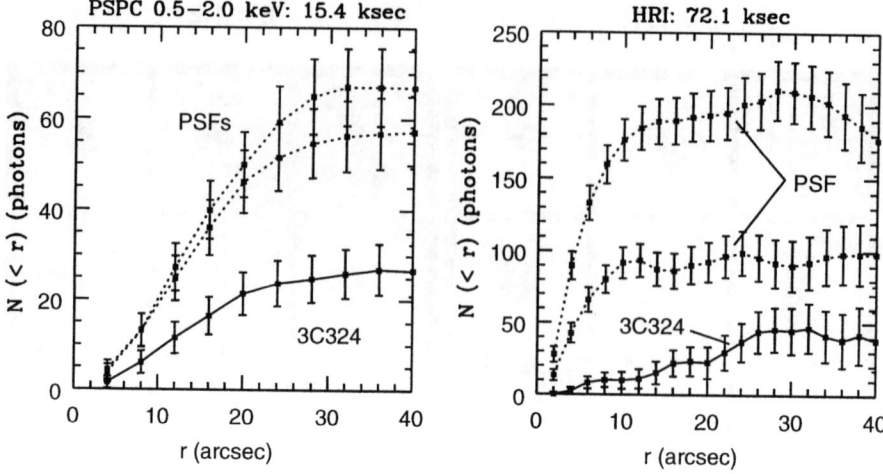

Figure 4: Radial curves of growth for x-ray emission around 3C 324 from the Rosat PSPC (left) and HRI (right). The dashed curves labeled PSF show the profiles of two point sources nearby in the field. In the low angular resolution PSPC data, 3C 324 is not convincingly resolved, but the HRI data shows that the x-rays are extended over a detectable diameter of $\sim 60''$, strongly suggesting the presence of a luminous x-ray cluster associated with the radio galaxy.

tant of our resolved HRI detections is for a radio galaxy at $z = 1.8$ with $L_X \approx 5 \times 10^{44}$ erg s^{-1}. Unless other explanations for highly extended x-ray emission with these luminosities can be concocted, it seems clear that some radio galaxies out to nearly $z = 2$ were situated in deep gravitational potential wells comparable to those of massive clusters today, providing a challenge for theories of cosmic structure formation.

3   Clusters at $z > 2$

Although the evidence for clusters at $z > 2$ is extremely heterogeneous, a number of observations have been presented recently which suggest that the idea is not pure fantasy. Most importantly, recent spectroscopic confirmation of galaxy redshifts in several examples (cf. Francis *et al.* 1996, Pascarelle *et al.* 1996, Malkan *et al.* 1996, Møller & Warren 1993, LeFèvre et al. 1996) has lent credence to earlier claims based on statistical overdensities or narrow band imaging studies alone. Table 1 below presents an almost certainly incomplete

Table 1: Cluster candidates at $z > 2$

z	Reference	Description
2.05	Dressler *et al.* 1993	QSO with local overdensity of peculiar objects in HST images
2.38	Francis *et al.* 1996	Multiple QSO absorbers at common z in separate sightlines; Lyα excess objects; some spectroscopic confirmation; red IR–IR color selected objects
2.39	Pascarelle *et al.* 1996	~18 Lyα selected compact galaxies in deep WFPC2 images; ~9 confirmed spectroscopically
~ 2.5	Aragón–Salamanca *et al.* 1996	Statistical K–band excess around radio loud quasars
2.50	Malkan *et al.* 1996	3 Hα selected (in infrared) galaxies at redshift of known damped Lyα QSO absorber
2.81	Møller & Warren 1993, Warren & Møller 1996	3 spectroscopically confirmed, Lyα selected compact galaxies (HST imaging) associated with $z_{abs} = z_{QSO}$ damped Lyα absorber and QSO.
3.14	LeFèvre *et al.* 1996	2 spectroscopically confirmed Lyα selected companions to a powerful radio galaxy
3.58	Giavalisco *et al.* (in prep.)	Excess of Lyman break selected galaxies in field of powerful radio galaxy
3.80	Lacy & Rawlings 1996	Excess of Lyman break selected galaxies in field of powerful radio galaxy
4.55	Hu & McMahon 1996	2 Lyα selected objects associated with QSO

summary of the more interesting examples of suggested (and in many instances confirmed) galaxy associations at $z > 2$.

Most impressively, Pascarelle *et al.* (1996) report spectroscopic redshifts for 5 objects out of 18 candidates in an apparent cluster at $z = 2.39$, selected by Lyα excess in intermediate bandwidth HST images. Windhorst (priv. comm.) reports that at least four further members have been confirmed with recent Keck spectroscopy. The cluster objects are very compact, like most very faint galaxies seen in deep WFPC2 images. Curiously, many of them show AGN–like spectral features (CIV, NV emission, etc.). Objects in several other $z > 2$ cluster candidates also show AGN features or radio emission (Malkan *et al.* 1996; LeFèvre *et al.* 1996; Francis *et al.* 1996) suggesting that AGN activity may have been quite prevalent in these environments and/or at these redshifts. In part, this may be a consequence of galaxy selection from narrow band emission line imaging, especially with Lyman α: evidence from high–z field galaxies indicates that Lyα emission is typically very weak or absent in star–forming galaxies, presumably due to effects of dust and resonant scattering.

At present, the relation of these $z > 2$ associations to nearby rich clusters is unknown. In particular, it is unclear whether these are massive, collapsed systems, groupings within unvirialized "sheets" of galaxies, or collections of "protogalactic" fragments destined to merge into single, more massive galaxies. But the range of new data is impressive and suggestive, and further investigations should be fruitful. With the dramatic progress of Keck spectroscopy on extremely faint high–z galaxies, and with the promise of infrared OH suppression spectrographs in the near future which will facilitate redshift measurement for galaxies at $z > 1.2$, we may fully expect this evidence to continue to accumulate.

Acknowledgements

I would like to extend special thanks to my collaborators (particularly Adam Stanford, Peter Eisenhardt, Hy Spinrad, Arjun Dey, Daniel Stern, Olivier LeFèvre, and Richard Mushotzky) for permitting me to present data prior to publication, and to the conference organizers for their invitation and travel support.

1. Aragón–Salamanca, A., Ellis, R.S., and O'Brien, K.S., 1996, MNRAS,
2. Bower, R.G., Lucey, J.R., and Ellis, R.S., 1992, MNRAS, 254, 601.
3. Crawford, C.S., and Fabian, A.C., 1996, MNRAS, 281, L5.
4. Dey, A., this volume.

5. Dickinson, M., 1995, in *Fresh Views on Elliptical Galaxies*, eds. A. Buzzoni, A. Renzini, & A. Serrano (ASP, San Francisco), p. 283.

6. Dressler, A., and Gunn, J.E., 1990, in *Evolution of the Universe of Galaxies* (ASP, San Francisco), p. 200.

7. Dressler, A., Oemler, A., Gunn, J.E., & Butcher, H., 1993, ApJ, 404, L45.

8. Dickinson, M., 1995, in *Fresh Views of Elliptical Galaxies*, ASP Conf. Series No. 86, eds. A. Buzzoni, A. Renzini, & A. Serrano, (ASP, San Francisco), p. 283.

9. Dunlop, J., Peacock, J., Spinrad, H., Dey, A., Jiminez, R., Stern, D., and Windhorst, R., 1996, Nature, 381, 581.

10. Francis, P.J., Woodgate, B.J., Warren, S.J., Møller, P., Mazzolini, M., Bunker, A.J., Lowenthal, J.D., Williams, T.B., Minezaki, T., Kobayashi, Y., and Yoshii, Y., 1996, ApJ, 457, 490.

11. Hu, E., & McMahon, R., Nature, 382, 231.

12. Lacy, M., and Rawlings, S., 1996, MNRAS, 280, 888.

13. LeFèvre, O., Deltorn, J.-M., Crampton, D., and Dickinson, M., 1996, ApJ, 471, L11.

14. Malkan, M.A., Teplitz, H., McLean, I.S., 1996, ApJ, 468, L9.

15. Møller, P., and Warren, S.J., 1993, A&A, 270, 43.

16. Pahre, M.A., Djorgovski, S.G., and DeCarvalho, R.R. 1996, ApJ, 456, L79.

17. Pascarelle, S.M., Windhorst, R.A., Keel, W.C., and Odewahn, S.C. 1996, Nature, 383, 45.

18. Schade, D., Carlberg, R.G., Yee, H.K.C., Lopez–Cruz, O., and Ellingson, E., 1996, ApJ, 464, L63.

19. Stanford, S.A., Eisenhardt, P.R.M., and Dickinson, M., 1997, in preparation.

20. Van Dokkum, P.G., and Franx, M., 1996, MNRAS, 281, 985.

21. Warren, S.J., and Møller, P. 1996, A&A, 311, 25.

NATURE OF E+A GALAXIES IN DISTANT CLUSTERS

P. BELLONI, R. BENDER, U. HOPP, R.P. SAGLIA, B. ZIEGLER

Universitätssternwarte München, Scheinerstrasse 1, 81679 München, Germany

Narrow band photometry of the galaxy populations of four galaxy clusters at z=0.4-0.5 (Cl0016+16, Cl0303+17, Cl0939+47, and Cl1447+26) has provided us with a sample of 73 E+A galaxies which are *secure cluster members* (Belloni et al.,1995, A&A 297,61; Belloni & Röser, 1996, A&ASuppl. 118,65). This is the largest sample of such galaxies found to date in distant clusters. For 34 of them HST images are available (WFPC2). From these we determined light profiles for all E+A galaxies but those showing highly irregular morphology. A fit with an exponential and an $r^{1/4}$ law allows us to classify galaxies on the basis of the bulge to disk ratio. We also derived *color gradients* for 10 E+A galaxies in Cl0016+16 (observed in both F555W and F814W filters). This allows us to study the spatial distribution of the starburst and thus, to distinguish among the physical mechanisms at its origin. In the case of an infalling or harrassed spiral the enhanced star formation would likely be a galaxy-wide phenomenon, or confined to the disk. Conversely, if a merger of two spirals has taken place or if the original galaxy was an elliptical that accreted gas from a dwarf galaxy, the burst signatures should be detectable as a bluing of the nuclear region (Mihos & Hernquist, 1994, ApJ 427, 112).

We found that *most of these galaxies are disk systems* judging from the exponential nature of their profiles. It is worthwhile to note that only one E+A galaxy has a regular $r^{1/4}$ surface brightness profile.

Preliminary results of the color profiles analysis indicate that most of the E+A galaxies in Cl0016+16 show little spatial variation in V-I colors (restframe U-V). Only one disk-like, young E+A galaxy (estimated post-starburst age 0.5 Gyr) has a starburst predominantly located in the nucleus. These results confirm the claim with a larger and homogeneous sample that *enhanced star formation in E+A galaxies is a galaxy-wide phenomenon.* (Caldwell et al.,1996, AJ 111, 78; Franx, 1993, ApJ 407, L5). Our results implicate that strong interactions like galaxy mergers play only a minor role for triggering the star formation in our sample of post-starburst cluster galaxies. It is more likely that non-disruptive interactions among galaxies (Moore et al., 1996, Nature 379, 613) or interactions between galaxies and the intra-cluster medium are the dominant mechanism.

Table 1: Results of the light profiles analysis of 34 E+A galaxies in intermediate redshift clusters. The E+A fraction is given with respect to the secure cluster members, about $100-150$ galaxies brighter than $m_R = 22.5$ per cluster. For mergers/interacting galaxies we indicate if possible whether they appear disk or bulge dominated. Cl0303+17 contains an irregular (Magellanic Cloud-like) E+A galaxy.

Cluster	E+A Tot.	E+A (HST)	Morph. Elliptical	Morph. S0-Sa	Morph. Sb-Sc	Merger Interact.
Cl0016+16 (z=0.54)	20 (21%)	11	1	3	6	1 Disk-domin.
Cl0939+47 (z=0.41)	35 (22%)	11			7	4
Cl0303+17 (z=0.41)	24 (25%)	7		1	5	1 Irregular
Cl1447+26 (z=0.38)	8 (9%)	4		1	1	2 Disk-domin.

Figure 1: Color profiles of two galaxies in Cl0016+16. The effect of 1σ errors in the mean sky level (the dominant error at large radii) on the colors are shown above and below the colors derived from the adopted sky level (continuous line). We remind that V-I=2.4 is the typical color for a passively evolving elliptical galaxy at z=0.54 (H_o=75 Km/sec/Mpc, q_o=0.5). Spiral galaxies have a larger spread depending on the Hubble type (V-I=1.3-1.5 for τ=3-10, Bruzual & Charlot,1996,in preparation). Galaxy #73 shows colors slightly bluer than those of an elliptical and no significant color gradient. On the other hand, the very blue central color of galaxy #54 (estimated post-starburst age 0.5 Gyr) indicates that the starburst is located in the nucleus.

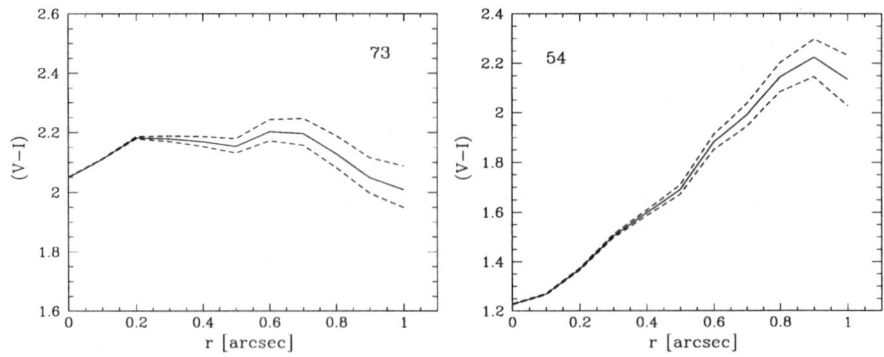

THE BUTCHER-OEMLER EFFECT IN THE CLUSTERS
0303+17 (z=0.42) AND 1447+26 (z=0.37)

P. BELLONI, B. VULETIC

Universitätssternwarte München, Scheinerstrasse 1, 81679 München, Germany

H.-J. RÖSER

Max-Planck Institut für Astronomie, Königstuhl 17, 69117 Heidelberg, Germany

Evidence that the galaxy population in the cores of rich clusters has evolved in the relatively recent past (z>0.2) has accumulated steadily since the seminal work of Butcher & Oemler (1978, ApJ. 219,18). Dressler (1987, *Nearly Normal galaxies from the planck Time to the Present*) recast the Butcher-Oemler effect in spectroscopic terms defining an "active" cluster population to include all galaxies with emission lines (AGN, starburst and spiral galaxies) and galaxies with strong Balmer absorption lines (and sometimes faint [OII]λ3727 in emission) superposed to an elliptical-type spectrum (so called E+A galaxies). Despite the progress made in multi-slit spectroscopy, up to date the evolutionary effects in the galaxy content of only a handful of rich clusters have been studied on the basis of statistically significant und unbiased samples of cluster members.

In the framework of our long-term project to analyse a statistically significant sample of intermediate redshift clusters (Thimm & Belloni, 1994, A&A 298, L27; Belloni & Röser,1996, A&AS 118, 65,) we report here the results obtained for Cl0303+17 and Cl1447+26. The clusters have been imaged with broad and narrow band (FWHM $\simeq$ 100-200 Å) filters, chosen to detect features characteristic of elliptical and E+A galaxies (4000 Å break, H$_\gamma$ and H$_\delta$ in absorption). Cluster membership and morphological types of the galaxies have been obtained by fitting the low-resolution spectral energy distributions with template spectra. Hubble type templates have been taken from Colemann et al. (1980, ApJS. 43, 393) whereas for E+A galaxies eight templates have been built up using Bruzual & Charlot population synthesis models. They represent the temporal evolution of a strong star formation episode in an elliptical galaxy or in a spiral galaxy with star formation truncated after the burst. Comparison with spectroscopic results demonstrated that for elliptical and E+A galaxies we reach a redshift accuracy of Δz=0.015.

The systematic analysis of all objects brighter than 23 R mag in the

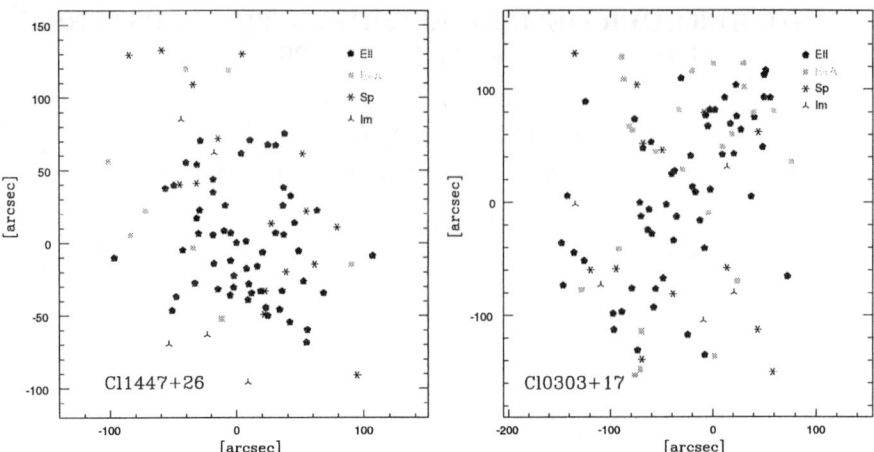

Figure 1: Spatial distribution of all member galaxies in the two clusters. Note that the few E+A galaxies in Cl1447+26 (active galaxies dominated by spirals/irregulars) avoid the central cluster region. Cl0303+17 shows an elongated structure that may be the result of the recent merger of two clusters. Support to this interpretation is given by the distribution of E+A galaxies concentrated around the north-west structure.

clusters' field resulted in a sample of 85 members in Cl0303+17 and 78 in Cl1447+26. This represents a statistical improvement of a factor of 3.5 and 9, respectively, over the past extensive studies of these clusters (Dressler & Gunn, 1992, ApJS. 78,1; Butcher & Oemler, 1984, Nature 310, 31).

Both clusters confirm *the presence of the Butcher-Oemler effect* with 40% (Cl0303+17) and 31% (Cl1447+26) of the whole cluster population being active galaxies. However, while in Cl0303+17 the fraction of active galaxies is dominated by E+A galaxies (18%), in Cl1447+26 E+A galaxies represent only 5% of the cluster members and the cluster is unusually rich in normal spirals (23%). In both clusters, spirals and E+A galaxies lack the central concentration that characterizes the elliptical population (Fig.1.)

By using only secure cluster members we calculated the concentration index for Cl1447+26 ($C=\log(R_{60}/R_{20})$ where R_n is the radius encompassing $n\%$ of the total cluster population). In contrast to previous determinations that classified Cl1447+26 as an open and irregular cluster we found that this cluster is quite concentrated ($C=0.48$). This has significant implications for any correlation between the concentration index (an indication of the dynamical state of the cluster) and other physical properties like the fraction of active galaxies or X-ray emission.

We also revealed that the elongated appearance of Cl0303+17 is due to the presence of two cluster substructures. They are clearly traced by the galaxy distribution concentrated around a cD and a bright E+A galaxy.

HIGH-RESOLUTION RICH CLUSTER SIMULATIONS

TEREASA G. BRAINERD

Boston University, Dept. of Astronomy, Boston, MA 02215, USA

DAVID M. GOLDBERG

Princeton University, Dept. of Astrophysical Science, Princetion, NJ 08544, USA

JENS VERNER VILLUMSEN

Max-Planck-Institut für Astrophysik, 85740 Garching bei München, Germany

Using large nested-grid particle-mesh simulations we investigate the formation of 3 large clusters of galaxies in a standard CDM universe.

1 The Simulations

The Hierarchical Particle Mesh (HPM) code (Villumsen, 1989) was used to run the simulations. To increase both spatial and mass resolution in a small region of the primary simulation volume, HPM nests small simulations ("subgrids") self-consistently within the main simulation volume (the "top grid"). By nesting subgrids within subgrids, very high resolution is progressively built up in a limited region of the overall simulation volume.

Three multi-grid simulations of cluster formation in a "standard" CDM universe were evolved from $z = 29$ to $z = 0$. All were 3-level calculations: a top grid of length $L_{top} = 200h^{-1}$Mpc, inside which was nested a subgrid of length $L_{sub1} = 33.3h^{-1}$Mpc, inside which was nested a subgrid of length $L_{sub2} = 8.3h^{-1}$Mpc. In all cases 256^3 grid cells were used in all 3 levels. The mass per particle in the top grid was $1.06 \times 10^{12}h^{-1}M_\odot$, in the largest subgrids it was $3.90 \times 10^{10}h^{-1}M_\odot$, and in the smallest subgrids it was $4.88 \times 10^9h^{-1}M_\odot$. In the highest-resolution (smallest) subgrid there were $\sim$400,000 particles within the Abell radii of the cluster centers at $z = 0$.

The clusters were sected as high peaks in the mass density field of the top grid at $z = 0$, smoothed on a scale of $1.5h^{-1}$Mpc. Cluster 1 was a 10-σ peak, while the other clusters were 6-σ peaks. These correspond to the largest clusters which would form in CDM universes and are not an unbiased, "average" sample of clusters, rather they are the "richest" clusters.

2 Results from Highest-Resolution Subgrid

Infall Distance: At $z = 29$, the average distance from the cluster centers of mass of those particles located within the Abell radii at $z = 0$ is $\sim 12h^{-1}$Mpc.

Particles initially located as far away as $18h^{-1}$Mpc from the centers of mass are present in the cluster at $z = 0$. Typical high-resolution cluster simulations utilizing excised peaks have "cut-out" radii of order the mean infall distance to our clusters. Our results suggest that, at least for rich clusters, a volume $\sim$3 times larger than that typically cut out from large simulations is necessary to follow all of the infall of mass into a cluster.

Mass Accretion: For particles within $3.0h^{-1}$Mpc of the cluster centers of mass, the mass accretion rate peaks between $z = 0.5$ and $z = 0.2$ for cluster 1 and between $z = 1$ and $z = 0.5$ for clusters 2 and 3. The peak accretion rate is $\sim$2 times the accretion rate averaged over the time interval from $z = 2$ to $z = 0$. The clusters gain $\sim$40% of their present-epoch mass between $z = 0.5$ and $z = 0$.

Density Profiles: Sperically-averaged density profiles were computed at $z = 0$. All three clusters have virtually identical density profiles, being roughly isothermal in the inner $0.5h^{-1}$Mpc, $\rho(r) \propto r^{-1.8}$, while at larger radii the density profiles are somewhat steeper, $\rho(r) \propto r^{-2.5}$.

Shapes: Using all particles within $3.0h^{-1}$Mpc of the cluster centers of mass, cluster shapes at were determined from a principal axis reduction of the moment of inertia tensors. A "triaxiality parameter", $T = (a^2 - b^2)/(a^2 - c^2)$, was computed where the convention $a > b > c$, $a \equiv 1$ is used. At $z = 0$ all three clusters are oblate (T =0.10, 0.19, and 0.37, respectively).

Galaxy Halos: A friends-of-friends algorithm was used to locate groups of particles with overdensities $\delta > 1000$ at $z = 0$. Of order 300 such halos with masses $\geq 10^{11}h^{-1}M_\odot$ were found within the Abell radius of each cluster. Most of these groups may be associated with the dark matter halos of individual galaxies. However, the gravitational force is only Newtonian on scales $> 65h^{-1}$ kpc, which can affect strongly the probability with which a halo will survive the cluster's deep central potential well. Because of this, the central $0.5h^{-1}$Mpc of each cluster is dominated by a single huge "halo" of mass $\sim 10^{15}h^{-1}M_\odot$. The size of this central "halo" is, however, somewhat artificial due to the simplicity of the halo-finding algorithm, which tends to link together close groups of particles in high density regions. Increasing the density contrast to $\delta > 10^5$, the massive central "halos" of the 3 clusters are found to consist of ~ 140 individual halos of mass $\geq 10^{11}h^{-1}M_\odot$, the largest having mass $\sim 10^{14}h^{-1}M_\odot$.

References

1. J.V. Villumsen ApJS, 89, 71, 407.

THE PROPERTIES OF THE GALAXY DISTRIBUTIONS OF FOUR NEARBY CLUSTERS

NOBUNARI KASHIKAWA

National Astronomical Observatory of Japan, 2-21-1 Osawa, Mitaka, Tokyo 181, JAPAN

Mosaic CCD group (NAOJ-Univ.Tokyo)

We have carried out a photometric survey of four nearby clusters of galaxies, A1656, A1367, A1644, and A1631 with mosaic CCD cameras. We have found strong segregations both in luminosity and in morphology for the total sample of all the clusters. In A1656, A1367, and A1631, early-type galaxies have a strong luminosity segregation while late-type galaxies show weak or no luminosity segregation, i.e., the strength of clustering of late-type galaxies does not depend on luminosity. Only in A1644 both early-type galaxies and late-type galaxies, have the luminosity segregation.

1 Introduction

The studies of clustering properties of distant faint galaxies are very important to see the evolutionary history of galaxies and large scale structures. In order to trace the clustering evolution without any biases, it is also necessary to know the clustering properties of nearby faint galaxies (dwarf galaxies), especially for its dependence on luminosity (luminosity segregation; hereafter LS) and morphology (morphology segregation; hereafter MS). In this study we present the angular two-point correlation functions of early- and late-type galaxies in four clusters based on large homogeneous samples with a faint limiting magnitude ($M_R \leq -16$). The luminosity functions for these samples as well as observation, data reduction, and sample selection are presented in Kashikawa et al.[1] We assume $H_0 = 100 \text{km/s/Mpc}$.

2 Angular two-point Correlation Function

To quantify the galaxy clustering, we use the angular two-point correlation function $\omega(\theta)$. We have found strong segregation both in luminosity and in morphology for the total sample of all the clusters. We extract subsamples with different limiting magnitudes from the early-type sample and the late-type sample, respectively. In order to see directly the degree of galaxy clustering and to quantify the LS, we introduce the parameter W_{int} as the integration of $\omega(\theta)$ from $0.12h^{-1}\text{Mpc}$ to $1.20h^{-1}\text{Mpc}$, which should be statistically more robust than $\omega(\theta)$ itself. The 1σ errors of the parameter W_{int}, and error bars of each bin

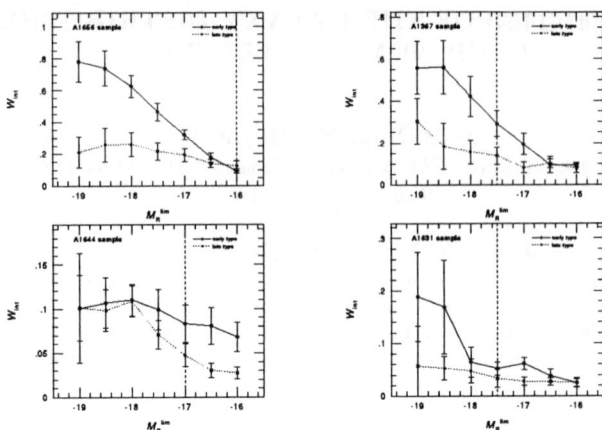

Figure 1: The W_{int} as a function of M_R^{lim} for each cluster sample. The solid line represents the early-type sample, dotted line represents the late-type sample respectively.

were estimated by the bootstrap resampling method. Figure 1 shows the W_{int} as a function of the limiting magnitude M_R^{lim}. The vertical dashed line denotes the limiting magnitude at which the estimate of the contamination by field galaxies exceeds 40%. Figure 1 demonstrates clearly that the LS is significant for early galaxies (otherwise, the MS is significant for bright galaxies) except in A1644. On the other hand in A1644, the clustering strength of early-type galaxies and that of late-type galaxies have almost the same dependence on the limiting magnitude. A1644 is the only cluster that lacks the population of "faint early-type galaxies" (Kashikawa et al.[1]). The LS observed in the total sample of these three clusters is caused by the fact that majority of the cluster members is of early type which shows a strong LS.

In order to reject this possibility that we have mis-classified many field galaxies as early types, we undertake the scaling test to evaluate statistically the effects of field galaxies on our results. The observed $\omega(\theta)$ of the sample with brighter limiting magnitude is scaled so that the scaled $\omega(\theta)$ would coincide with the observed $\omega(\theta)$ of the sample with fainter limiting magnitude if the difference between the two $\omega(\theta)$s is totally due to the contamination of field galaxies. We found that the scaled $\omega(\theta)$s of A1656 still show a large difference from the observed $\omega(\theta)$ of the sample with the faintest limiting magnitude. This demonstrates that the observed strong LS is not due to the contamination of field galaxies at least in A1656. The difference is statistically significant in A1644 and A1631, but marginal in A1367.

1. Kashikawa, N., Shimasaku, K., Yagi, M., Yasuda, N., Doi, M., Okamura, S., and Sekiguchi, M., ApJ, 452, L99, 1995.

COLOR EVOLUTION IN MODERATE REDSHIFT GALAXY CLUSTERS

S.A. STANFORD,

IGPP/LLNL, 7000 East Avenue, Livermore, CA 94550, USA

PETER EISENHARDT,

JPL, 4800 Oak Grove, Pasadena, CA 91109, USA

MARK DICKINSON

STScI, 3700 San Martin, Baltimore, MD, 21218, USA

We present results from a large–scale optical–IR imaging survey of moderate red-shift galaxy clusters.

1 Results

We have obtained optical–IR imaging using NOAO facilities of $\sim$40 galaxy clusters from Coma out to $z \sim 0.9$. 14 of these clusters have $WFPC2$ imaging available from the HST Archive. We used the public data to morphologically choose E/S0 galaxies in the 14 clusters which are at $0.3 \leq z \leq 0.9$. We compare transformed Coma cluster colors with the observed optical–IR colors of the moderate redshift early-type galaxies to determine the amount of color evolution in each cluster. We also measure the scatter in the colors of the E/S0 galaxies within each cluster.

We find color evolution consistent with a passive Bruzual–Charlot elliptical model formed at high redshift, as has been seen in similar studies such as Aragón–Salamanca et al.[1], Rakos & Schombert[2], and Lubin[3]. Furthermore, we find that the intrinsic scatter in each color is small and roughly constant with redshift, also indicative of an early and coeval formation for the early–type galaxies of clusters.

Acknowledgments

S.A.S. acknowledges financial support from IGPP/LLNL.

References

1. Aragon-Salamanca, A. *etal.*, MNRAS, 262, 764, 1993.
2. Rakos,K.D. & Schombert, J., ApJ, 439, 47, 1995.
3. Lubin, L.M., AJ, 112, 23, 1996.

COLOR EVOLUTION IN MODERATE REDSHIFT GALAXY CLUSTERS

S. A. STANFORD
IGPP/LLNL, 7000 East Avenue, Livermore, CA 94550, USA

PETER EISENHARDT
JPL, 4800 Oak Grove Dr., Pasadena, CA 91109, USA

MARK DICKINSON

THE X-RAY STRUCTURE OF THE BUTCHER-OEMLER CLUSTERS AC114 AND AC118

A.I. TERLEVICH, R.G. BOWER, I. SMAIL

Physics Department, University of Durham, South Road, Durham DH1 3LE, U.K.

A.J. BARGER AND R.S. ELLIS

Institute of Astronomy, Madingley Road, Cambridge CB3 0HA, U.K.

We present ROSAT HRI data of the $z \sim 0.31$ Butcher Oemler clusters AC114 and AC118. Using a wavelet technique to examine the morphology of these systems we show that they contain significant substructure, especially AC118 which is composed of two distinct elements. By comparing the structure of the clusters with the position of the galaxies currently undergoing starburst activity we show that their triggering cannot be associated with the infall of the large structures into the cluster.

1 Discussion

In order to quantify the structure present in the cluster gas, we have developed a new algorithm based both on wavelet analysis [1] and likelihood ratio techniques [2] to produce maps of significant structure in the x-ray images. Figure 1 shows that AC118 and AC114 have very different morphologies. AC118 appears to be a system where two largely undisturbed subcomponents are in the process of merging, whereas AC114 is not only highly elliptical, but also has a second peak to the SE of the main one.

These clusters were originally targeted by Couch & Sharples [3] (1987, CS) who found that the large number of abnormally blue 'Butcher Oemler' galaxies had spectra which could be explained by a strong 0.1 to 0.7Gyr starburst after which star formation ceases and the stellar population evolves passively. In this timescale a galaxy moving at $700Kms^{-1}$ would only move $70 - 490h^{-1}Kpc$ (the arrow in figure 1 represents a distance of $350h^{-1}Kpc$ in both clusters). Because of the short timescale of the burst, if the starbursting was triggered by some localised event, i.e. a massive merger between clusters or the infall of groups of galaxies into the main cluster, we would expect to see some coherence in the positions of the currently starbursting galaxies.

We have used the position, $(b - r)$ colour and $H\delta$ line strengths from CS to classify the galaxies in these clusters according to the classification scheme set out in Barger *et al.* [4]. The distribution of the SB galaxies in figure 1 shows that they cannot be associated with the merging of substructures. In the case of AC118 the only SB galaxy in the CS data lies on the other side of the main

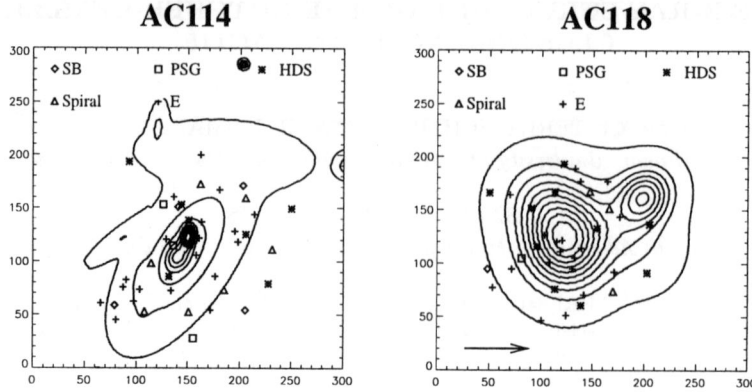

Figure 1: The positions of the galaxies found in these clusters by CS are shown superposed upon the models for the distribution of x-ray gas in AC118 & AC114 produced using our wavelet algorithm. AC118 clearly has a large subcomponent falling into the North West and AC114 has possibly undergone a merger in the recent past leaving it highly elliptical and with a slight double peak towards the SE of the main one. The large arrow in the figure of AC118 represents the possible distance travelled by a galaxy with a velocity of $700 \mathrm{kms}^{-1}$ in $0.5 \mathrm{Gyr}$ ($350 h^{-1} \mathrm{Kpc}$) ($q_0 = .5$). Different symbols represent different stages in the evolution of the starburst and post starburst galaxies (Barger *et al.* and CS). The diamonds represent the starburst galaxies.

cluster from where the subcomponents are merging, while in AC114 the 3 SB galaxies are distributed seemingly randomly around the cluster, indicating that they can't originate from the same infall event.

Our x-ray data is only sensitive enough to detect major subcomponents. Any infall of small groups or field galaxies would affect the overall x-ray morphology so little as to be undetectable and the increased frequency of such events would explain the even distribution of SB galaxies in AC114. Galaxy harassment[5] would also produce galaxies evenly distributed around the cluster.

References

1. S.A. Grebenev, W. Forman, C. Jones & S. Murray, ApJ, 445, 607, 1995.
2. Webster Cash, ApJ, 228, 939, 1979.
3. W.J. Couch & R.M. Sharples, MNRAS, 423, 229, 1987.
4. A.J. Barger, A. Aragón-Salamanca, R.S. Ellis, W.J. Couch & R.M. Sharples, MNRAS, 279, 1, 1996.
5. B. Moore, N. Katz, G. Lake, A. Dressler & A. Oemler Jr, Nature, 613, 379, 1996.

INFRARED IMAGING OF A GALAXY CLUSTER AT z=2.39

I. WADDINGTON, J.S. DUNLOP

Institute for Astronomy, University of Edinburgh, Royal Observatory,
Edinburgh EH9 3HJ, Scotland.

J.A. PEACOCK

Royal Observatory, Edinburgh EH9 3HJ, Scotland.

R.A. WINDHORST

Dept. of Physics & Astronomy, Arizona State University, Tempe,
Arizona 85287-1504, USA.

We present results from deep J,H,K imaging of the field around the $z = 2.39$ radio galaxy 53W002. Our observations are designed to identify potential evolved galaxy cluster members by the presence of a 4000Å break in their spectra, producing a red $J - H$ colour. We find ten galaxies that are redder and fainter than the radio galaxy, and consistent with a spectral break at this redshift. Two of these reddest galaxies have been confirmed spectroscopically to lie at $z = 2.4$.

An important recent breakthrough in the search for high-redshift galaxies has been the discovery [1,2] of a substantial population of star-forming galaxies at $3.0 < z < 3.5$, selected using a custom set of broadband filters designed to identify the presence of a Lyman limit break at these redshifts. In a complementary study,[3] we have used a similar technique to search for red, passively-evolving galaxies around the radio galaxy 53W002 at $z = 2.39$. At this redshift, the spectral break at 4000Å falls between the infrared J and H bands and thus an evolved galaxy with a strong break will have a red $J - H$ colour.

From our UKIRT observations, we find ten galaxies with $J - H > 1.0$ and $K > 18.8$, redder and fainter than the radio galaxy, and consistent with a spectral break at this redshift. The reddest and third reddest of these sources were also identified from *HST* imaging [4,5] as having excess Ly-α emission and have been confirmed spectroscopically to lie at $z = 2.4$. Further comparison with the *HST* data shows that all our infrared selected galaxies and most of the Ly-α sources are clustered within an area of $\sim$1.5 arcmin diameter. Our red cluster candidates are similar in size to 53W002, which has a scale-length of $\sim$10 kpc, typical of present day ellipticals and significantly larger than the $\sim$1 kpc-sized Ly-α emitters.[5] Comparing the strength of the 4000Å break in the reddest galaxy with spectrophotometric models of elliptical galaxy evolution,[6] we find that it is consistent with a 1 Gyr burst model at an age of 2 Gyr, at a redshift when the age of the universe is only $1.0 h_{100}^{-1}$ Gyr ($\Omega_0 = 1$).

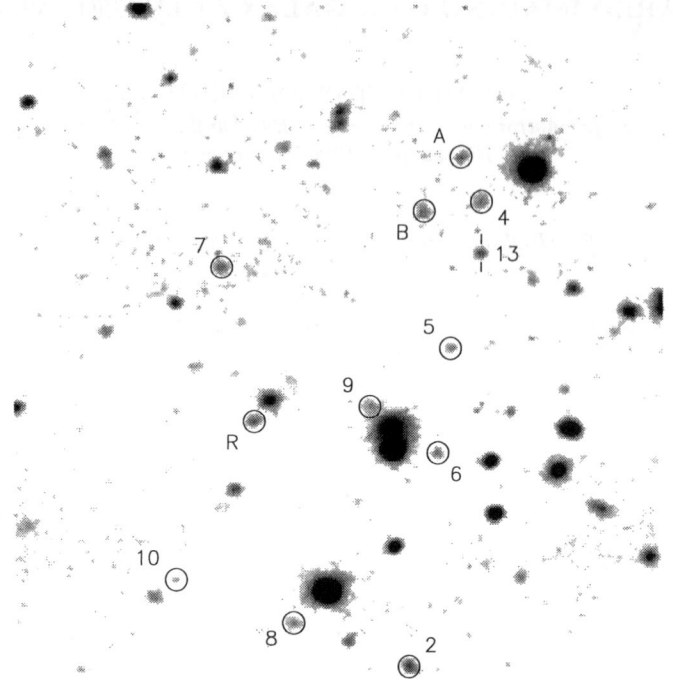

Figure 1: The field around 53W002. Circles identify the radio galaxy (R), and infrared-selected potential cluster members (A,...,10). Sources A and B have spectroscopic redshifts of 2.4. Of the eighteen sources that show excess Ly-α emission in narrow-band imaging,[5] sources R, A, B and 13 are the only ones also detected at K. Note how most of the encircled objects are of similar size to the radio galaxy. This grey-scale plot is the sum of the sky-subtracted J, H & K images, scaled such that the radio galaxy has equal counts in each filter, and smoothed by a 0.3-arcsec gaussian. North is to the top, east to the left, and the image is 103×105 arcsec^2.

References

1. Steidel C.C., Pettini M., Hamilton D., AJ, 110, 2519, 1995.
2. Steidel C.C., et al., ApJ, 462, L17, 1996.
3. Waddington I., Dunlop J.S., Peacock J.A., Windhorst R.A., MNRAS, in preparation, 1996.
4. Pascarelle S.M., et al.,, ApJ, 456, L21, 1996.
5. Pascarelle S.M., Windhorst R.A., Keel W.C., Odewahn S.C., Nature, 383, in press, 1996.
6. Guiderdoni B., Rocca-Volmerange B., A&A, 186, 1, 1987.

CLUSTERS OF GALAXIES AT INTERMEDIATE REDSHIFTS: A SAMPLE SELECTED AT RADIO WAVELENGTH

A. ZANICHELLI[1,2], R. SCARAMELLA[1], M. VIGOTTI[2], G. VETTOLANI[2], G.GRUEFF[3,2], S. BARDELLI[4], G. ZAMORANI[5], M. MIGNOLI[5], J.V. WALL[6], C.R. BENN[6]

[1] *Osservatorio Astronomico di Roma, Monteporzio Catone, Italy*

[2] *Istituto di Radioastronomia - C.N.R., Bologna, Italy*

[3] *Dipartimento di Astronomia, Università di Bologna, Bologna, Italy*

[4] *Osservatorio Astronomico di Trieste, Trieste, Italy*

[5] *Osservatorio Astronomico di Bologna, Bologna, Italy*

[6] *Royal Greenwich Observatory, Cambridge, UK*

In order to study the status and the evolution of clusters of galaxies at intermediate redshifts ($z \sim 0.1 - 0.3$), as well as their spatial correlation and relationship with the local environment, we built a sample of candidate clusters using radiogalaxies as tracers of dense environments. This selection technique is complementary to purely optical (e.g. Postman, *et al.*, 1996) or X-ray cluster selection methods, which suffer from both limited sensitivity of instrumentation and evolutionary effects (Gioia *et al.*, 1990), and preferentially select the richest clusters at large distance (or the most massive). Being radiosources luminosity uncorrelated with cluster richness (Ledlow & Owen, 1996), this selection is a powerful tool to the selection of clusters of all richness at moderate redshift.

¿From the NRAO VLA Sky Survey maps (Condon *et al.*, 1994) we extracted a catalogue of 11922 pointlike and 3371 extended or double radiosources over an area of ~ 550 square degrees at the South Galactic Pole.

Optical identifications of pointlike radiosources were made with galaxies brighter than $b_j = 20.0$ in the Edinburgh-Durham Southern Galaxy Catalogue (Heydon-Dumbleton, Collins & MacGillivray, 1988) inside a search radius of $7''$ centered on the radio position, thus obtaining a sample of 609 identified radiogalaxies with contamination from chance coincidence $< 5\%$. We searched for excesses in surface galaxy density near each identified radiogalaxy fainter than $b_j = 17.5$. After having discarded 45 known ACO or Edimburgh-Durham Cluster Catalogue clusters, we obtained a sample of 59 candidate clusters.

The same selection procedure applied to the extended or double identified radiosources will complete our sample of cluster candidates.

R-Gunn photometry and multislit spectroscopy for a sub-sample of 10 candidates was acquired at the ESO 3.6m telescope and led to the detection of 8 clusters in the range $z \sim 0.13 - 0.3$, corresponding to a positive cluster identification rate of $\sim 80\%$. Velocity dispersions for the detected clusters, evaluated by means of the ROSTAT (RObust STATistical) package (Beers, Flynn & Gebhardt, 1990), range from $241 km/sec$ to $846 km/sec$ so it seems that our selection technique is not biased towards the selection of very rich clusters, but also small clusters are picked up.

Assuming $H = 100 Km/secMpc$ and $q_0 = 0.5$, the radiogalaxy power for the observed clusters range from $P_{1.4GHz} = 4.87 \times 10^{22} Watt/Hz$ to $P_{1.4GHz} = 1.92 \times 10^{24} Watt/Hz$, thus representing FR I population of extragalactic radiosources. We expect that the bulk of FR II population will be represented by the cluster candidates that we are selecting through optical identifications of the 3371 radiosources that our algorithm classified as double or extended, whose fluxes are sensibly higher than for pointlike ones.

Acknowledgments

We would like to thank Harvey MacGillivray and Chris Collins for having provided us the Edimburgh-Durham Southern Galaxy Catalogue.

References

1. T.C. Beers, K. Flynn, K. Gebhardt, AJ, 100, 32, 1990..
2. J.J Condon, W.D. Cotton, E.W. Greisen, Q.F. Yin, R.A. Perley, J.J. Broderick in *Astronomical data analysis software and systems III.*, Astron.Soc.Pac. Conf. Ser. vol. 61, p. 155, ed D.R. Crabtree, R.J. Hanisch, J. Barnes, 1994.
3. I.M. Gioia, J.P. Henry, T. Maccacaro, S.L. Morris, J.T. Stocke, A. Wolter, ApJ, 356, L35, 1990..
4. N.H. Heydon-Dumbleton, C.A. Collins, H.T. MacGillivray in *Large-Scale Structure in the Universe - Observational and Analytical Methods*, p. 71, eds W.C. Seitter, H.W. Duerbeck& M. Tacke, Springer-Verlag, Berlin, 1988.
5. M.J. Ledlow, F.N. Owen, AJ, 112, 9, 1996..
6. M. Postman, L.M. Lubin, J.E. Gunn, J.B. Oke, J.G. Hoessel, D.P. Schneider, J.A. Christensen, AJ, 111, 615, 1996..

MEASURING THE EVOLUTION OF ELLIPTICAL GALAXIES WITH THE Mg_b–σ TEST

B. L. ZIEGLER, R. BENDER

Universitätssternwarte, Scheinerstr. 1, D–81679 München, Germany

The evolution of elliptical galaxies is investigated by comparing their Mg_b–σ relations at different redshifts. Locally, a very tight relationship between the Mg_b index ($\lambda_0 \approx 5170$ Å) and the velocity dispersion σ is observed (e.g., Bender *et al.* 1993, ApJ 411, 153). For a given σ, both the metallicity and the age are constrained to $< 20\%$. Therefore, relative ages can be deduced from the observed reduction of Mg_b at higher z. Using Worthey's population synthesis (1994, ApJS 95, 107) we derive the following dependence of Mg_b on age and metallicity: $\log Mg_b = 0.37 + 0.20 \ \log t + 0.31 \ \log Z/Z_o$. The Mg_b–σ test is a very robust estimator of the evolution (no K–correction, no correction for extinction, very little dependence on the IMF; see Bender *et al.* 1996, ApJ 463, L51).

So far, we have established the Mg_b–σ relation at $z = 0.37$ in three clusters (*Abell 370, CL 0949+44* and *MS 1512+36*) using the 4m-class telescopes on Calar Alto and at ESO (*fig. 1*). For any given σ, the Mg_b of the distant ellipticals is significantly lower than the mean value of the nearby sample (Coma and Virgo ellipticals). The difference of Mg_b between the two samples is very small and can be fully attributed to the younger age of the distant stellar populations in accordance with the passive evolution model for ellipticals. The low reduction of Mg_b at a look–back time of about 5 Gyrs requires that the bulk of the stars in cluster ellipticals have formed at very high redshifts of $z_f > 2$; for the most massive galaxies, where the reduction is even lower, z_f probably exceeds 4. In *fig. 1* the expected Mg_b–σ relations of ellipticals at $z = 0.37$ are given for different z_f. In quantitative agreement with the reduction of the Mg_b absorption we find an increase of the B magnitude at fixed σ from the Faber–Jackson relation using the formula $\Delta M_B = 1.2 \Delta Mg_b$ (*fig. 2*).

From our HST images of *Abell 370* and *MS 1512+36* we have derived effective radii r_e and effective surface brightnesses SB_e of the ellipticals with measured σ allowing us to construct the fundamental plane at $z = 0.37$. Applying the evolutionary corrections derived from the Mg_b–σ test, we can calibrate ellipticals as standard candles and compare the FP at $z = 0$ with that at $z = 0.37$ to constrain the value of q_o (Bender *et al.* 1996, ESO workshop).

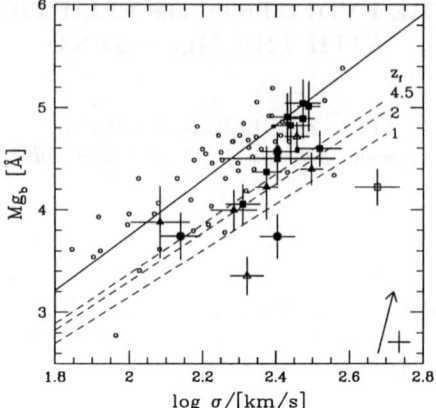

Figure 1: Mg_b–σ relations at $z = 0$ (small dots with no error bars; typical error bar in lower right corner; straight line: linear χ^2–fit) and at $z = 0.37$ (big symbols with errorbars; arrow in lower right corner: aperture correction applied). Dashed lines: theoretical Mg_b–σ relations at $z_{obs} = 0.37$ for redshifts of formation $z_f = 1$, 2 and 4.5.

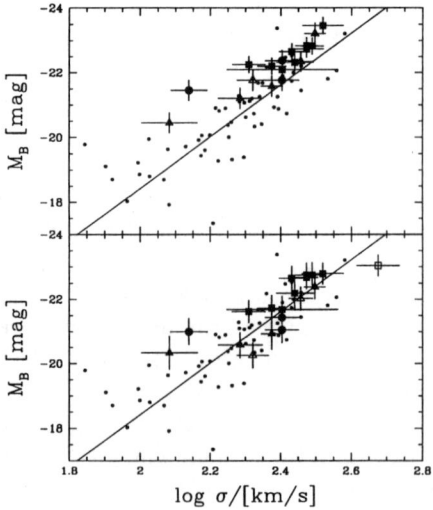

Figure 2: Faber–Jackson relations: top: rest–frame M_B ($q_o = 0.5, H_o = 50\,km/s/Mpc$), bottom: M_B of $z = 0.37$ ellipticals corrected for evolution according to their individual reduction in Mg_b.

SECTION 3

GRAVITATIONAL LENSING

GRAVITATIONAL LENSING AND THE REDSHIFT DISTRIBUTION OF B> 25 GALAXIES

Y. MELLIER

Institut d'Astrophysique de Paris, 98bis Boulevard Arago
75014 Paris, France

In addition to the determination of the mass distribution of lenses (see P. Schneider, this conference), strong and weak lensing can also be powerfull tools to analyse the redshift distribution of faint galaxies. In this review, I summarize the present status of redshifts of galaxies beyond $B = 25$ as they are inferred by spectroscopy of magnified galaxies, lensing inversion and magnification bias.

1 Introduction

During the last decade, outstanding results have been obtained on the redshift distribution of faint galaxies up to $I = 22.5$ (see O. Le Fèvre, this conference) or $I = 23$ (Cowie et al. 1996). With the coming of 10 meter class telescopes equipped with wide field multi-object spectrographs these surveys will be extended to thousands of galaxies. The leading goals of these programme is to explore the evolution of clustering of galaxies, their physical and stellar evolution with redshift up to $B = 24 - 25$ or $I = 22.5 - 23.5$ as well as to study very high redshift galaxies. The galaxies with magnitudes $B > 25$ are also important for the models of galaxy formation since we do not know whether they are all at large redshift or if there is a significant fraction of faint nearby dwarfs galaxies. Furthermore, the weak lensing inversion uses the grid of faint distant sources with magnitudes between $B = 25$ to $B = 28$ for which the redshift distribution is unknown. While this information is not important for nearby lenses, it is crucial for for those having redshifts larger than $z = 0.5$ and can be a major source of uncertainty in the mass determination (see Luppino & Kaiser 1996). In the perspective of the new surveys to study the large-scale mass distribution by using weak lensing, the need of the redshift distribution of the faintest galaxies is thus esssential.

Unfortunately, while we do expect considerable informations in the magnitude range $B < 25$, beyond this limit even 10 meter class telescopes are still too small and redshifts of a complete sample of $B > 25$ galaxies cannot be secured in a reasonable amount of observing time. The possibility of using photometric redshifts has been proposed as early as the beginning of eighties and has been studied in great details. Observations as well as reliability tests are still underway (Pelló private communication). However, they are based on

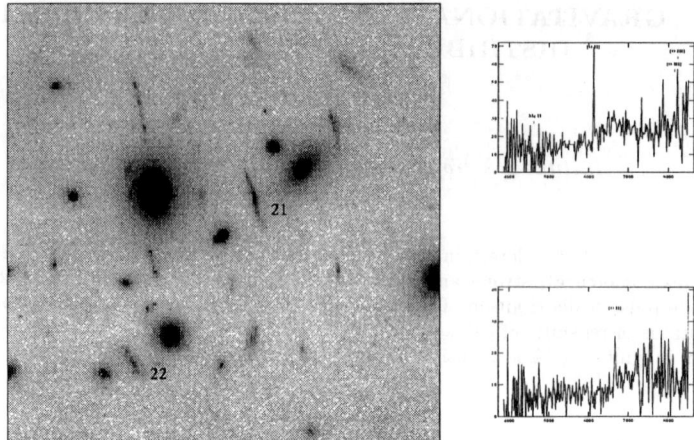

Figure 1: Redshift of faint arclets in lensing-clusters. The left panel is a deep HST image of A2390. Arclets are visible with some of them showing multiple images with image parity changes. The arclets 21 and 22 have been observed by Bézecourt and Soucail (1996). As expected, they show the [OII]λ3727 emission line from which the redshifts are easily measured ($z = 0.643$ and $z = 0.790$ respectively. Courtesy J. Bézecourt)..

theoretical evolution scenarios of galaxies whose predictions about faint distant galaxies are not confirmed yet. Furthermore, there is no hope to calibrate the photometric redshifts of the faint samples with spectroscopic data.

The most attractive alternative is the use of the deviation, magnification and distortion effects induced by gravitational lensing on extended objects. In this review, I discuss the recent advances in the redshift distribution of $B > 25$ galaxies by using spectroscopic samples of arc(let)s, the lensing inversion and finally the magnification bias. Though many works in these fields are still underway, they are new promising approaches that must be tested jointly with photometric redshifts in order to cross-check their reliablity and the consistency of their predictions.

2 Spectroscopic surveys of arc(let)s

Spectroscopic redshifts of arc(let)s are indispensable to calculate the angular distances D_d, D_{ds} and D_s and to get the absolute scaling of the projected mass density. The redshifts of a large number of arc(let)s in each individual lensing-clusters provide the positions of many critical lines and allow to probe locally the mass distribution. In practice, the redshifts of arc(let)s check the lens modelling obtained from giant arcs and can be used to refined it. It is also possible to obtain information on the cosmological parameters if one could have enough redshift to constrain both the deflector and the geometry of the Universe.

More recently, with the development of the lensing inversion technique (see next section), the need of spectroscopic confirmations of its predictions led to intensive observations of arclets. Spectroscopic surveys of the "brightest" arclets in many clusters are now underway and first results have been obtained in A2390 (Bézecourt & Soucail 1996) and A2218 (Ebbels et al. 1996; Ebbels this conference). Since most of these objects are very faint, only arclets showing bright spots of stars forming regions on HST images are selected in order to detect an emission line and to optimize the chance to get reliable redshift (see figure 1).

About 20 redshifts of arcs and 20 redshifts of arclets have been measured. However, the use of this sample to recover the redshift distribution of $B > 25$ galaxies is difficult because it is biased. First, only arclets with star forming emisson lines are selected. Second, beyond $B = 25$ the magnification bias favours observations of blue galaxies rather than red. So, even if the spectroscopy of arclets is crucial for the lens modelling and eventually to obtain the spectral energy distribution of high-redshift galaxies, the redshift distributions obtained from these methods are still questionnable.

3 Redshifts of arclets from lensing inversion

When it is possible to recover the lensing potential with a good accuracy, the lensing equation can be inverted to reconstruct the lensed images back to the source plane. This is basically the procedure of the lensing inversion which searches for each arclet the source plane where its distortion is minimum, assuming it gives its most *probable redshift*. The obvious interest of this method is that it does not depend on the magnitude of the arclet but on its position and its shape in the image plane. Potentially, it provides redshift of galaxies up to $B = 27$.

The lensing inversion has been developped by the Toulouse/Paris group and was first applied on A370 (Kneib et al. 1994) from the lens modelling of the giant arc. Though the (intrinsic)magnitude-redshift found for these arclets shows a good continuity with the faint spectroscopic surveys, there are still some uncertainties. In fact, as it is shown in figure 2, the X-ray isophotes and the arclet positions do not follow the expectations of the lens modelling in the eastern region. This is an indication that while the modelling is excellent in the cluster center, the mass distribution has not a simple geometry beyond the

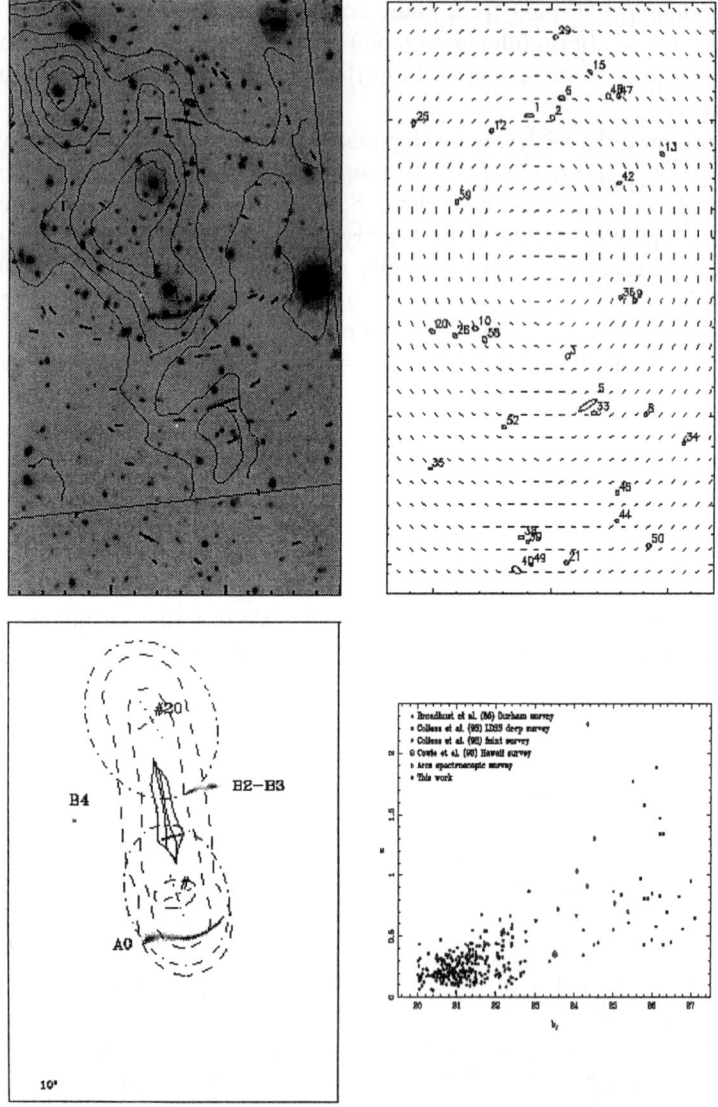

Figure 2: The top left pannel shows a B image of A370 with the ROSAT isocontours and the position of arclets superimposed (dark segments). They are consistent with the lens modelling (bottom left) and we can expect good redshift predictions of arclets from the reconstructed shear map (top right). Kneib et al. inverted the arclets in A370 and found the magnitude-redshift diagramme plotted in the bottom right panel. However, the ROSAT isophotes and the arclets positions show some discrepancies with the modelling on the eastern side (left) and in this region the predicted redshifts could be wrong.

giant arc. Furthermore, the lensing inversion is also sensible to the accuracy of the shape measurements of each arclet, and for so faint objects errors can be large.

There are two solutions to solve these issues: first, it is highly preferable to use HST images instead of ground based images. The results obtained by Kneib et al. (1996) on A2218 show the efficiency of the lensing inversion when applied on excellent images. Second, it is important to have lensing-clusters with simple geometry. In this respect, though A370 and A2218 are rather well modelled, they are not the simplest and clusters like MS0440 or MS2137-23 are better candidates.

4   The distribution of faint galaxies from the magnification bias

The projected number density of galaxies through a lens results from the competition between the gravitational magnification that increases the detection of individual objects and the deviation of light beam that increases the area and thus decreases the apparent number density. Therefore the amplitude of the magnification bias depends on the slope of the galaxy counts as a function of magnitude and on the magnification factor of the lens (Broadhurst et al. 1995): when the slope is higher than 0.4 the number density increases, whereas below 0.4 is decreases and the radial distribution shows a typical depletion curve (see figure 3).

When the slope is lower than 0.3, a sharp decrease of the number of galaxies is expected close to the critical radius of the lens corresponding to the redshift of the background sources. For a broad redshift distribution, it can result a shallower depletion between the smallest and the largest critical line which depends on the redshift distribution of the galaxies (Figure 3). Therefore, the analysis of the shape of the depletion curves provide a new way to sort out their redshift distribution. As the lensing inversion, this is a statistical method which can also infer redshift of very faint sources (up to $B = 28$) but does not need anymore information on the shapes of arclets. However, the need of a good lens modelling is still necessary.

This method was first used by Fort et al (1996) in the cluster Cl0024+1654 to study the faint distant galaxies population in the extreme range of magnitude $B = 26.5 - 28$ and $I = 25 - 26.5$. For these selected bins of magnitude they found on their CFHT blank fields that the counts slope was near 0.2, well suited for the study of the effect. After analysis of the shape of the depletion curve (figure 4), 60% ± 10% of the B-selected galaxies were found between $z = 0.9$ and $z = 1.1$ while most of the remaining 40% ± 10% galaxies appears to be broadly distributed around a redshift of $z = 3$. The I selected popula-

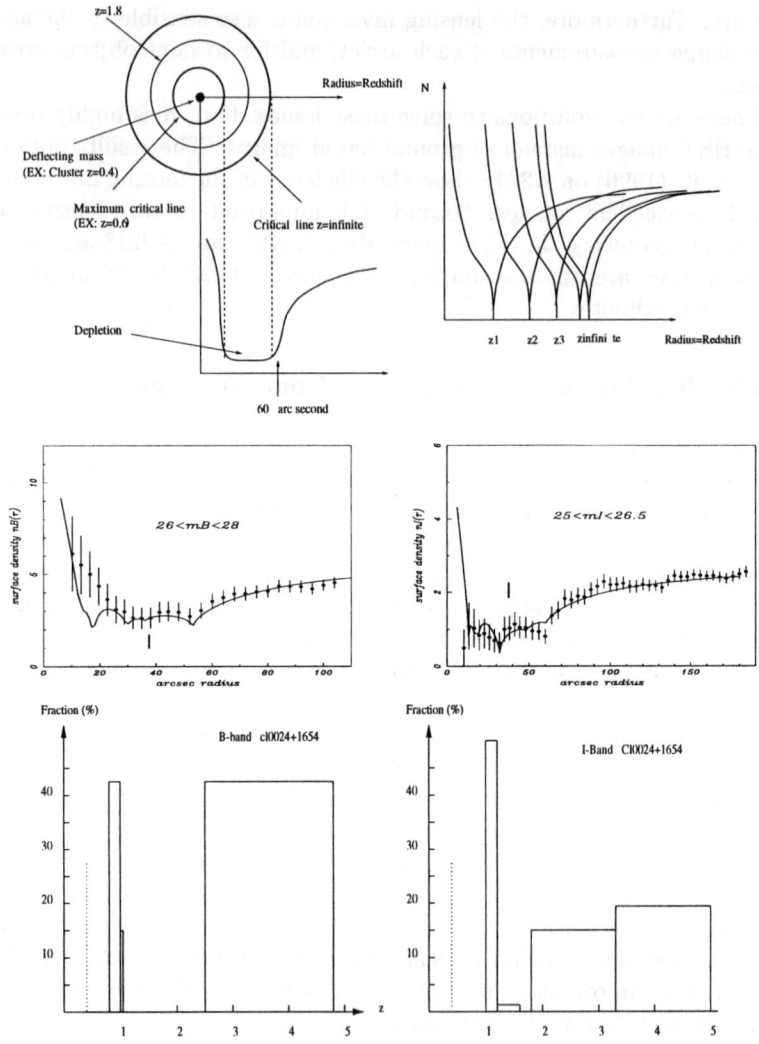

Figure 3: Depletion by a singular isothermal sphere as it would be observed on the sky and radial density of galaxies (top left). For a given redshift, the minimum of the depletion is sharp and its radial position is equivalent to a redshift (top right). The miminum increases with the redshift of sources but the depletion curves tighten and converge towards the curve corresponding to sources at infinity. In a realistic case, the redshift distribution is broad and the individual curves must be added. In this case, instead of the single peaked depletion we expect a more pronounced minimum between two radii (i.e. two redshifts; top left). The middle panels show the depletion curves observed in B and I in Cl0024. Since the mass distribution of this lens is well known, one can recover the redhsift of the sources for the B and I populations (bottom panels: note that this is a fraction of galaxies. The width of boxes is the redshift range, not a total number of galaxies).

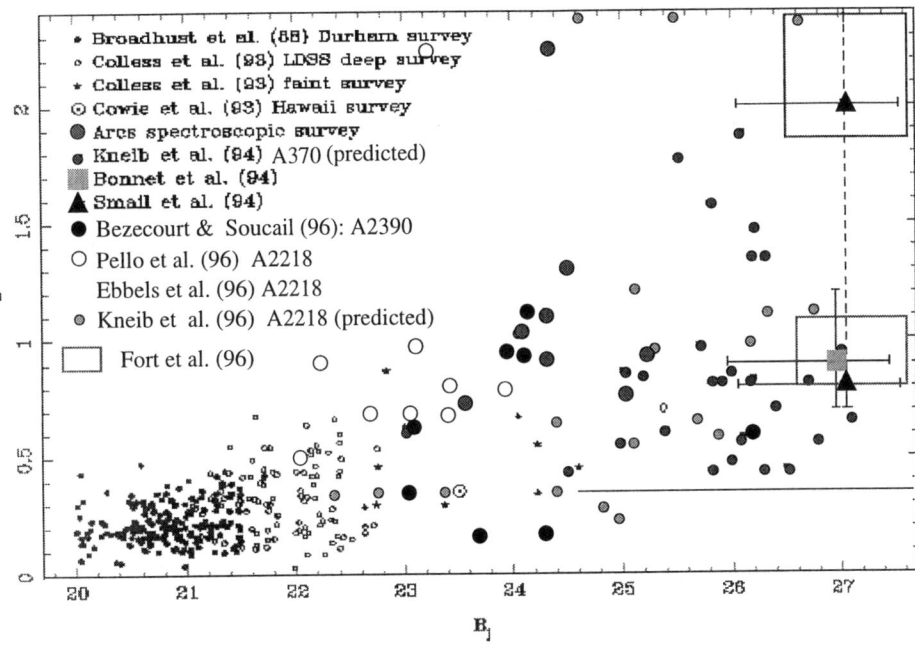

Figure 4: A magnitude-redshift diagramme showing the positions of the redshift surveys (dark symbols on the left), the arc(let)s spectroscopic surveys (large circles. Those concerning A2218 have been kindly provided by Pelló prior to publication), the predictions of lensing inversions for A370 and A2218 (small circles), of weak lensing studies by Bonnet et al. and Smail et al. (triangles) and finally, of the depletion curves in Cl0024 (large boxes). The spectroscopic redshift of Cowie et al. (1996) with Keck would be between $B = 22.5$ and $B = 24.5$. We see the potential interest of gravitational lensing which provide redshifts up to $B = 28$. The straight line on bottom right is the redshift of A370 which is a limit of the lensing inversion in this cluster.

tion present a similar distribution with two maxima, but spread up to a larger redshift range with about 20% above $z > 4$ (figure 3).

This very first tentative must be pursued on many lensing clusters in order to provide significant results on the redshift distribution of the faintest distant galaxies. Though it is a very promising approach, it also need to be applied on clusters with simple geometry. Furthermore, the detection procedure demands ultra-deep exposures with subarcsecond seeing.

5 Conclusions

The redshift distribution of galaxies beyond $B = 25$ is a crucial scientific question for galaxy evolution and weak lensing inversions. I have discussed three innovative ways which can go as faint as $B = 28$. They must be considered jointly with photometric redshifts which will need confirmations from others observations. But whatever the method, how and when will we be sure that these redshifts are correct from spectroscopic data? This key issue may be solved with ultra-deep CCD spectroscopic exposures with the VLTs. This is a major challenge for the coming years that will be possible to match with the systematic use of gravitational telescopes.

Acknowledgments

I thank B. Fort, R. Ellis, R. Pelló, P. Schneider and L. Van Waerbeke for stimulating discussions on lensing and on distances of faint galaxies.

References

1. J. Bézecourt, G. Soucail. Preprint astro-ph/9606064, 1996.
2. H. Bonnet, Y. Mellier, B. Fort, apj, 427, L83, 1994.
3. T. J. Broadhurst, A. N. Taylor, J. Peacock, apj, 438, 49, 1995.
4. L. L. Cowie, A. Songaila, E. M. Hu, J. G. Cohen. Preprint astro-ph/9606079, 1996.
5. T. Ebbels, J.-F. Le Borgne, R. Pelló, R. S. Ellis, J.-P. Kneib, I. Smail, B. Sanahuja. Preprint astro-ph/9606015.
6. B. Fort, Y. Mellier, M. Dantel-Fort. Preprint astro-ph/9606039.
7. J.-P. Kneib, G. Mathez, B. Fort, Y. Mellier, G. Soucail, P.-Y. Longaretti, aap, 286, 701, 1994.
8. J.-P. Kneib, R. S. Ellis, I. Smail, W. J. Couch, R. M. Sharples. Preprint astro-ph/9511015, 1995.
9. G. Luppino, N. Kaiser. Preprint astro-ph/9601194, 1996.

WEAK GRAVITATIONAL LENSING

PETER SCHNEIDER

Max-Planck-Institut f. Astrophysik,
Postfach 1523, D-85740 Garching, Germany

The distortion of images of faint high-redshift galaxies by the tidal gravitational field of mass concentrations allows to investigate the mass distribution of individual galaxy clusters, the investigation of the statistical properties of the mass distribution in galaxy halos, and the detection of dark halos without any reference to their luminosity. In addition, the statistical properties of the image distortion field on large scales can be used to infer directly the power spectrum of cosmological density fluctuations. I will outline the basic methods of this new research field in extragalactic astrophysics, and mention several recent results. The important role played by the HST for weak lensing studies is highlighted.

1 Introduction

Light bundles are not only deflected as a whole by an intervening mass concentration, but distorted by the tidal gravitational field of the deflector. This image distortion can be quite weak and can then not be detected in individual images. However, since we are lucky to live in a Universe where the sky is full of faint distant galaxies, this distortion effect can be discovered statistically. This immediately implies that weak lensing requires excellent and deep images so that image shapes (and sizes) can be accurately measured and the number density be as high as possible to reduce statistical uncertainties. Weak gravitational lensing can be defined as using the faint galaxy population to measure the mass and/or mass distribution of individual intervening cosmic structures, or the statistical properties of their mass distribution, or to detect them in the first place, independent of the physical state or nature of the matter, or the luminosity of these mass concentrations. In addition, weak lensing can be used to infer the redshift distribution of the faintest galaxies. After introducing the necessary concepts, I will list the main applications of weak lensing as isolated today and discuss some of them in slightly more detail, stressing the need for very deep images of the sky taking with instruments of excellent image quality.

2 Gravitational lensing of small sources

A gravitational lens provides a map from the observer's sky to the undistorted sky, $\vec{\beta} = \vec{\theta} - \vec{\alpha}(\vec{\theta})$, where $\vec{\theta}$ is the angular position of a (point) source as seen by the observer, $\vec{\beta}$ the angular position of the source in the absence of the deflector, and $\vec{\alpha}(\vec{\theta})$ is the (scaled) deflection angle the light ray undergoes near the lens (see Schneider, Ehlers & Falco [1] for details). The deflection angle is a linear functional of the dimensionless surface mass density $\kappa(\vec{\theta}) = (4\pi G D_d D_{ds})/(c^2 D_s)\Sigma(\vec{\theta})$ where $\Sigma(\vec{\theta})$ is the physical surface mass density, and D_d, D_{ds} and D_s are the angular diameter distances to the lens and source, and from the lens to the source, respectively. The distance factors in κ, and thus in the lens equation, imply that sources at different redshifts are affected differently; this fact will allow to estimate the redshift distribution of faint galaxies. If one considers sources whose angular extent is small compared to the typical angular scale over which the deflection angle changes appreciably, the distortion of such sources can be described by the locally linearized lens equation, $\delta\vec{\beta} = A(\vec{\theta})\,\delta\vec{\theta}$ where $A(\vec{\theta})$ is the local Jacobian matrix. Its trace depends only on the local surface mass density κ, and its trace-free contribution describes the local tidal field or shear $\gamma = \gamma_1 + i\gamma_2$, which is again linearly related to κ through $\gamma(\vec{\theta}) = \pi^{-1} \int d^2\theta' \, \mathcal{D}(\vec{\theta} - \vec{\theta}') \, \kappa(\vec{\theta}')$ with the kernel $\mathcal{D}(\vec{\phi}) = -(\phi_1^2 - \phi_2^2 + 2i\phi_1\phi_2)/\left|\vec{\phi}\right|^4$. Together with the fact that the surface brightness is unchanged by gravitational light deflection, the locally linearized mapping fully specifies the imaging of small sources; a circular source is mapped onto an ellipse, with axis ratio depending on the strength of the shear [or more precisely, on the strength of the reduced shear $g = \gamma/(1 - \kappa)$], and the size determined by the magnification $\mu = \left[(1 - \kappa)^2 - |\gamma|^2\right]^{-1}$.

2.1 Shear (or distortion) effects

Thus, if one had a sample of circular sources, the matrix elements of A, and thus the local reduced shear, could be determined from the shape of the lensed images. But galaxies are not circular intrinsically, and so individual images do not constrain the local shear. However, making the basic assumption that the intrinsic orientations of an ensemble of galaxies taken from a large cosmic volume are randomly distributed one can statistically infer the local distortion from an ensemble of galaxy images. Thus, any net orientation (or net ellipticity) of this ensemble in excess of Poisson noise is due to light propagation. One component of this light propagation occurs in the telescope and camera, so that an anisotropic PSF can cause a net ellipticity of the images. The great advantage of imaging with HST is that its PSF is well determined and can be corrected for. The remaining net ellipticity is due to gravitational light deflection. This implies that the tidal gravitational field can be measured locally: if ϵ denotes the (complex) image ellipticity, then its expectation value $E(\epsilon) = g(\vec{\theta}) = \gamma(\vec{\theta})/[1 - \kappa(\vec{\theta})]$. Identifying the local mean of ϵ with the expectation value, the reduced shear can be determined locally. Here, 'local' means a region around a point $\vec{\theta}$ which is small enough so that the reduced shear does not vary strongly over this region, and which is still large enough to contain a sufficiently large number N of images. As a rough estimate, the magnitude of the reduced shear is several percent in clusters of galaxies, rising to order unity close to their centers, whereas the typical magnitude of the shear caused by the density fluctuations of the large-scale structure is of order one percent, depending on the cosmological model (see, e.g., Villumsen [2] and references therein). A coherent shear signal in a cluster has first been discovered by Fort et al. [3] and Tyson, Valdes & Wenk [4].

2.2 Magnification effects

In addition to changes of the image shape, the size of an image relative to the size of the source is affected. Surface brightness conservation then implies that the flux of the image is changed by the magnification μ. This latter effect changes the local source counts: if $n_0(> S)$ denotes the number density of sources with flux greater than S in the absence of lensing, the local number density is changed to $n(> S) = n_0(> S/\mu)/\mu$. Thus, if the counts deviate from an S^{-1} form, comparing the local number density to n_0 yields an estimate of the local magnification [5]. Alternatively, if $\omega_0(s)$ denotes the (appropriately defined) solid angle subtended by an unlensed source with surface brightness s, then in the presence of a lens, $\omega(s) = \mu\omega_0(s)$, and the magnification can locally be determined [6] from $\mu = \langle \omega(s) \rangle / \langle \omega_0(s) \rangle$.

2.3 Practical considerations

In order to obtain good angular resolution and/or high accuracy on the local determination of the shear and the magnification, one has to take very deep exposures to be able to work with a high number density of galaxy images. The images are affected by any residual anisotropic PSF which mimics a shear. In order to correct for these instrumental effects, a stable PSF is needed, and good sampling of the PSF is required. It is also obvious that the seeing is crucial in this game: seeing circularizes small elliptical images and thus significantly reduces the shear signal. In order to regain the image ellipticities 'before seeing', correction factors have to be applied, which are determined by simulating images with the same PSF and comparing the input shear with that estimated from the convolved image with pixelization and noise added. These correction factors can be quite large and reduce the accuracy with which the shear can be measured significantly [7,8]. These problems are significantly reduced if HST images are used.

2.4 Digging in the noise

These 'traditional' methods to determine the shear and magnification use the properties of (isolated) galaxy images; for each one has to determine a center and the tensor of second brightness moments, and the rest of the CCD is unused. Alternatively, one can use the two-point auto-correlation function (ACF) of the light distribution on the CCD, $\xi(\vec{\theta})$. This is related to the unlensed ACF $\xi^s(\vec{\theta})$ by $\xi(\vec{\theta}) = \xi^s(A\vec{\theta})$. Since the unlensed ACF can be assumed to be isotropic, the anisotropy of the observed ACF immediately yields the reduced shear g. One can calculate the ACF locally and determine g locally. In addition, since the ACF is caused by very many faint galaxies per solid angle, one might suppose that it is a universal function (which can be determined from deep HST exposures); in that case, also the magnification can be determined locally. To avoid being dominated by just the brighter objects on the frame, they can be cut out, so that one works on a field with the topology of a Swiss Cheese. Eventually, if all objects are cut out which are significantly detected, one works in the noise limit. If the ACF of the noise is caused by faint high-redshift galaxies, the value of g determined from the noise should agree with that determined from the images, but gives independent information. This method was proposed and successfully tested both on synthetic images as well as on real data; in the latter case, the shear field obtained from individual galaxy images has been reproduced by the ACF of the noise[9]. The ACF method is also a sensitive diagnostics for testing image quality; improper data reduction shows up immediately as artificial features in the ACF.

3 Main applications of weak lensing

3.1 Reconstruction of cluster mass profiles

The pioneering paper by Kaiser & Squires[10] paved the way for a non-parametric two-dimensional mass reconstruction, by obtaining the inversion γ-κ-relation,

$$\kappa(\vec{\theta}) = \frac{1}{\pi}\mathcal{R}e\left(\int_{\mathbb{R}^2} \mathrm{d}^2\theta'\, \mathcal{D}^*(\vec{\theta} - \vec{\theta}')\, \gamma(\vec{\theta}')\right) + \kappa_0 , \qquad (1)$$

where the asterisk denotes complex conjugation, $\mathcal{R}e(z)$ is the real part of the complex number z, and κ_0 is an undetermined additive constant. The shear γ in the weak lensing regime ($\kappa \ll 1$) can be obtained from the local image ellipticities, as described above, and thus from an ensemble of images, the surface mass density can be evaluated by replacing the integral in (4) by a sum over images. This method was first applied to the cluster MS1224[11], and quite a large lower limit for the mass-to-light ratio of this cluster was obtained. Since then, several more clusters have been investigated with that method. The method has been modified to allow the inclusion of strong lensing[12,13,14] and to replace the integral over the $\mathbb{R}^2$ by one over a finite region represented by the data field (CCD)[15,16,17]. Using these generalizations, together with explicitly accounting for a broad redshift distribution of the galaxies, Seitz et al.[18] have reconstructed the mass profile of the inner part of the cluster Cl 0939+4713 from a deep image taken with the WFPC2. The resulting detailed two-dimensional mass map, when compared with the distribution of bright cluster galaxies, shows that the light traces the mass very well in this cluster. Also, the number density effect caused by the magnification has been discovered in this cluster. The mass-to-light ratio is only moderate ($\sim$ 200, depending on the mean redshift of the galaxies), but that should be no surprise: Cl 0939 is the highest-redshift cluster in the Abell catalog (A851) and therefore expected to have a very high optical luminosity. A low-resolution X-ray map[19] indicates that also the X-ray emission traces the (dark) mass; this will be checked in more detail once a HRI map of this cluster becomes available.

The prospects of this method are simply excellent: deep images taking under good conditions will allow to study the dark mass distribution in clusters (e.g., the radial density profile, detection of substructure and ellipticity), independent of assumptions about symmetries or dynamical or thermal equilibrium of the matter. It therefore provides the least prejudiced mass distributions, and can be used to calibrate other methods, e.g., those using the X-ray profile and temperature (for example, see Squires at al.[20]). As stressed before, the accuracy of this method depends sensitively on the data quality, and on the available number density of galaxy images – thus on the depth of the observations. The combination of distortion and magnification effects, using maximum-likelihood techniques[21], will increase the efficiency and accuracy of the reconstructions. Probably, the most promising observational strategy is to combine HST images of the cluster center with wide-field ground-based images to compensate for the small field-of-view of WFPC2.

3.2 Statistical properties of the (dark) mass distribution in galaxies

Individual galaxies are not massive enough to produce a significant shear signal, but statistically combining the signals from many (foreground) galaxies can yield a detectable 'relative alignment' of background images relative to the direction of the nearest foreground galaxy. First attempted by Tyson et al.[22], this effect has now been discovered by Brainerd, Blandford & Smail[23]. Fitting a parametrized model to the alignment data, they have shown that the characteristic velocity dispersion (or rotational velocity) of galaxies is in the range expected from other investigations. In addition, they were able to obtain an interesting lower bound on the spatial extent of the dark halos in galaxies. This study was carried out with a relatively small number of galaxies; Schneider & Rix [24] have shown that even with moderately-sized samples of galaxies, one can obtain very accurate determinations of model parameters such as σ_* or the characteristic size s_* of an L_* galaxy. In addition, the Tully-Fischer exponent can be probed, as well as the evolution of the mean redshift with apparent magnitude. All that is needed is a collection of wide-field images taking in excellent seeing conditions. This method can be applied to a combined data set of HST and ground-based images; in fact, galaxy-galaxy lensing has been detected in the Groth-strip [25]. Our attempt to obtain a wide-field image covering the strip in March 1996 failed – remember the bright comet then, and guess where is was located ...

3.3   Detection of 'dark' mass concentrations

On wide-field images, one can search for (dark) mass concentrations by looking for statistically significant alignements of faint galaxy images. Based on the aperture densitometry [14], I have investigated the statistical properties of the appropriately-defined aperture mass calculated from the image ellipticities in annular regions [26]. The expectation is to detect isothermal halos with velocity dispersion in excess of $\sim 600 \, \mathrm{km/s}$, without any reference to the optical or X-ray luminosity of these halos. Depending on the cosmological model, one expects about 10 such halos per square degree for a standard CDM model, increasing by a factor of order 10 in a COBE-normalized CDM model. This method will thus allow for the first time to investigate the statistics of dark halos without any assumption about bias factors, so that these results can be directly compared to numerical LSS simulations. In fact, dark halos have already been discovered by their shear effects: the 'dark' lens in the double QSO 2345+007 was discovered by the shear field it creates [27], and significant shear fields have been discovered around several high-redshift radio-loud quasars [28], supporting the magnification bias hypothesis for the associations of these QSOs with foreground galaxies [29].

3.4   Constraints on the redshift distribution of very faint galaxies

The dimensionless surface mass density κ depends on the source redshift through the factor $D_{\mathrm{ds}}/D_{\mathrm{s}}$; hence the lensing strength increases with increasing source redshift. This yields the possibility to obtain information about the redshift distribution of the faintest detectable galaxies, as proposed by Smail, Ellis & Fitchett [30], Bartelmann & Narayan [6] and others. In particular, the fact that significant shear was observed [31] in the high-redshift ($z_d = 0.83$) cluster MS 1054−03 shows that a large fraction of the galaxies used in this study ($21.5 < I < 25.5$) must have a redshift significantly larger than 1. For a different study of source redshifts from weak lensing, using the magnification effect, see Fort et al.[32].

3.5 Determination of the power spectrum of cosmic density fluctuations

The density fluctuations of the mass inhomogeneities in the Universe distort light bundles from distant sources and can produce an observable effect. It has been shown in several papers (see, e.g., Villumsen [2] for references) that the statistical properties of the distortion field are directly related to the power spectrum of the density fluctuations. For example, the two-point correlation function of the image ellipticity caused by the LSS is obtained by a convolution of its power spectrum with a known kernel function. Whereas the expected magnitude of the shear is quite small (of order 1%), its detection and quantitative investigation will allow to study the statistical properties of the density field in the Universe, on (comoving) scales much smaller than those achievable with CMB experiments, again without any assumption about bias factors.

4 Conclusions

Weak lensing has become a quantitative tool for cosmology, though we have only started to apply it. The theoretical tools needed are all in place, although they certainly will be refined, but the bottleneck is the availability of deep and high-quality data. HST will continue to play a dominant role in this field, with its best use probably being in combination with ground-based data. A major problem, the small field-of-view of WFPC2, will be cured one the Advanced Camera is installed, from which substantially improved data can be expected.

Acknowledgments

This work was supported by the "Sonderforschungsbereich 375-95 für Astro–Teilchenphysik" der Deutschen Forschungsgemeinschaft.

References

1. Schneider, P., Ehlers, J. & Falco, *Gravitational Lenses*, Springer-Verlag, New York, 1992.
2. Villumsen, J.V., MNRAS 281, 369, 1996.

3. Fort, B., Prieur, J.L., Mathez, G., Mellier, Y. & Soucail, G., A&A 200, L17, 1988.
4. Tyson, J.A., Valdes, F. & Wenk, R.A., ApJ 349, L1, 1990.
5. Broadhurst, T.J., Taylor, A.N. & Peacock, J.A., ApJ 438, 49, 1995.
6. Bartelmann, M. & Narayan, R., ApJ 451, 60, 1995.
7. Bonnet, H. & Mellier, Y., A&A 303, 331, 1995.
8. Kaiser, N., Squires, G. & Broadhurst, T., ApJ 449, 460, 1995.
9. Van Waerbeke, L., Mellier, Y., Schneider, P., Fort, B. & Mathez, G. A&A, in press, 1996.
10. Kaiser, N. & Squires, G., ApJ 404, 441, 1993.
11. Fahlman, G., Kaiser, N., Squires, G. & Woods, D., ApJ 437, 56, 1994.
12. Schneider, P. & Seitz, C., A&A 294, 411, 1995.
13. Seitz, C. & Schneider, P., A&A 297, 287, 1995.
14. Kaiser, N., ApJ 439, L1, 1995.
15. Schneider, P., A&A 302, 639, 1995.
16. Seitz, S. & Schneider, P., A&A 305, 383, 1996.
17. Squires, G. & Kaiser, N., preprint, 1996.
18. Seitz, C., Kneib, J.-P., Schneider, P. & Seitz, S., A&A, in press, 1996.
19. Schindler, S. & Wambsganss, J., preprint, 1996.
20. Squires, G. et al., ApJ 461, 572, 1996.
21. Bartelmann, M., Narayan, R., Seitz, S. & Schneider, P., ApJ 464, L115, 1996.
22. Tyson, J.A., Valdes, F., Jarvis, J.F. & Mills Jr., A.P., ApJ 281, L59, 1984.
23. Brainerd, T.G., Blandford, R.D. & Smail, I., ApJ, in press, 1996.
24. Schneider, P. & Rix, H.-W., ApJ, in press, 1996.
25. Griffiths, R.E., Casertano, S., Im, M. & Ratnatunga, K.U., preprint, 1996.
26. Schneider, P., MNRAS, in press, 1996.
27. Bonnet, H., Fort, B., Kneib, J.-P., Mellier, Y. & Soucail, G., A&A 280, L7, 1993.
28. Fort, B., Mellier, Y., Dantel-Fort, M., Bonnet, H. & Kneib, J.-P., A&A 310, 705, 1996.
29. Bartelmann, M. & Schneider, P., A&A 284, 1, 1994.
30. Smail, I., Ellis, R.S. & Fitchett, M.J., MNRAS 270, 245, 1994.
31. Luppino, G. & Kaiser, N., preprint 1996.
32. Fort, B., Mellier, Y. & Dantel-Fort, M., A&A, submitted, 1996.

PROBING GALAXY HALOS IN CLUSTER-LENSES FIRST RESULTS FOR AC114

PRIYAMVADA NATARAJAN
*Institute of Astronomy, Madingley Road,
Cambridge CB3 0HA, U.K.*

Weak shear maps of the outer regions of clusters have been successfully used to map the distribution of mass at large radii. The signature of galactic-scale substructure on the measured weak lensing signal has not yet been systematically studied. Two new methods to study the effect of bright cluster galaxies on the cluster weak shear field are proposed - aperture averaging of the local shear and a maximum likelihood method. We demonstrate using simulations that the observed local weak-shear on galaxy scales within the cluster can be used to statistically constrain the mean M/L of cluster members and their fiducial parameters. Preliminary results from the HST image of the cluster AC114 are presented.

1 Introduction

The observed gravitational lensing of faint background galaxies by foreground clusters is a promising probe of the detailed mass distribution within a cluster as well as on larger scales. Lens modeling of the mass distribution in clusters[1,2,3] suggests that the dark matter associated with individual galaxies is of consequence for consistency with the observed multiple images and the geometry of arcs. Galaxy evolution in clusters also indicates that redistribution of dark matter on smaller scales is likely to occur. Therefore, the issue of the possible existence and survival of dark halos associated with individual bright galaxies is an important one as it addresses the key question of whether the mass-to-light ratio (M/L) of galaxies is a function of the environment, and if it is indeed significantly different in the high density regions like cluster cores as opposed to the field.

2 Aperture averaging of the shear

In order to quantify the lensing distortion produced by the individual galaxy-scale components, a minimum number of parameters are used to characterize cluster galaxy halos. The newtonian cluster potential is modelled as a linear sum of a large scale (> 20 arcsec) component and a superposition of smaller scale 'perturbers'. In the 'weak' regime, the average shear is computed in an aperture centred on each bright cluster member assuming that the large-scale shear contribution from the global potential is constant or atmost linear over

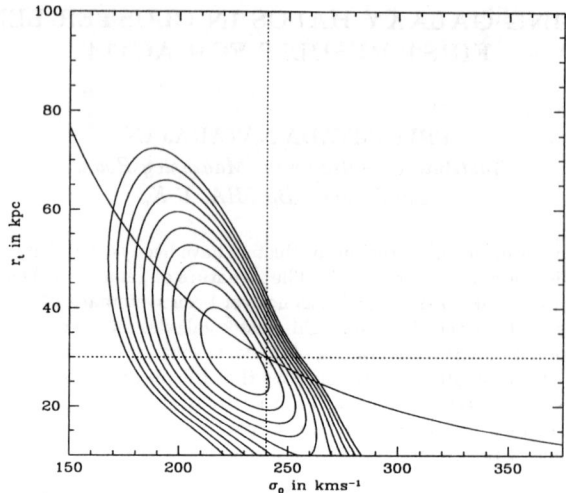

Figure 1: Log-likelihood contours for the retrieval of the fiducial parameters σ_{0*} and r_{t*} - the input values are indicated by the intersection of the dashed lines. The solid line maps the constant M/L curve.

the area of integration and which allows its efficient subtraction.

The strength of the extracted signal varies with (i) the input M/L of cluster galaxies (increasing with higher M/L), (ii) the shape of the redshift distribution of the sources, (iii) the redshift of the lens (with an optimum range for detection $0.1 < z_{\text{lens}} < 0.35$) and (iv) the total number of background sources. For a given value of the M/L (~ 4 in the R-band), a minimum number (~ 5000) of background galaxies are needed for a significant (3.5σ) detection. While the direct method does not require a precise knowledge of the smooth cluster mass distribution, it yields a bound on the statistical mean M/L ratio. It is however, not efficient in decoupling the dynamical parameters (σ_0 - the velocity dispersion and r_t - the truncation radius) that characterize the typical cluster galaxy, or in taking the over-critical region into account.

3 Maximum Likelihood Method

To break the degeneracy of the fiducial parameters and to take into account the ellipticity of the mass distribution of individual galaxies as well as the over-critical region, a maximum likelihood method is proposed. The likelihood of the source ellipticity distribution (measured from observations of faint field galaxies in the Medium Deep Survey[4]) is maximized. An example of the retrieved parameters from a numerical simulation are plotted in Fig. 1. The input M/L is recovered reliably, the halo size however, is not as well constrained as the velocity dispersion. This procedure requires however a precise knowledge of the strong lensing model. In terms of feasibility of application to data - the prospects are promising: stacking a minimum of 20 WFPC2 deep cluster fields would furnish the requisite number of background galaxies to infer down to an M/L of 4.

4 Application to HST data: a first attempt for AC114

The analysis developed in the previous sections was applied to HST data of the cluster AC114 (Smail, Ellis and Kneib 96) The extracted image catalog (from a nearly 5 arcmin X 3 arcmin field) contains ~3500 objects. Background galaxies were selected solely on the basis of their magnitudes; all objects with $23 < R < 26$ were classified as lensed background galaxies (~1500 objects satisfy the criterion). In order to delineate the cluster galaxies, a magnitude limit ($R < 21$) coupled with a luminosity cut-off was used ($L > 3L_*$). The results of the maximum-likelihood analysis are plotted in the Fig. 2. The log likelihood does have a well-defined albeit broad maximum at $\sigma_0^* = 160$ km/s ($\sigma_0^* > 250$ km/s is severely excluded) and $r_t = 85$ kpc. These observed maximum values translate into a typical mass for an L_* galaxy of $\sim 2.\,10^{10}\,M\odot$ or an average M/L ~ 10 in the R-band.

5 Conclusions and Prospects

Although the direct averaging cannot decouple the velocity dispersion and the halo size it yields a statistical mean M/L reliably and is easy to implement as it does not require knowledge of any details of the smooth mass distribution. The maximum likelihood approach permits independent estimation of the fiducial σ_{0*} and r_{t*} parameters, but requires knowledge of the smooth cluster mass distribution rather accurately. The methods presented here can be feasibly applied to real cluster data as demonstrated by the success of the first attempt with AC114. The extraction of the signal is reliable on stacking a minimum of 20 clusters, and this programme will be specially suited to using the Advanced Camera for Survey due to be installed on HST in 1999.

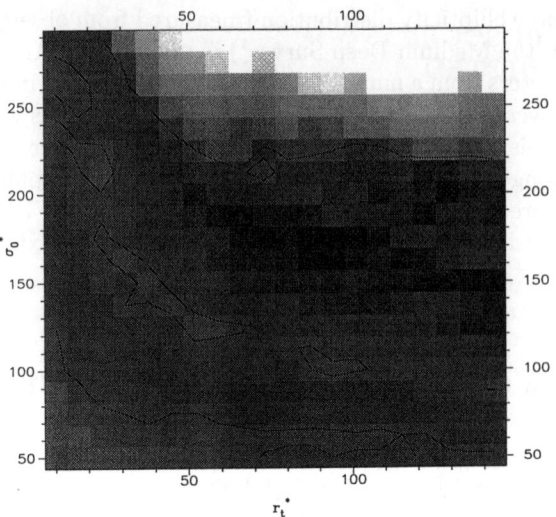

Figure 2: Maximum-likelihood retrieval of the fiducial parameters for the HST image of AC114. The maximum value is found at $r_t^* = 85$ kpc and $\sigma_0^* = 160$ km/s; and the solid curves correspond to the 2σ confidence limits.

Acknowledgments

Jean-Paul Kneib is acknowledged for a fruitful collaboration that helped develop the ideas into an implementable form and Martin Rees for support and encouragement during the enterprise. Richard Ellis, Ian Smail and Jean-Paul Kneib are thanked for allowing use of the HST AC114 data. Financial support from the Isaac Newton Studentship and Trinity College is gratefully acknowledged.

References

1. A. Kassiola and I. Kovner,ApJ, 400, 41, 1993..
2. I. Smail, A. Dressler, J-P. Kneib, R. S. Ellis, W. Couch, R. Sharples and A. Oemler, preprint,1996.
3. J-P. Kneib, R. S. Ellis, I. Smail, W. Couch and R. Sharples,ApJ in press, 1996.
4. T. Ebbels et. al, in preparation, 1996.
5. P. Natarajan, J-P. Kneib and I. Smail, in preparation, 1996.

SPECTROSCOPIC VERIFICATION OF REDSHIFTS PREDICTED BY GRAVITATIONAL LENSING

TIM EBBELS

Institute of Astronomy, Madingley Road,
Cambridge, CB3 0HA, U.K.

We present spectroscopic observations of 21 faint arcs seen in the HST image of the cluster Abell 2218 for which redshifts have been predicted using a lensing inversion technique based on the cluster mass model discussed by Kneib et al. (1996)[1]. This mass model is constrained by multiply-imaged sources seen in the HST image, 2 of which have known redshifts. Preliminary results indicate that the spectroscopic redshifts obtained agree well with predicted values and this gives us confidence that the lens inversion redshifts of even fainter arclets are similarly well constrained. These arclets are well beyond reach of current spectrographs. We conclude that the technique remains a highly promising method of studying the statistical properties of the faint field population.

1 Introduction and Observations

Multiply-imaged arcs in the cores of rich clusters offer a promising route to breaking the degeneracy between the mass distribution of the cluster and the distances of the background population. The location and orientation of these multiple images may be used to tie down the mass model in the core which can in turn be used predictively to investigate the properties of the faint population. For any faint galaxy within the core region, one may compare the predicted shear to the observed shape and orientation of the image, and derive a 'most likely' redshift for the object. The inversion technique was developed in detail by Kneib et al.[1] (hereafter KESCS) on the basis of an impressive image of Abell 2218 ($z=0.175$) obtained with the Hubble Space Telescope (HST). The superlative HST resolution was able to confirm the presence of no less than 7 sets of multiply images. Further, the fainter sheared images were individually identified with much greater confidence than was possible using ground-based data, enabling accurate inversion for a greater number of faint sources. Predicted redshifts were determined by for ~ 80 arclets in Abell 2218 to $R=25$, the brightest of which are just within reach of deep spectroscopy on 4m telescopes.

To verify the inversion proceedure, we have undertaken just such a program, and although only the bright end of the whole sample may be reached, this is a perfectly valid test, since the method relies only on the shape and position of each arclet. Objects were selected to have orientations close to the local shear direction, total magnitudes of $B < 25$ and $B - R$ colour at least one magnitude bluer than the cluster sequence. The observations were

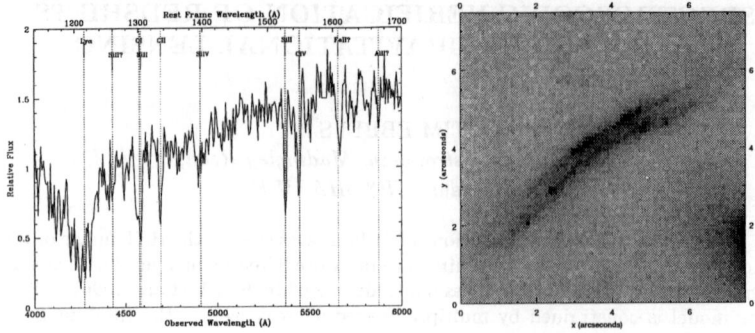

Figure 1: *left:* The LDSS-2 spectrum of the arc #384 showing both strong and more tentative line identifications, giving a redshift of $z = 2.515$. *right:* WFPC-2 F702W image of arc #384 (after KESCS) clearly showing mirror symmetry characteristic of multiple images.

conducted with the LDSS-2 multi-object spectrograph on the 4.2m William Herschel Telescope during 1995 May/June and 1996 June under varying conditions. Further limitations were imposed by a our adoption of a minimum slit length of 10 arcsec to enable good sky subtraction. We exposed 6 multi-slit masks, each containing 5-10 arclet candidates. First results show that 19 of the arclets surveyed gave satisfactory results, of which one (arc #384) was particularly interesting.

2   Arc #384

This arc had been inferred to be a multiple image on the basis of ground-based imaging, and the associated lensing model predicted a redshift of $z_{lensing} = 2.8 \pm 0.3$. The spectrum obtained from the 1995 May observing run is shown in figure 1 from which a redshift of $z_{spec} = 2.515$ was determined in spectacular confirmation of the lensing prediction. The spectrum reveals UV absorption lines characteristic of hot young stars and the ISM, as well as a particularly broad Lyman α absortion feature suggesting neutral gas is abundant within the galaxy. Both the spectral and photometric properties[3] show the source to be similar to that of sources found at higher redshift by Steidel et al. [2]

3   Uncertainties in Lens Inversion

Lens inversion does not suffer many of the biasses inherent in traditional proceedures. Its sources of error may be divided into four groups [1]. Firstly, an

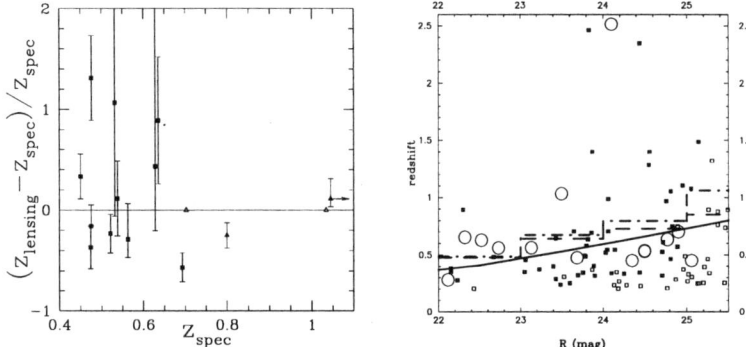

Figure 2: *left:* Fractional error in inversion redshifts for the arclets with determined spectroscopic redshifts. Model constraints and predictions are denoted by open and filled points while triangles and squares indicate multiple and single imaged systems respectively.*right:* The magnitude-redshift diagram of the sample (see figure 11 of KESCS) with our new spectroscopic results (open circles. Magnitudes have been corrected for lensing amplification.)

intrinsic uncertainty is that deriving from the unknown shape of the source; the inverted redshift corresponds only to the *most likely* source shape (derived from field survey images.) Secondly, the depth of the survey is limited by the extent to which image shape parameters can be accurately measured. Our survey extends to $R \simeq 25.5$ from an exposure of only $6.5ksec$ in R. Beyond this, errors in shape determination severely limit the accuracy of the method. Thirdly, uncertainties in the mass distribution will affect the redshift predictions. For A2218 it is estimated that residuals are present only on scales of $\lesssim 75kpc$ due to the high degree of constraints applied by the multiple images. Finally, contamination of the arclet sample by foreground and cluster galaxies dilutes the measured shear field. This leads to a reduction in the mean inverted redshift of the background galaxies. In the absence of any colour information from our single filter WFPC2 exposure to eliminate cluster galaxies in the sample, a statistical method was developed to account for contamination and correct the derived mean redshift accordingly.

4   Comparison with lensing predictions

Figure 2 (*left*) shows the fractional error in inverted redshift plotted against spectroscopic redshift from the preliminary data reduction. We attribute the lack of points above a redshift of about $z \simeq 0.7$ to the difficulty of retrieving the

$[OII]$ emission line from those areas of the spectra with high sky contamination and to the lack of emission features entering the optical window above a redshift of ~ 1. Due to the statistical nature of the method, we would naturally expect a distribution of points around the zero error line. This is certainly consistant with the figure, although the scatter is somewhat larger than might be expected. The inversion technique is an iterative one, so that as more redshifts are obtained, residuals can be used to refine the initial mass model, thus increasing the accuracy of the redshift predictions. We are therefore confident that the observed scatter may be reduced in this way (particularly by including the new redshifts of 2 multiply imaged systems.) Finally, we show the preliminary spectoscopic results on the magnitude-redshift plot of KESCS (figure 2 (*right*). It can readily be seen that the spectroscopic redshifts sample well the brighter end of the distribution.

5 Conclusions

First results from our campaign to investigate the accuracy of lens inversion show it to be a very promising method to investigate the properties of the faint field population. Due to the geometrical nature of this technique, the statistical properties of faint galaxies may be obtained to far fainter limits with deep imaging than with deep spectroscopy, even on 10m telescopes. It benefits from a more even selection function in redshift than approaches based on colour selection. However, in order to take out the effect of clustering in the background population one must average results from several cluster lenses.

Acknowledgments

Other collaborators in this project are: Richard Ellis, Jean-Paul Kneib, Roser Pelló, Jean-Francois Le Borgne, Ian Smail and Blai Sanahuja. We acknowledge Karl Glazebrook for the use of his *REDSHIFT* software, and Mike Breare at La Palma for ensuring a smooth run with LDSS-2.

References

1. Kneib, J.-P., Ellis, R.S., Smail, I.R., Couch, W.J. & Sharples, R., ApJ, , 1996, in press.
2. Steidel, C., Giavalisco, M., Pettini, M., Dickinson, M., Adelberger, K. L. ApJ, 462, L17, 1996.
3. Ebbels T.M.D., Le Borgne J.-F., Pelló R., Ellis R.S., Kneib J.-P., Smail I. & Sanahuja B., MNRAS, 281, L75, 1996.

REDSHIFT SURVEY OF GRAVITATIONAL ARCLETS IN ABELL 2390

J. BEZECOURT, G. SOUCAIL

Observatoire Midi-Pyrénées, 14 avenue E. Belin, 31400 Toulouse Cedex, France

1 Redshift and spectroscopic content of the arclets

Spectroscopic observations of gravitational arclets in cluster Abell 2390 were performed on August, 1995, at the CFHT with the MOS spectrograph, 7 exposures of 1 hour each were obtained with an average seeing of 0.9".

New redshift determinations are based mostly on the identification of the [O II] emission line at λ=3727Å in addition to the previous results of Pelló et al. (1991). The 9 arclets redshifts range from 0.398 to 1.268 confirming the trend observed in the whole sample of cluster lenses. Only one spectrum (#19) has no redshift identification, giving a success rate of 90% . This is important to note, as we can exclude any strong bias towards the high redshift tail of galaxies, regardless the fact that between $z \simeq 1.4$ and $z \simeq 2.2$, no strong emission line falls in the optical band for "normal" galaxies.

The redshift distribution of arclets (combined with the results of Le Borgne et al. 1991) indicate that clustering is present behind Abell 2390 in a plane at z=0.64 (4 objects) and in another one at z =0.90 (5 objects). The two planes are quite significant as they are spatially coherent: in both cases, all the identified objects fall inside a radius smaller than 1'.

2 Star formation rates

In the spectroscopic content of our sample of distant galaxies, among the 12 background objects listed in the table, 10 emit the [O II] line characteristic of star forming H II regions. A rough estimate can be made for high redshift objects using the relation (Kennicut, 1992): $SFR\,(M_\odot/yr) = 7\,10^{-12}L_BW_\lambda$, where L_B is the blue luminosity in solar units on the continuum ($\lambda_0 \simeq 4400$Å) and W_λ is the rest frame [O II] equivalent width. Corrections from the magnification factor were taken from the modelling proposed in Pierre et al. (1996).

The mean star formation rate among our sample of distant galaxies is $3\,M_\odot/yr$, with a factor 2 of uncertainty, a value of the same order of magnitude as for nearby blue galaxies. Starburst activity in high redshift galaxies seems to be more frequent than in the local universe but its intensity is comparable to nearby spiral galaxies, at least up to $z \sim 1$.

Table 1: Spectroscopic properties of arclets candidates in A2390, W_λ: equivalent width of [O II] in Å, B: blue magnitude, SFR: star formation rate in $M_\odot/yr$.

#	identified features	z	W_λ	B	SFR
14	[O II]	0.886	12		
16_1	[O II]	1.268	22	24.21	4
16_2	[O II]	1.082	20	24.13	3
17	[O II]	0.859	38	24.04	3
19		?		23.39	
20	[O II], no continuum detected	0.647	<100	> 26	
21	Mg II, [O II], [O III] 4959Å and 5007Å	0.643	44	23.48	3
22	[O II]	0.790	17	24.11	1
24_1	Ca II K, Ca II H, Hγ	0.631	0	} 23.00	
24_2	[O II], [O III] 5007Å	0.398	29		

3 Gravitational lensing by two planes at different redshifts

Two pairs of objects with different redshifts have been discovered and were used to derive an upper limit for the mass of the less distant one in each pair. The slit #16 contains two objects at $z = 1.082$ and $z = 1.268$ and in the slit #24 two objects at $z = 0.398$ and $z = 0.631$ are overlapping. By assuming a singular circular isothermal potential of velocity dispersion σ for the lensing galaxy (respectively #16_2 and #24_2) and using the modelling of Pierre et al. (1996) for the cluster A2390, we can derive an upper limit for σ (and consequently M/L_V) with the hypothesis that the axis ratio of the source b_S/a_S is greater than 0.5. Hence, for the object #16_2 $M/L_V < 29h_{50}$ which is not a strong constraint but for the object #24_2 $M/L_V < 12h_{50}$ which falls within standard values for spiral galaxies.

References

1. Kennicut R.C. 1992, ApJ, 388, 310
2. Le Borgne J.F., Mathez G., Mellier Y., Pelló R., Sanahuja B., Soucail G. 1991, A&A Supp. Ser., 88, 133
3. Pelló R., Le Borgne J.F., Soucail G., Mellier Y., Sanahuja B., 1991, ApJ, 366, 405
4. Pierre M., Le Borgne J.F., Soucail G., Kneib J.-P., 1996, A&A, 311, 413

SUBMILLIMETRE-WAVE GRAVITATIONAL LENSING

A.W. BLAIN

Cavendish Laboratory, Madingley Rd., Cambridge, CB3 0HE, UK

Gravitational lensing of distant sources, both by galaxies and clusters of galaxies, is an important phenomenon in observational cosmology,[1] and it can be used to investigate the properties of both the lensing galaxies/clusters and the distant lensed galaxies. Recent spectacular *Hubble Space Telescope (HST)* images of clusters[2][3] have shown the giant arcs and multiple images of distant galaxies in great detail. In the submillimetre (sub-mm) waveband, distant dusty star-forming galaxies are relatively much brighter, as compared with normal galaxies in the optical waveband[4], because of the strong negative K-corrections predicted by redshifting the steep spectrum of dust emission in the far-infrared waveband. The increased intrinsic brightness of distant sources in the sub-mm waveband produces a large increase in the surface density of lensed images as compared with observations in the optical waveband.

The gravitationally-lensed images of distant galaxies are predicted to be both much brighter and more numerous in the sub-mm waveband than in the optical waveband, independent of whether field galaxies[5] or clusters provide the lensing mass (Blain in prep.). In the case of galaxy–galaxy lensing, a significant fraction, of order 10% or more, of distant sub-mm–wave sources could be magnified strongly by galaxies along the line-of-sight.[5] The dramatic effects of the sub-mm K-corrections on cluster lensing are illustrated in Fig. 1, which contrasts the appearance of a cluster in the optical waveband with that at $850\,\mu$m in the sub-mm waveband. At $850\,\mu$m the surface density of bright lensed images is much larger, and the contribution from galaxies within the cluster is much smaller than in the R-band. The mass and redshift of the simulated cluster are chosen to match those of A2218[3], $\sigma_{\mathrm{v}} = 1370\,\mathrm{km\,s^{-1}}$ and $z = 0.171$. New large sub-mm interferometer arrays[6] could image the fields of clusters at the same angular resolution as the *HST*, but could detect a much larger number of arcs and multiple images.

1. R.D. Blandford and C.S. Kochanek in *Dark matter in the universe*, eds J. Bahcall *et al* (World Scientific, Singapore, 1987).
2. W.N. Colley, J.A. Tyson and E.L. Turner, ApJ, 461, L59, 1996.
3. J.-P. Kneib *et al*, ApJ, in press, 1996.
4. A.W. Blain and M.S. Longair, MNRAS, 279, 847, 1996.
5. A.W. Blain, MNRAS, submitted.
6. D. Downes, in *Astronomy with Millimeter and Submillimeter Wave Interferometry*, eds W. Wamsteker *et al* (Kluwer, Dordrecht, 1994).

R-band   850 μm

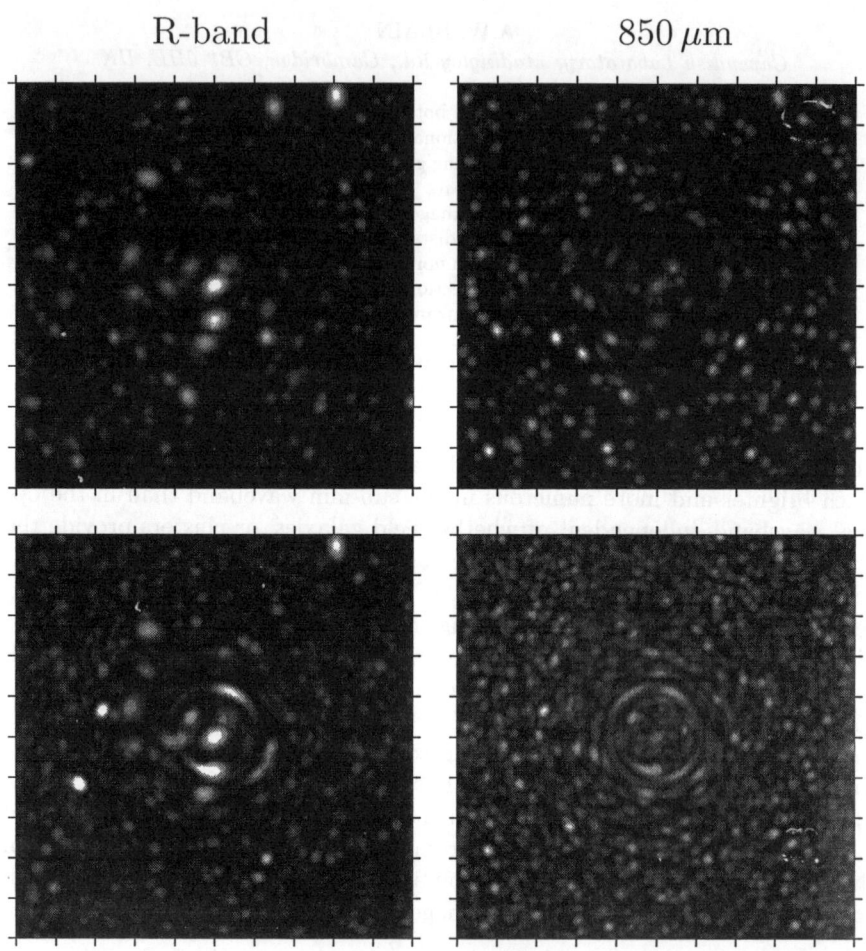

Figure 1: Simulated surface brightness distributions of a rich cluster of galaxies, such as Abell 2218 at a redshift $z = 0.171$, in the optical R-band (left) and at $850\,\mu$m in the sub-mm waveband (right). Both the emission from galaxies within the cluster and from lensed images is shown. Two models of the population of distant lensed galaxies are shown[4]: one described by an evolving *IRAS* luminosity function (top), and the other by an hierarchical model (bottom). The angular resolution of all the frames is 1 arcsec, and the axes are graduated at 20 arcsec intervals.

DISTRIBUTION OF MASS AND LIGHT IN HIGH-REDSHIFT E (LENS) GALAXIES

J. HJORTH

Institute of Astronomy, Madingley Road,
Cambridge CB3 0HA, England

J.-P. KNEIB

Observatoire Midi-Pyrénées, 14 Av. E. Belin,
F-31400 Toulouse, France

I. JØRGENSEN

McDonald Observatory, The University of Texas at Austin,
Austin, TX 78712, USA

We probe the distribution of mass out to more than 2 effective radii in two elliptical (E) galaxies at $z = 0.5$ and $z = 0.8$ that act as gravitational lenses and compare the ellipticity, extent and amount of dark matter (DM) inferred from detailed modeling with the observed distribution of luminous matter.

1 Introduction

Whereas the existence of DM in nearby spiral galaxies is well-established from their observed rotation curves, the problem of dark matter in E galaxies is still largely an unresolved issue. Gravitational lensing is a promising independent tool for probing the distribution of mass in galaxies that produce multiple images by acting as deflectors of light emitted from background quasars. HST provides images with sufficient spatial resolution for studying the lens galaxies and so allows detailed modeling. We here report on a study of two such systems—the E galaxies responsible for producing the 'double quasar' UM673[1] and the quadruple lens system HST14176+5226=CFRS14.1311[2,3].

2 Elliptical Lens Galaxies

The gravitational lens system UM673 consists of a quasar at $z = 2.727$ split into two images separated by $2\rlap{.}''2$ by an intervening galaxy at $z = 0.493$. The images were obtained with the WFPC2 camera in two filters, F555W and F675W. The two point sources were subtracted using a model for the point spread function (PSF). The galaxy is oriented almost perpendicular to the line joining the two images. The morphology (elliptical), surface photometry ($R^{1/4}$ law) and colour ($V - R = 1.25$) suggest that the lens is an E galaxy.

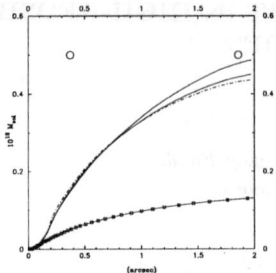 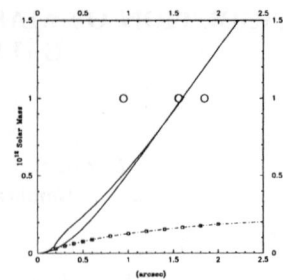

Figure 1: Enclosed mass (in units of $10^{12} M_\odot$ ($h = 0.5, q_0 = 0.5$) for various models; solid or dashed-dotted curves) and light (in units of $10^{12} L_{\odot B}$; curves through points) for UM673 (left) and HST14176+5226 (right). The slope of the mass profile is well constrained at the indicated locations (circles) of the multiple images.

The derived effective (half-light) radius is $0.''75$ compared to the distance from the lens center to image A of $1.''9$. Thus, A constrains the mass out to 2.3 R_e.

The HST quadruple lens system consists of an $I = 19.5$ E galaxy at $z = 0.81$ lensing a blue source at $z = 3.4$, presumably an AGN. The effective radius has been determined to $1.''20$, compared to the Einstein radius of $1.''6$. We find that the surface-brightness profile is well-fit by an $R^{1/4}$ law.

3 Modeling

We have recently proposed a mass model [4] which reproduces the $R^{1/4}$ law for any ellipticity, and is analytically tractable in its lensing properties. For UM673, the centres, ellipticities, orientations, effective radii, and profile shapes of the light and mass are found to be consistent (Fig. 1a), i.e., mass follows light in this system with a small inferred mass-to-light ratio of $6h$ in solar units (in restframe B). For HST14176+5226 we also find that the orientation and ellipticity of the matter distribution is consistent with the distribution of the light. However, in Fig. 1b we see that the DM is more extended than the light, and that the mass-to-light ratio increases with radius. Thus we have detected a significant amount of DM out to a radius of 1.5 R_e in this galaxy.

References

1. J. Surdej et al., Nature, 329, 695, 1987.
2. K. A. Ratnatunga et al., ApJ, 453, L5, 1995.
3. D. Crampton et al., A&A, 307, L53, 1996.
4. J. Hjorth and J.-P. Kneib, ApJ, submitted, 1996.

STRONG AND WEAK GRAVITATIONAL LENSING IN THE HUBBLE DEEP FIELD

DAVID W. HOGG, ROGER BLANDFORD, C. D. FASSNACHT, TOMISLAV
KUNDIĆ
California Institute of Technology, mail code 130-33, Pasadena CA 91125

T. G. BRAINERD
Boston University

SANGEETA MALHOTRA
Infrared Processing and Analysis Center; and Jet Propulsion Laboratory

Cosmologically distant galaxies ought to act as multiply imaging (i.e., "strong") gravitational lenses for a fraction $\sim$ 0.002 to 0.005 of background sources[1,2,3,4] so in the Hubble Deep Field[5] (HDF), $\sim$ 3 to 10 cases of multiple imaging are expected. The number of multiple image systems constrains the redshift distribution of the very faint sources observed in the HDF relative to other populations for which both the lensing rate and the redshift distribution are better known, such as flat-spectrum radio sources[6,7]. The lens rate can be used in this comparative way to constrain the redshift distribution of sources at much fainter levels than the limits of current spectroscopic surveys. We have begun a partially automated search derived from that used in the CLASS survey[7] for multiply imaged sources. The two best candidates so far, selected by eye, are red ellipticals with adjacent, tangentially sheared arcs, HDF J123652+621227 and HDF J123656+621221 (the names specify the positions in the field). The former may be multiply imaging multiple sources at different redshifts[8]. The total lensing rate will be known once the candidate selection procedure has been automated and all candidates which are bright enough have been followed up spectroscopically.

A measurement of a coherent polarization among faint galaxy images due to "weak" gravitational lensing by large-scale structure can be used to constrain both the redshift distribution of the galaxies and the size, nature and mass of structure in the Universe[9]. From a catalog of 1500 objects detected in the $F606W$-band images of the HDF we place an upper limit on any mean image polarization in the HDF consistent with a previous null result[10]. Splitting the sample by median apparent flux and by median color, apparently significant polarizations are found for the fainter and bluer halves of the catalog. If the polarization is due to lensing, this suggests that fainter and bluer objects are, on average, at higher redshifts than the brighter or redder objects.

The measurements are on the right order of magnitude for standard structure formation scenarios and source redshifts around 1 to 2^9, but there may be systematic distortions in the HDF images which could produce the signal.

The masses of field galaxies can be measured to large radius, in a statistical way, by measuring the average tangential shear among background sources induced by gravitational lensing[11]. An apparently significant detection of this shear is found in the HDF, at the level expected, when spectroscopic redshifts (Hogg et al, this proceedings) are used to identify luminous foreground objects. However, concerns about systematic errors remain. Similar results are reported when photometric redshifts (Hudson, this proceedings) or fluxes[12] are used to differentiate foreground and background, although in the latter case a positive signal can only be found at radii $<$ 5 arcsec where contamination by tidally distorted companions is possible.

The HDF in itself is too small for extensive weak lensing projects but shows what might be accomplished with much larger imaging surveys with HST in the future.

Acknowledgements

We thank the Hubble Deep Field team, led by Bob Williams, for planning, taking, reducing, and making public the images of the HDF. We benefited from helpful conversations with and assistance from Lee Armus, Judy Cohen, Len Cowie, Mark Dickinson, Mauro Giavalisco, Richard Hook, Jordi Miralda-Escudé, Lexi Moustakas, Gerry Neugebauer, Peter Schneider and Chuck Steidel.

1. Turner, E. L., Ostriker, J. P. & Gott, J. R., 1984, ApJ, 284, 1
2. Blandford, R. D. & Narayan, R., 1992, ARA&A, 30, 311
3. Miralda-Escudé, J. & Lehar, J., 1992, MNRAS, 259, 31P
4. Schneider, P., Ehlers, J., and Falco, E. E., 1992, *Gravitational lenses,* Springer-Verlag, New York
5. Williams, R. et al, 1996, AJ, in press
6. Patnaik, A. R. et al, 1992, MNRAS, 254, 655
7. Myers, S. T. et al., 1995, ApJ, 447, L5
8. Hogg, D. W. et al, 1996, ApJ, 467, L73
9. Blandford, R. D. et al, 1991, MNRAS, 251, 600
10. Mould, J. et al, 1994, MNRAS, 271, 31
11. Brainerd, T., Blandford, R. & Smail, I., 1996, ApJ, 466, 623
12. Dell'Antonio, I. P. & Tyson, J. A., 1996, ApJ, submitted

GALAXY–GALAXY GRAVITATIONAL LENSING IN THE
HUBBLE DEEP FIELD

MICHAEL J. HUDSON & STEPHEN GWYN

Dept. of Physics & Astronomy, University of Victoria,
P.O. Box 3055, Victoria V8W 3P6 Canada

We have detected in the *Hubble Deep Field* the gravitational lensing of background galaxies by foreground galaxy haloes at the 99.97% confidence level. We use photometric redshifts to separate foreground and background galaxies.

A mass concentration, such as a cluster, will distort the shapes of background galaxies by gravitational lensing. Recently, Brainerd, Blandford & Smail (1996, preprint) detected lensing by individual galaxies using deep, ground-based, two-colour photometry. However, the precision of the method is limited because the distances to individual lens and source galaxies are unknown.

We have examined galaxy-galaxy lensing in the *Hubble Deep Field*, making use of photometric redshifts derived from four-colour photometry (Gwyn & Hartwick 1996, preprint; see also Gwyn, this volume). Photometric redshifts are available for 541 galaxies with $I_{ST} < 28$. The accuracy of the photometric redshifts (as determined from a direct comparison with spectroscopic redshifts) is approximately 0.5 in z for $z > 1.5$, which is sufficient to discriminate foreground and background galaxies and to determine angular diameter distances.

The signature of galaxy-galaxy lensing is an excess of tangentially-oriented source galaxies around massive foreground lens galaxies. Figure 1a shows the histogram of position angles of the major axis of the source galaxies (measured with respect to the line joining the lens and the source). The selection criteria for the lens and source samples are given in the caption. For the case of no lensing a uniform distribution of angles is expected; this is rejected at the 99.97% confidence level. The control histogram of the orientation of the major axis of lens galaxies shows no preference for tangential orientation, as expected. This lack of signal in the control sample rules out a number of potential systematics, e.g. uncorrected geometric distortion in the HDF. Furthermore, it shows that the photometric redshifts are sufficiently accurate to separate foreground and background galaxies.

We are currently performing a maximum-likelihood analysis similar to that proposed by Schneider and Rix (1996, preprint) in order to determine galaxy halo circular velocities and the Tully–Fisher/Faber–Jackson relation of galaxies at $z \sim 2$.

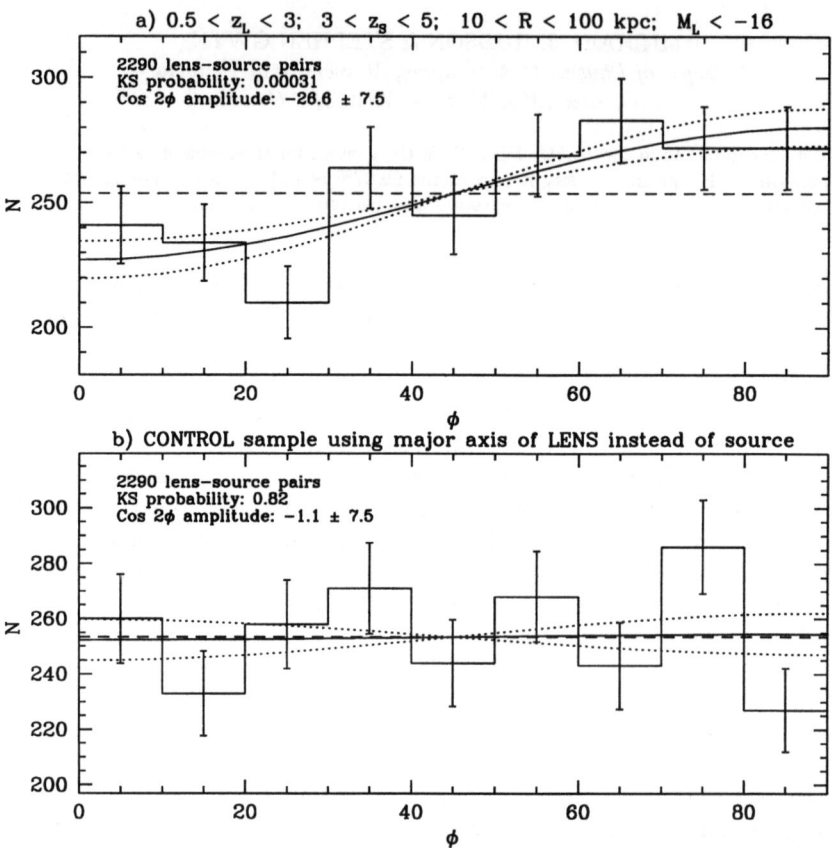

Figure 1: a) Histogram of the angle, ϕ, between the major axis of the source galaxies and the line joining the lens and source galaxies for all lens-source pairs. Lens galaxies are defined as all galaxies with $M < -16$ and $0.5 < z_{phot} < 3$. Source galaxies are all galaxies with $3 < z_{phot} < 5$ within a projected distance of $10 - 100$ kpc of a lens and with $z - z_{lens} > 0.5$. There are 2290 lens-source pairs. If there were no lensing signal we would expect a uniform distribution of ϕ. This is rejected by a KS test at the 99.97% confidence level. The solid and dotted curves show the $\cos 2\phi$ fit and errors, respectively. b) The same pairs are used but now we use the major axis of the lens galaxies as a control sample. There is no $\cos 2\phi$ variation and it is perfectly consistent with a uniform ϕ distribution.

OPTICAL EINSTEIN RINGS AS PROBES OF THE HIGH-REDSHIFT UNIVERSE

[1]G.F. LEWIS, [1]P.C. HEWETT, [2]S.J. WARREN, [3]P. MØLLER AND [1]J.P. WILLIS

[1] *Institute of Astronomy, Madingley Road, Cambridge. UK*
[2] *Imperial College, London. UK*
[3] *Space Telescope Institute, Baltimore. USA*

Recently, a z=3.6 star-forming galaxy was identified, lensed into a ring by a z=0.5 elliptical galaxy. Further studies have demonstrated that these Einstein rings are powerful probes of the properties of high-redshift, star-forming systems, and provide a non-dynamical determination of the mass in the deflector population at $z \sim 0.5$. We present a review of current study and outline future projects.

The UKIRT spectrum shown here confirmed the high redshift nature of the $1''.35$ radius ring structure observed in a $z = 0.485$, massive elliptical galaxy[3]. Simple modelling reveals this ring structure is consistent with the action of gravitational lensing. The background source is amplified by a factor of ~ 17 and, without the action of gravitational lensing, would have a scale-size of $0''.5$, and $m_v \sim 27$. The Ly_α flux implies a star-formation rate of $\sim 4M_\odot$ yr^{-1}, comparable to other high redshift systems[2].

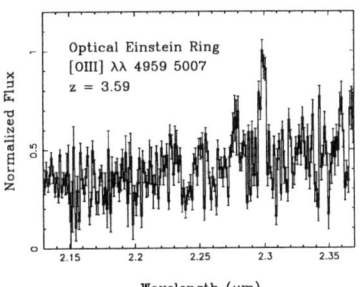

HST cycle 6 observations are scheduled, allowing a more accurate determination of the lensing model. This will be coupled with observations of spatially resolved spectra using ARGUS on the CFHT, left, to probe the kinematics in the source galaxy. The resolution of such observations will increase by ×10 with the advent of STIS on HST. The sky surface density of optical Einstein rings is calculated to be $\sim 20/\Box'$, down to a B$\sim 24^l$. These should be readily identifiable in multi-fibre spectroscopic surveys, such as 2dF,

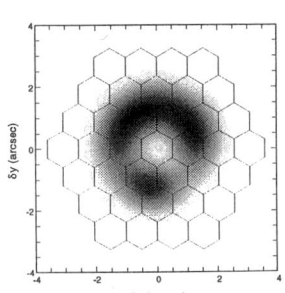

providing a valuable probe of high redshift star-forming systems, and a non-dynamical mass estimate of galaxies at $z \sim 0.5$.

1. J. Miralda-Escudé and J. Lehár, MNRAS, 259, 31, 1992.
2. C. C. Steidel et al, ApJ, 462, 17, 1996.
3. S. J. Warren et al, MNRAS, 278, 139, 1996.

KECK SPECTRA OF "LENSES" IN THE HUBBLE DEEP FIELD

L.A. MOUSTAKAS, S.E. ZEPF, & M. DAVIS

Department of Astronomy, University of California, Berkeley, CA 94720

The frequency of gravitational lensing by individual galaxies can set constraints on models of galactic evolution and on cosmological parameters. With LRIS on Keck, we have obtained spectra of three of the strongest lensing candidates in the Hubble Deep Field (HDF). We conclude that two of the three candidates are almost certainly not lenses, and that the lensing frequency is extremely low.

1 The Project

Gravitational lensing by individual galaxies provides constraints on the mass of the lensing galaxy, and on the numbers of background galaxies. We selected the three likeliest lens-like candidates based on morphology (Zepf et al. 1996), and included them in a modest Keck/LRIS[a] program on 1996 March 14 UT in which we targeted more than thirty galaxies in the HDF+Flanking Fields (Moustakas et al. 1996). We used two slitmasks, oriented at PAs of 61° (PA61; 80 minutes) and 19° (PA19; 60 minutes).

2 J123641+621204 (L4.1)

This system was observed in both orientations. We confirm the Lyα identification of Steidel et al. (1996) at $z \approx 3.22$. Our data show two distinct Lyα components ($\Delta v_{rest} \approx 800$ km s^{-1}) which are spatially separated by about 0.5″ in the PA19 spectra. *This rules out gravitational lensing* as the cause of the characteristic morphology of this system. The detection of strong NV is suggestive of an AGN. Therefore, L4.1 appears to be composed of two or more starforming galaxies with AGN, separated by roughly 10 kpc and 800 km s^{-1}.

3 J123652+621227 (L3.1)

From **W** to **E**, our slit crossed the "arc," the "elliptical," and the "counterimage." At the position of the counterimage we detect an emission line at 5301Åwith no continuum. Assuming this to be Lyα, which is consistent with the object's colors, the redshift is $z = 3.36$. At the position of the arc, a very

---

[a] Based on observations obtained at the W. M. Keck Observatory, which is operated jointly by the California Institute of Technology and the University of California.

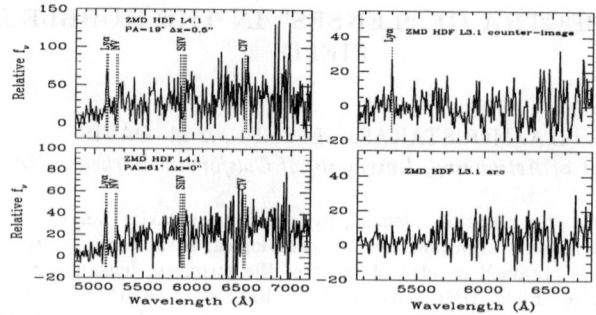

Figure 1: Spectra of the lens-like J123641+621204 (L4.1) (left column); and spectra of the "counter-image" and the "arc" of the lens-like system J123652+621227 (L3.1) (right column).

faint blue continuum is detected with *no emission line*. Furthermore, the counterimage is a F300W dropout (ie $z \gtrsim 2.5$), while the arc is not. Most likely, then, they are *at different redshifts*, making a lensing hypothesis implausible.

4   Conclusions

Based on our spectra, it appears that two of the three targets we observed are *almost certainly not* gravitational lens systems. The third target remains a viable lens candidate. These results imply that to a limiting magnitude of roughly $F814W_{AB} < 27$, there are one, or possibly zero, cases of multiply-imaged lens systems in the HDF. With ~ 750 objects with $F814W_{AB} < 27$, the lensing rate is $\sim 0.000 - 0.002$. Models where the large numbers of faint blue galaxies are explained as a normal galaxy population in a cosmology with a significant cosmological constant are at odds with this measurement. With continuing work in the HDF, the statistics on the lensing probability may better constrain other models for the faint blue galaxies.

References

1. L.A. Moustakas, S.E. Zepf and M. Davis 1996, in preparation
2. C. Steidel, et al., AJ, 112, 352, 1996.
3. S.E. Zepf, L.A. Moustakas and M. Davis, Astrophys. Lett., , , 1996.

HST WFPC & NEAR-IR STUDY OF THE CLUSTER-LENS ABELL 2390

R. PELLÓ[1], J.-P. KNEIB[1], A. ARAGÓN-SALAMANCA[2], R.S. ELLIS[2], B. FORT[3], J.-F. LE BORGNE[1], Y. MELLIER[13], J. MIRALDA-ESCUDÉ[4], J.-M. MIRALLES[1], J.-P. PICAT[1], I. SMAIL[5], G. SOUCAIL[1]

[1] *UMR 5572, Observatoire Midi-Pyrénées, 14 Av. Edouard-Belin, 31400 Toulouse, France*

[2] *Institute of Astronomy, Madingley Road, Cambridge CB3 0HA, UK*

[3] *Observatoire de Paris, 61 Avenue de l'Observatoire, 75014 Paris, France*

[4] *Institute for Advanced Study, Olden Lane, Princeton NJ 08540, USA*

[5] *Physics Department, University of Durham, South Road, Durham DH1 3LE, UK*

We present a highly-constrained model for the mass distribution in A2390, based on HST-WFPC2 images in two filters, F814W (I_W) and F555W (V_W), and also near-IR J and K images. Striking new features have been detected in the cluster core, leading to the identification of multiple-image and counter-image candidates. The brightest arcs are resolved and show complex internal features. The tangential shear-map has been obtained for different populations of background galaxies, selected according to their colors. Figures can be found at $http://www.obs-mip.fr/omp/umr5572/cosmolens/pub.html$

1 Summary of relevant observations

A2390 ($z = 0.231$) was observed by the HST WFPC-2 camera in filters V_W and I_W. Exposures were shifted relative to each other by 1".4 to 3" providing a good overlap between the chip fields. After IRAF/STSDAS pipeline processing, the final frames have an effective resolution of 0".14, with a common field of 2'.5 (see Fig.). Photometry and morphological parameters were obtained with both the *Amaphot* and *Sextractor* packages. Besides, deep near-IR J and K' images were obtained with the *Redeye* camera at the CFHT (Miralles et al., in prep.) (sampling 0".5/pixel). The spectroscopic redshifts of arc(let)s comes from Pelló et al. (1991) (the main "straight arc" and the D object, both at z=0.913) and Bézecourt & Soucail (1996) and Bézecourt (this Conference). X-ray ROSAT/HRI observations can be found in Pierre et al. (1996).

275

2 Modelling the mass-distribution and results

The method employed to compute the mass-distribution is similar to that used by Kneib et al. (1996) (KESCS) for the HST-based model of A2218. The idea is to use all the possible constrains coming from multiple-images and the $r \leq 500kpc$ shear-map to fit the shape parameters of the different mass-components (position, ellipticity, orientation, core radius and central velocity dispersion) through a χ^2 minimization procedure. The cluster-scale mass-components correspond to the brightest cluster galaxies, leading to a bimodal mass-distribution in this case: the main one is associated to the central cD and a secondary one corresponds to an enhancement in the 2D density of galaxies. Galaxy-scale components are also included, whith shapes following the distribution of light in the parent galaxies, to take into account the effects of the brightest cluster-members ($I_W \leq 20$). The other mass parameters (central velocity dispersion and truncature radius) are scaled to the luminosity of the galaxy and included in the minimization procedure.

A large scale mass-distribution is needed to reproduce the lensing features, in addition to the galaxy-scale components. This result is similar to previous modelling of cluster-lenses containing cDs.

The $V_W - I_W$ color gives a first discrimination between cluster and field objects. We have computed the tangential shear-map for 3 different samples: i) red population ($V_W - I_W \geq 1.7$ and $I_W \geq 22.0$, with predicted $0.23 \leq z \leq 1.1$), ii) the blue population ($V_W - I_W \leq 1.0$ and $I_W \geq 22.0$, with $z \geq 1$ or foreground), and iii) the "cluster-compatible" population ($1.0 \leq V_W - I_W \leq 1.7$, $I_W \geq 23.0$). The blue one exhibits a stronger shear than the other two at any distance, whereas the red sample tends to decrease steeply than the bluer with the radial distance. This result could be interpreted as a difference in the averaged redshift of these samples, the bluer being more distant than the redder. Using the mass model and the inversion-method by Kneib et al. (1994), we have derived the lensing-redshift for a large number of arclets as well as the most likely redshift distribution for the faint galaxies down to $I_W=25$. In a forthcoming paper (Kneib et al. in preparation) we compare these results with those already obtained by KESCS in the field of A2218.

3 References

1. Kneib J-P., Ellis R.S., Smail I.R., Couch W.J., Sharples R., ApJ in press.
2. Pelló R, Le Borgne JF, Soucail G, Mellier Y, et al. ApJ, 366, 405, 1991.
3. Pierre M, Le Borgne JF, Soucail G, Kneib JP, A & A, 311, 414, 1996.

THE GIANT PROTO-GALAXY cB58; AN ARTIFACT OF GRAVITATIONAL LENSING?

L.L.R. WILLIAMS

Institute of Astronomy, Madingley Rd., Cambridge CB3 0HA, UK

G.F. LEWIS

State University of New York at Stony Brook, Department of Earth and Space Sciences, Stony Brook, NY 11794-2100, USA

The proto-galaxy, cB58, was discovered in the Canadian Network for Observational Cosmology (CNOC) survey of cluster redshifts. Its high absolute magnitude, $M_V = -26$, implies a large star-formation rate of several 100 $M_\odot$ yr^{-1}, making it the most "active" star-forming galaxy. This proto-galaxy lies $\sim 6''$ away from a cD galaxy in a foreground cluster. The X-ray properties of the cluster suggest that its mass, and therefore its lensing potential, could be greater than that found using a virial analysis. We argue that the phenomenal properties of this proto-galaxy could be due to the gravitational lensing effect of the foreground cluster, and the unlensed properties of the source are typical of high-redshift star-forming systems.

We show that if the lensing cluster is just subcritical, and has a core, then the single lensed image of a background extended source, if projected roughly within the cluster core, will be highly magnified, yet almost completely undistorted. We used a non-singular isothermal sphere (see Schneider et al. 1992) model for the galaxy cluster. Figure 1 plots contours of image magnification and distortion as a function of two cluster parameters, central normalized surface mass density, and velocity dispersion. Note that the definition of e, and hence Δe are different from conventional (see Williams & Lewis 1996). Figure 2 shows an example of highly magnified, undistorted image. Even though our model is highly non-unique, the parameters assumed here could be similar to those of cB58.

References

1. Carlberg, R. G., Yee, H. K. C., Ellingson, E., Abraham, R. G., Gravel, P., Morris, S., and Pritchet, C. J., ApJ, 462, 32, 1996.
2. Schneider, P., Ehlers J. and Falco E. E. , 1992, Gravitational Lenses, Springer-Verlag Press
3. Williams, L. L. R., & Lewis, G. F., MNRAS, 281, L35, 1996.
4. Yee, H. K. C., Ellingson, E., Bechtold, J., Carlberg, R. G. & Cuillandre, J. C., AJ, 111, 1783, 1996.

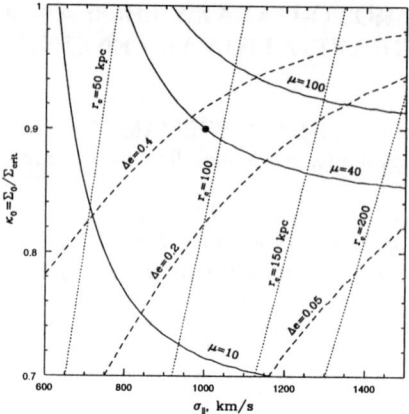

Figure 1: Contours of constant magnification μ (solid lines), ellipticity change Δe (dashed lines), and cluster core radii, r_c (dotted lines) for a range of cluster central surface mass densities, κ_0, and line of sight velocity dispersion, $\sigma_{\parallel}$. The redshift of the source and cluster, and the cluster-image separation are same as observed in the cB58 case. Images with ellipticity changes $\Delta e \lesssim 0.6$ appear undistorted. The dot at $\sigma_{\parallel} = 1000 \mathrm{km\ s^{-1}}$ and $\kappa_0 = 0.9$ gives the cluster parameters used in Figure 4.

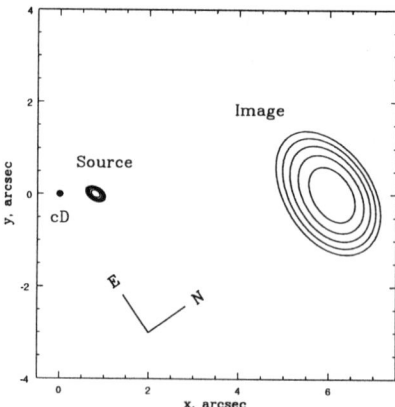

Figure 2: An example of a source/image geometry with $\sigma_{\parallel} = 1000 \mathrm{\ km\ s^{-1}}$, and $\kappa_0 = 0.9$. Cluster core radius r_c is 92 kpc, while its mass is $1.6 \times 10^{15} M_{\odot}$. Note that the isophotes are undistorted ($\Delta e = 0.35$) in spite of a rather large magnification $\mu = 38.5$. The source is located 0.8″ from the central cD and has an orientation of $\sim 115°$, semi-major axis of 0.24″, and axis ratio of 1.6. The resultant image is aligned eastward, has a semi-major axis of 1.5″, and an axis ratio of 1.5. Compare to Figure 1, which is taken from Yee et al. (1996).

A SEARCH FOR OPTICAL EINSTEIN RINGS: GALAXY SELECTION

J.P. WILLIS, P.C. HEWETT, G.F. LEWIS

Institute of Astronomy, Madingley Road, Cambridge, CB3 OHA, England

S.J. WARREN

Imperial College, Prince Consort Road, London SW7 2BZ, England

Optical Einstein rings are a probe of the mass distribution of lensing galaxies and the spatial emission line structure of lensed galaxies. Warren *et al*[1] reported the discovery of the first optically selected Einstein ring system from a sample of 154 early-type galaxies . The aim of the current search is to extend the spectroscopic sample to ~2,000 early-type galaxies. Based on the results of the original survey, up to 10 new ring systems may be identified. This contribution considers the effectiveness of the survey's photometric selection criteria at selecting early-type galaxies in the desired redshift interval.

The survey is designed to select bright early-type galaxies in the redshift interval $0.3 < z < 0.5$. Such galaxies, with large central mass concentrations, possess large cross-sections for strong gravitational lensing, and their space density is high enough to produce a large sample. Deflectors in the redshift interval $0.3 < z < 0.5$ provide the most advantageous geometry for lensing background sources at $z > 1.5$[2].

Candidate galaxies are selected from APM scans of UKST plates taken in B_J, R and I. The photometric selection criteria applied to these data are $16.4 < i < 18.85$, $b_j - r > 2.15$, $r - i < 1.05$. The survey to date covers 5 Schmidt fields, a total area of $\sim 150\,\mathrm{deg}^2$. Spectra of the candidate galaxies may then be most efficiently obtained using a wide-field, multi-fibre spectrograph such as the 2dF instrument on the AAT.

A simulation was developed to test the effectiveness of the photometric parameters employed in the survey. A model universe, characterised by the cosmological parameters $h = 0.5$ and $q_0 = 0.5$ was populated by galaxies of three morphological classes (E/S0, Sab, Scd), each class with its own SED, luminosity function and evolutionary behaviour. The results of applying the selection criteria to this simulation indicate that ~80% of the galaxies selected are E/S0s in the specified redshift interval.

1. S.J.Warren *et al*, MNRAS, 278, 139, 1996.
2. S.J. Warren *et al*, *'Identifying Optical Einstein Rings'*, p329 IAU Symposium 173 'Astrophysical Applications of Gravitational Lensing' Eds. C.S Kochanek & J.N. Hewitt, Kluwer 1996.

SECTION 4

QUASAR ABSORBERS

QUASAR ABSORPTION LINES: REFLECTIONS AND VIEWS

ALEC BOKSENBERG

Institute of Astronomy, Madingley Road, Cambridge, CB3 0HA, UK

Building on the considerable foundation work of the seventies and eighties, current progress in understanding the quasar absorption lines through the recent advent of facilities yielding real and virtual data of superb quality has catapulted the topic to the vanguard of observational cosmology.

1 Introduction

I was introduced to quasar absorption lines by Wal Sargent at the Hale Telescope in 1973, on the first of many observing runs together. Although I had been working as an astronomer, in ultraviolet astronomy, for some years this run with Wal was a watershed for me – he made clear what astronomy truly was about and it is only from that time, the start of a long and (at least for me) happy association, that I could begin to call myself an astronomer. The first object we observed was PKS 0237-23, known to show absorption lines in low dispersion spectra. We had set up the IPCS at the coudé focus and this was its first real trial. Among its electronics racks was a display screen indicating the photon counts accumulating in the computer memory. It was then a very slow display, with a single spot repetitively stepping through the high dispersion spectrum, taking some minutes for each pass. This both seemed interminable and made it difficult to see any pattern in the spectrum; Wal reserved his position on what he thought we were getting. Later I sent him a graphical printout showing well-resolved complex structure in the hydrogen and metal absorption lines. His telexed response "I am tremendously impressed" (from Wal I took this to be the ultimate in praise) is something I have treasured ever since.

The question of whether the majority of the narrow absorption lines observed in quasar spectra represent cosmologically distributed objects then was controversial, but today has been long settled. Realising their initial promise, quasar absorption lines now are established as important indicators of the content and structure of the Universe at observed redshifts up to nearly 5. In recent years research in this field has been given an enormous boost through the use of the Hubble Space Telescope and the Keck Telescope and from advances in theoretical studies and detailed numerical simulations; to quote David Tytler "these are superb times to be an observational or experimental astronomer". This new research has revealed, not surprisingly, that some of the earlier models for the absorbing objects are somewhat naive. Here I will touch on a few aspects which currently interest and impress me but avoiding some which are dealt with by others at this Conference.

2   The Lyman Forest

The dense array of high redshift H I absorption lines in systems having no detectable metal lines, first noted by Lynds (1971) and now known as the Lyman forest, were studied in detail by Sargent et al. (1980) who concluded that these lines arise from cosmologically distributed intervening material. An important finding was no significant clustering on all scales from 300 to 30000 km s^{-1}, contrasting with metal line systems which were seen to be strongly clustered on small scales. They attributed the Lyman forest lines to a population of intergalactic clouds not associated with galaxies, distinct from the metal line systems which probably arose in galactic haloes (Bahcall 1975), and determined that the forest lines are photoionised by the integrated quasar light and are pressure confined by a warm general intergalactic medium. From an expanded sample of Lyman α lines Young, Sargent & Boksenberg (1982) found that the population was evolving in the sense that there are more absorption lines at higher redshifts.

The Lyman forest systems (Figure 1) provide an excellent probe of the high-redshift Universe, being observable relatively without bias, outnumbering other detectable tracers of structure and indicating their initial conditions more sensitively and probably more directly than highly non-linear objects like galaxies. Following the early idea of pressure-confined clouds (Sargent et al. 1980; Ikeuchi & Ostriker 1986), among several other suggestions convincing analytical attempts were made to describe the formation of the absorbers as a natural consequence within the general gravitational instability theory describing the formation of cosmic structure (Rees 1986; Ikeuchi 1986; Bond, Szaley & Silk 1988). Recently, massive and detailed high-resolution numerical hydrodynamic simulations of gravitational collapse on small scales in normalised cold dark matter dominated models with a photoionising ultraviolet background which yield direct quantities for comparison with observations have been performed by several groups (Cen et al. 1994; Zhang, Anninos & Norman 1995; Hernquist et al. 1996; Miralda-Escudé et al. 1996; Weinberg et al. 1996). In these simulations it is straightforward to compute the neutral hydrogen absorption that would be produced in the light of a background quasar along an arbitrary line of sight through the simulation volume. Although different numerical methods were used by the different groups the results are very similar in character and predict (with an appropriately assumed radi-

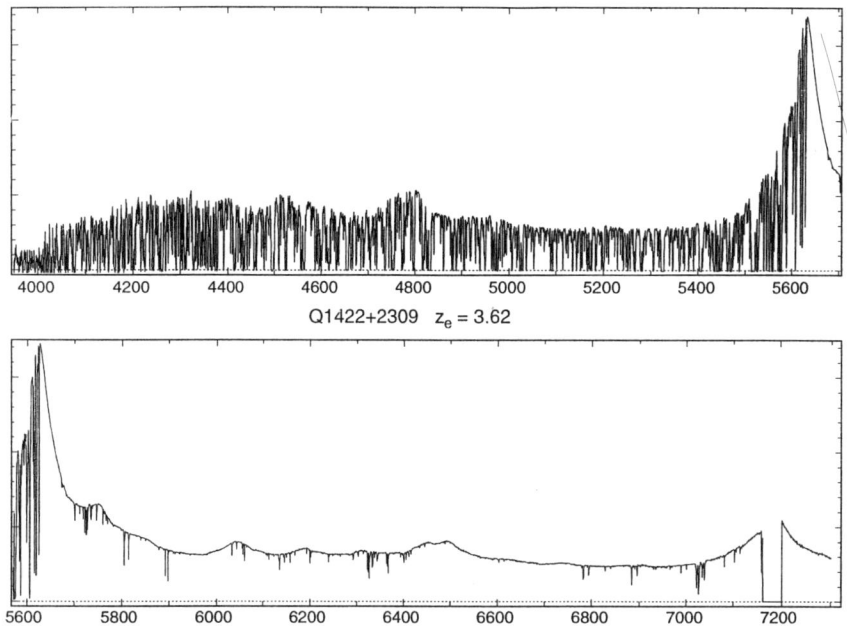

Figure 1: Upper panel shows the spectrum of the quasar Q1422+2309 (in Angstroms) short-ward of the Lyman α emission line (at right), exhibiting the dense forest of H I absorption known as the Lyman forest. Lower panel shows the spectrum longward of Lyman α emission with the apparently sparser forest of weak metal absorption lines, mainly C IV, which fall shortward of the C IV emission line (at right). The apparent deep well in the C IV emission line is a gap in coverage of the spectrum. Keck HIRES spectrum courtesy of Wal Sargent.

ation background) distributions of column densities, Doppler parameters and equivalent widths, and their evolution, which agree remarkably well with the observed ones. In the simulations these quantities arise from an interlocking structure of evolving sheets, filaments and clumps surrounding voids of growing size. Both the characteristic spectrum as observed from absorption along a single line of sight and large-scale spectral correlations between separated lines of sight as found in absorption studies of quasar pairs (Bechtold et al. 1994; Dinshaw et al. 1994, 1995; Impey in this volume) are produced by the simulations.

The good agreement between the simulated and observed absorption line properties suggests that the adopted scenario provides a realistic general picture of the origin of the Lyman forest, even if the physical assumptions and numerical realisations are not correct in all details. An important insight into the physical world gained from the simulations is that manifestations of both galaxies and H I absorption systems develop naturally together in the hierarchial formation of structure. High column density lines (N(H I) $\gtrsim 10^{17}$ cm^{-2}) arise from radiatively cooled gas associated with forming galaxies in collapsed, high density, compact regions. Low column density absorption (N(H I) $\lesssim 10^{15}$ cm^{-2}) arises in photoionised gas as it falls into the shallower dark matter potential wells, occurring in physically diverse systems spread over a wide range of scales and in various stages of gravitational infall and collapse. The typical low column density absorbers are flattened or filamentary structures of mild overdensity ($\rho/\bar{\rho} \sim 1-10$) with Doppler parameters that are often set by peculiar motions or Hubble flow rather than thermal broadening. Gravitational, pressure and ram-pressure confinement all play significant roles but most of the absorbers are far from dynamical and thermal equilibrium. An interesting common feature is the production of caustics in frequency space produced by converging velocity flows (McGill 1990) which locally distorts the velocity-distance relation for the forest lines; this demands comparison with simulations for the physical interpretation of the observations.

Traditional searches for the Gunn-Peterson (1965) effect look for continuum absorption by an assumed pervading uniform intergalactic medium in which discrete clouds are embedded. The simulations however reveal a smoothly fluctuating medium containing a network of complex structures having a continuous spectrum of clumpiness, with no sharp distinction between a diffuse 'background' and 'Lyman forest clouds' and this is borne out by high quality observations (Tytler *et al.* 1995). In this sense the observed Lyman forest *is* the Gunn-Peterson absorption, with the lowest column density lines arising from low-contrast structure in minivoids – small regions with densities below the cosmic mean (Zhang *et al.* 1996). These may be expanding faster than the Hubble flow (Reisenegger & Miralda-Escudé 1995).

It is worth mentioning here that the standard technique for characterising properties of the observed absorption lines involves fitting a spectrum with a superposition of Voigt profiles. The implicit physical model is of a collection of discrete, compact clouds each characterised by a single Doppler parameter (including temperature and Gaussian turbulence broadening). In the real world, as indicated by the simulations, the absorbing regions merge continuously on all scales, often contain gas at a range of temperatures and are usually influenced by coherent velocity flows that diverge or converge and do not resemble Gaussian turbulence. Nonetheless, any spectrum can be described with clustered Voigt profile fits to a level consistent with the available signal-to-noise ratio in the data. While this process has limited physical meaning, it provides a useful statistical basis for comparing simulations and observations (Davé *et al.* 1996).

3 The C IV Forest

It has been of continuing interest to determine whether the Lyman forest truly is chemically primordial or has some finite metallicity. The C IV $\lambda\lambda 1548,1550$ doublet is the most common heavy element feature found in the strong metal line systems in the absorption spectra of high redshift quasars. Most of these C IV systems must arise in highly ionised and strongly clustered gas clouds (Bergeron & Stasinska 1986; Sargent, Boksenberg & Steidel 1988; Steidel 1990). Because of the limited sensitivity of past spectroscopic observations the search for metals associated with the weak members of the Lyman forest was very difficult. Some associated metal lines were found in individual clouds (Meyer & York 1987), and stacking of spectra corresponding to many Lyman α lines has been used to try to determine a mean metallicity but with mixed results (Norris et al. 1983; Lu 1991; Tytler & Fan 1994). With the much higher signal-to-noise ratio coupled with the exceptionally high resolution (R=36000) of the spectra available from the HIRES spectrograph on the Keck Telescope it has now become possible to detect individual metal features related to the high redshift Lyman forest clouds for a large fraction of the stronger lines (Tytler et al. 1995; Cowie et al. 1995; Womble, Sargent & Lyons 1996; Rauch et al. 1996; Songaila & Cowie 1996; and see Figure 1). Clouds with N(H I) $> 10^{15}$ cm^{-2} are more or less all metal-enriched, typically having carbon abundances very approximately 10^{-2} solar and Si/C about three times solar. These are values very close to those of Galactic halo stars. More than half of the clouds with N(H I) $> 3 \times 10^{14}$ cm^{-2} show measurable C IV absorption with N (C IV) $\geq 10^{12}$ cm^{-2}, again with abundances very approximately 10^{-2} solar. In all cases the estimate of relative abundance is highly dependent on the (unknown) ionising conditions in the clouds.

Figure 2, from Womble, Sargent & Lyons, of the C IV redshifts two-point correlation function demonstrates that the C IV systems are clustered on velocity scales less than 500 km s^{-1} and lumped on scales of several thousand km s^{-1}, corresponding to comoving distances of $\leq 2.5h^{-1}$ Mpc and $\sim 25h^{-1}$ Mpc, respectively. These clustering properties are similar to those of the traditional metal line systems (Sargent et al. 1988). The latter scale length at z ~ 3 is comparable to the local cluster-cluster correlation length. Rauch et al. (1996) found that the kinetic energy of each individual C IV component is dominated by thermal broadening with average temperature ~ 38000 K although temperatures possibly $> 3 \times 10^5$ K occur occasionally . To produce line widths as small as those measured the individual clouds must be rather quiescent. The shape of the two-point correlation function on scales up to 500 km s^{-1} then can be understood if the C IV systems represent an ensemble of absorbing ob-

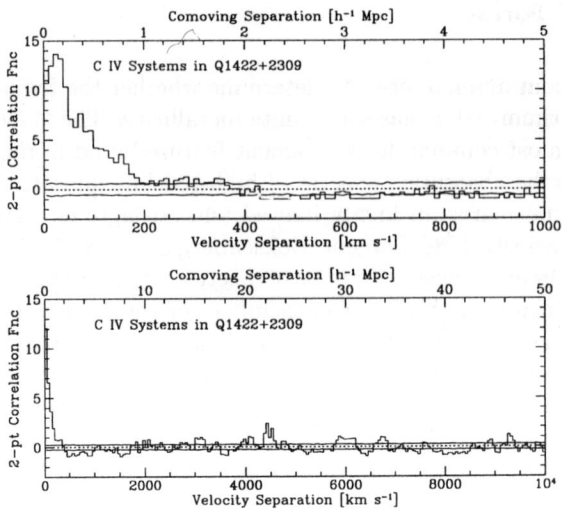

Figure 2: Two-point correlation function for 146 C IV redshifts in the spectrum of Q1422+2309; pair separations are denoted in both velocity and comoving distance for small and large scales (from Womble, Sargent & Lyons 1996).

jects with small internal velocity dispersion embedded in flattened large scale walls expanding with the Hubble flow (Rauch *et al.* 1996). While clustering of the Lyman forest lines is not evident on velocity scales $v \geq 300$ kms^{-1}, as found by Sargent *et al.* (1980), weak but significant clustering now has been found on small scales (Webb 1987; Crotts 1989; Chernomordik 1995; Cristiani *et al.* 1995; Hu *et al.* 1995; Irwin, Storrie-Lombardi & McMahon 1995; and see Fernández-Soto *et al.* 1996 for a different view), with a progression from virtually no clustering for lines of N(H) $\sim 10^{13}$ cm^{-2} to quite strong clustering for N(H) $\sim 10^{15}$ cm^{-2} (Cristiani *et al.* 1995). Note that it is in the column density range where significant clustering is seen that the Lyman forest lines also show detectable C IV absorption. (Incidentally, in the standard approach to investigating the correlation properties of the absorption lines, based on a numerical assessment of resolved lines, much information is lost because this process takes no account of the column densities. Measuring the pixel-to-pixel fractional transmission without resolving any single line (Press, Rybicki & Schneider 1993; Press & Rybicki 1993; Irwin, Storrie-Lombardi & McMahon 1995) is one, more satisfactory alternative technique).

Such results overturn the traditional view of the Lyman forest objects purely as intergalactic clouds of primordial composition. There might still be essentially two separate populations: lower column density primordial systems and higher column density (N(H I) $\geq 10^{14}$ cm^{-2}) evolved systems with different clustering properties. If carbon exists only in those Lyman forest clouds with higher N(H I), strictly as observed, then these absorbers possibly are associated with galaxies, which would have to be extended to about $200h^{-1}$ kpc in radius to account for the number density of N(H I) $> 10^{15}$ cm^{-2} clouds at redshift ~ 3, but it seems unlikely that star formation in the galaxy could enrich this whole region. Alternatively, if the lack of detection of carbon in clouds of lower N(H I) is due to limitation of signal to noise, then all Lyman forest clouds could contain metals. The metals then may originate in the clouds themselves, perhaps produced in subgalactic clumps at an early stage in the process of forming galaxies, or in Population III stars which have formed before galaxies. The latter development might result in a more uniform metallicity than *in situ* star formation where enrichment might be strongest in higher density clouds.

However it originates, the enrichment of the Lyman forest clouds must be understood within the same cosmological context of hierarchial structure formation as the existence of the clouds themselves. In the numerical simulations the low column densities of the high redshift Lyman forest clouds typically correspond to gas densities lower than those expected for fully collapsed objects, so the enrichment of the clouds appears to occur outside virialised regions. Rauch, Haehnelt & Steinmetz (1996) used hydrodynamical simulations with gas assumed to be homogeneously contaminated by metals and studied the mock line formation process as a function of the characteristic physical properties of the absorbing structures. While their emphasis was on hierarchial galaxy formation, conditions in the gas over the low density range appropriate for the Lyman forest lines were well covered in the analysis. They found that in the diffuse gas of the extensive regions exhibiting the lowest H I column densities (10^{12}–10^{14} cm^{-2}) high ionisation species are prevalent and O VI $\lambda 1031$ is often the strongest metal absorption line. Towards higher H I column densities the lines of sight pass through filamentary structures embedded in large scale sheets and in these regions low column density C IV lines from infalling gas dominate the metal absorption features. C IV remains the most readily observable metal ion in the yet unvirialised regions still quite far from protogalactic clumps which are later to merge into present-day galaxies. Strong C II and Si IV absorption occurs for lines of sight passing through the denser regions close to the centres of the protogalactic clumps. The observed column density ratios of the different ionic species overall are well reproduced in the mock spectra from these simulations if a mean metallicity [Z/H] = -2.5, relative abundances as found in metal-poor stars and an ultraviolet background with either a power law spectrum ($\propto \nu^{-1.5}$) or the spectral shape proposed by Haardt and Madau (1996) are assumed. This points to the broad validity of an imposed homogeneous metal distribution. The simplest implication is that much of the gas which is yet unvirialised has been subject to a widespread phase of star formation well before redshift ~ 3. Nevertheless, the observed scatter in [C/H] is substantially larger than predicted by the simulations, which may indicate that some of the metal enrichment took place *in situ* with incomplete mixing thereafter. The observed distribution of Doppler parameters and the relative contributions to the line widths from thermal and non-thermal motions (Rauch *et al.* 1996) also are well reproduced by these simulations. Rauch, Haehnelt & Steinmetz find that the C IV optical depth arises mostly in narrow, post-shock regions where the shocked gas already has come to rest and is cooling rapidly. This explains the observed narrow width, typically only 8 to 10 km s^{-2}, of the C IV lines in spite of the large peculiar velocities (~ 100 kms^{-1}) of the infalling gas apparent in the simulations. At the same time the contribution of bulk motions to the Doppler parameters of the low column density H I lines are predicted to be greater because much of the absorption arises at larger impact parameters where infall has a more important broadening effect.

Recently, ultraviolet spectroscopic observations with the Hubble Space Telescope have opened up the low end of redshift space for study (e.g. Bahcall *et al.* 1993, 1996). Relatively strong clustering of Lyman forest lines at $<z> = 0.7$ now has been found on scales of 250–500 km s^{-1} (Ulmer 1996). A reasonably strong correlation function might be expected for the low redshift Lyman forest systems because structural amplitudes grow with time. Evidence has been found for clumping of Lyman α lines near strong metal line systems (Bahcall *et al.* 1996), and low redshift observations of galaxies near quasar lines of sight show that a substantial fraction of Lyman forest absorbers are associated with galaxies (Lanzetta *et al.* 1995; le Brun, Bergeron & Boissé 1996). The latter authors find, however, that these absorbers are not clouds that belong in a strict sense to galaxies but instead are distributed in the local large scale structure. Moreover, the density of Lyman forest lines is much lower and increases much more slowly with redshift at low redshifts than in the high redshift region observed from the ground (Bahcall *et al.* 1996; Jannuzi in this volume). This all is indicating (and shows in the simulations too) that in contrast to the wide distribution of absorbing material observed at high redshifts, by late epochs intergalactic space has been rather thoroughly cleared of gas except for concentrations near galaxies.

4 Concluding Remarks

Quasar absorption lines have assumed an important role in illuminating the formation of cosmic structure. The observations now are firmly placed in the broad intellectual context of galaxy formation and of the evolution of the intergalactic medium as a whole. The Lyman forest absorption allows us to observe gas at lower densities than in any other observational method, ranging from the most underdense regions in the voids to the mildly overdense complex regions, as well as up to the dense central regions in dark matter haloes where galaxies form. This gives an enormous advantage, for example, over the use of galaxies which only indicate regions where the gas densities become high enough for the massive onset of star formation.

Clearly, much more work is needed to determine which cosmological models can best match the observed characteristics of the absorption. The success of rather different models in reproducing the broad statistical characteristics of the absorption lines as normally catalogued, now known from high quality observations, suggests that quite sophisticated tests are needed to form a more discriminatory basis for comparison. For such tests correlated observations along multiple lines of sight (Charlton *et al.* 1996; Hernquist *et al.* 1996; Miralda-Escudé *et al.* 1996; Fang *et al.* 1996; Impey in this volume) will give vastly more information than conventional data sets from single lines of sight. For the models, to be more realistic it will be necessary properly to include the effects of star formation and feedback (Katz, Hernquist & Weinberg 1992) and to contain a photoionising ultraviolet background self-consistently (Miralda-Escudé *et al.* 1996). But it is certain that, overall, significant further progress can be made only if observations and representative simulations are taken forward together.

Acknowledgements

I was enormously pleased that so many came to this 37th Herstmonceux Conference which has provided extraordinary science and entertainment in equally large measure. I particularly thank the organisers for the compliment of identifying me with the Conference, for their inspiration and hard work in arranging it all and for the not insignificant work to come in publishing the Proceedings.

References

1. J.N. Bahcall, ApJ **200**, L1 (1975).
2. J.N. Bahcall *et al.*, ApJS **87**, 1 (1993).

3. J.N. Bahcall *et al.*, ApJ **457**, 19 (1996).

4. J. Bechtold, A.P.S. Crotts, R.C. Duncan & Y. Fang, ApJ **437**, L83 (1994).

5. J. Bergeron & G. Stasinska, A&A **169**, 1 (1986).

6. J.R. Bond, A.S. Szalay & J. Silk, ApJ **324**, 627 (1988).

7. R. Cen, J. Miralda-Escudé, J.P. Ostriker & M. Rauch, ApJ **437**, L9 (1994).

8. J.C. Charlton, P. Anninos, Y. Zhang & M.L. Norman, ApJ , in press (1996).

9. V.V. Chernomordik, in *QSO Absorption Lines*, ed. G. Meylan (Berlin: Springer-Verlag, p. 343, 1995).

10. L.L. Cowie, A. Songaila, T.-S. Kim & E.M. Hu, AJ **109**, 1522 (1995).

11. S. Cristiani, S. D'Odorico, A. Fontana, E. Giallongo & S. Savaglio, MNRAS **273**, 1016 (1995).

12. A.P.S. Crotts, ApJ **336**, 550 (1989).

13. R. Davé, L.H. Hernquist, N. Katz & D.H. Weinberg, ApJ , in press (1996).

14. N. Dinshaw, C.B. Foltz, C.D. Impey, R.J. Weymann & S.L. Morris, Nature **373**, 223 (1995).

15. N. Dinshaw, C.D. Impey, C.B. Foltz, R.J. Weymann & F.H. Chaffee, ApJ **437**, L87 (1994).

16. Y. Fang, R.C. Duncan, A.P.S. Crotts & J. Bechtold, ApJ **462**, 77 (1996).

17. A. Fernández-Soto, K.M. Lanzetta, X. Barcons, R.F. Carswell, J.K. Webb & A. Yahil, ApJ **460**, L85 (1996).

18. J.E. Gunn & B.A. Peterson, ApJ **142**, 1633 (1965).

19. F. Haardt & P. Madau, ApJ **461**, 20 (1996).

20. L. Hernquist, N. Katz, D.H. Weinberg & J. Miralda-Escudé, ApJ **457**, L51 (1996).

21. E.M. Hu, T.-S. Kim, L.L. Cowie & A. Songaila, AJ **110**, 1526 (1995).

22. S. Ikeuchi, Ap Space Sci **118**, 509 (1986).

23. S. Ikeuchi & J.P. Ostriker, ApJ **337**, 609 (1986).

24. M.J. Irwin, L.J. Storrie-Lombardi & R.G. McMahon, in *QSO Absorption Lines*, ed. G. Meylan (Berlin: Springer-Verlag, p. 363, 1995).

25. N. Katz, L. Hernquist & D.H. Weinberg, ApJ **399**, L109 (1992).

26. K.M. Lanzetta, D.V. Bowen, D. Tytler & J.K. Webb, ApJ **442**, 538 (1995).

27. L. Lu, ApJ **379**, 99 (1991).

28. C.R. Lynds, ApJ **164**, L73 (1971).

29. C. McGill, MNRAS **242**, 544 (1990).

30. D.M. Meyer & D.G. York, ApJ **315**, L5 (1987).
31. J. Miralda-Escudé, R. Cen, J.P. Ostriker & M. Rauch, ApJ , in press (1996).
32. J. Norris, F.D.A. Hartwick & B.A. Peterson, ApJ **273**, 450 (1983).
33. W.H. Press & G.B. Rybicki, ApJ **418**, 585 (1993).
34. W.H. Press, G.B. Rybicki & D.P. Schneider, ApJ **414**, 64 (1993).
35. M. Rauch, W.L.W. Sargent, D.S. Womble & T.A. Barlow, ApJ **467**, L5 (1996).
36. M. Rauch, M.G. Haehnelt & M. Steinmetz, ApJ , in press (1996).
37. M. Rees, MNRAS **218**, 25P (1986).
38. A. Reisenegger & J. Miralda-Escudé, ApJ **449**, 476 (1995).
39. W.L.W. Sargent, P.J. Young, A. Boksenberg & D. Tytler, ApJS **42**, 41 (1980).
40. W.L.W. Sargent, A. Boksenberg & C.C. Steidel, ApJS **68**, 539 (1988).
41. A. Songaila & L.L. Cowie, AJ **112**, 335 (1996).
42. C.C. Steidel, ApJS **72**, 1 (1990).
43. D. Tytler & X.-M. Fan, ApJ **424**, L87 (1994).
44. D. Tytler, X.-M. Fan, S. Burles, L. Cottrell, C. Davis, D. Kirkman & L. Zuo in *QSO Absorption Lines*, ed. G. Meylan (Berlin: Springer-Verlag, p. 289, 1995).
45. A. Ulmer, ApJ , in press (1996).
46. J.K. Webb in *Observational Cosmology, Proc. of IAU Symposium No.124*, eds. A. Hewitt, G. Burbidge & L.-Z. Fang (Dordrecht: Reidel, p. 803, 1987).
47. D.H. Weinberg, L. Hernquist, N.S. Katz & J. Miralda-Escudé in *Cold Gas at High Redshift*, eds. M. Bremer, H. Rottgering, C. Carilli & P. van de Werf (Dordrecht: Kluwer, 1996), in press.
48. D.S. Womble, W.L.W. Sargent & R.S. Lyons, in *Cold Gas at High Redshift*, eds. M. Bremer, H. Rottgering, C. Carilli & F. van de Werf (Dordrecht: Kluwer, 1996), in press.
49. P. Young, W.L.W. Sargent & A. Boksenberg, ApJ **252**, 10 (1982).
50. Y. Zhang, P. Anninos & M.L. Norman, ApJ **453**, L57 (1995).
51. Y. Zhang, A. Meiksin, P. Anninos & M.L. Norman, Nature , in press (1996).

QUASARS, DEUTERIUM AND COSMOLOGY

DAVID TYTLER and SCOTT BURLES

CASS 0424, University of California San Diego,
La Jolla, CA 92093-0424, USA

QSO absorption lines encode information on the density, temperature, abundances and motions of cosmologically distributed gas. The deuterium to hydrogen ratio (in gas with low metal abundances) was established in the first minutes of the big bang, and gives the most direct and accurate measure of the cosmological baryon density. Today many of those baryons are in hot gas, detected in individual Ly-alpha lines, and in Gunn-Peterson absorption from intergalactic H I and He II. We review the connection of QSO absorption to the rest of astronomy and then discuss measurements of deuterium in QSO spectra.

1 The Beginning of Quantitative Astrophysics on Faint Objects

I first met Alec Boksenberg in September 1977, in the Physics Building of University College London. His research groups were unusually active and I was looking for an undergraduate research project. I asked to work on interferometry, but he recommended Quasar absorption lines and gave me a strip chart graph showing a 60 km s^{-1} resolution spectrum of the Lyman alpha forest of the bright QSO 0424–388 which was vastly better than any published.

Alec was the inspiration for three dynamic instrument groups: one working on ultraviolet spectroscopy from balloons, a second building the superb detectors for the International Ultraviolet Explorer, and the third was the Flying Circus, the group which I joined. In all there were 22 people, huge by todays standards, but these were very productive times. The Flying Circus was a very happy group, in many ways the model of a modern self contained team, including people specialized in mechanical, electronic and optical engineering, computers, data analysis and theory. The Flying Circus designed, build and used the Image Photon Counting System (IPCS). We used several enormous 4x4x6 foot wooden crates to transport the whole system to major telescopes around the world. One create held the computer, which had flashing lights on its front to show activity. This robust giant was re-booted using switches to enter a sequence of long hexadecimal numbers.

The Image Photon Counting System revolutionized astronomical spectroscopy by permitting the first quantitative astrophysics on faint objects. Some of the best spectra before that time, obtained by my colleagues in San Diego[4] using a Carnegie Image Tube and IIIaJ photographic emulsion, were severely limited by non-linearity and sky background. The IPCS spectra[7] had four huge

Table 1: CLASSIFICATION OF QSO ABSORBERS BY H I COLUMN DENSITY

Name	log N(H I) (cm^{-2})	Number per z	Size (kpc)	Feature
Gunn-Peterson	< 12	...	...	H I, He II troughs
Ly-α Forest	12 – 14	500	>200	H I, He II lines
Ly-α Forest	14 – 17	10	100	H I, C IV lines
Ly Limit	17 – 22	2	40	Lyman edge
Damped Lyα	20 – 22	0.2	10	Damped Lyα

advantages: they were digital, linear, allowed sky subtraction, and reached very faint targets because of the extremely low noise. Although detectors and telescopes have improved gradually since then, not until the HIRES spectrograph on the W. M. Keck 10-m telescope in January 1996 was there a comparable leap in data quality.

2 General Properties of Quasar Absorption

Gas along the line of sight to QSOs produces multiple spectral absorption lines, from the most abundant element: H, C, N, O, Si, Mg. The absorbing gas is usually transparent, so that we can see multiple absorbers towards high z QSOs. We see more absorbers at high z because the universe was denser and less ionized. Many of the absorbers, and all lines of all elements except H and He, arise in discrete entities which each occupy an insignificant fraction of the distance to a QSO (40 kpc/10 Gpc), but we a also see H I and He II absorption from gas which is distributed nearly continuously at high z.

The IPCS allowed the first accurate measurements of equivalent widths and column densities, revolutionizing the study of QSO absorption, and leading to the modern classification given in Table 1.

3 Significance of Quasar Absorption

Quasar absorption samples nearly all gas in the universe, but with biases which can be calibrated. There are strong bias towards observing absorption in three cases: (1) the absorbing object is large; (2) the ionization is low enough that the dominant ions have strong lines at $\lambda > 900$ Å, and preferably > 1216 Å;

and (3) gas near the QSO is more likely to be seen.

The absorption samples gas in a variety of environments: the diffuse intergalactic medium (IGM), the outer regions of galaxies, the disk of galaxies, and the immediate surroundings of QSOs. The properties of the absorbers tell us about galaxy formation and the intergalactic spectrum of ionizing radiation [9].

QSO absorption now connects to the rest of astronomy and astrophysics in six areas. (1) The cosmological baryon density – using D/H to measure primordial density, which is comparable to the density of baryons at recent times in the Lyα forest and IGM. (2) The distribution of cold and warm gas in the IGM and galaxies. (3) The formation of structure, seen in the internal velocity dispersion and clustering of absorbing gas. (4) The formation and cosmological distribution of the elements. (5) The ionizing radiation field – its spectrum and origin, and how and when the IGM was re-ionized. (6) The opacity of the universe, from H I, He II and dust.

4   The Significance of the Two Primordial Deuterium Measurements

There are seven reasons why it is important to measure the primordial D/H ratio accurately. (1) D/H is the best measure of η = (baryon density) / (photon density), a fundamental cosmological constant, which can be converted into baryon density Ω_b because the photon density is well determined from the cosmic microwave background. (2) The η ratio depends on high energy physics, and may one day be calculated. (3) Comparison of the measurement of Ω_b from D/H and from the cosmic microwave background will provide a new type of test of the cosmological model (Michael Turner). (4) A precise value for Ω_b removes a degree of freedom from cosmological models. (5) We can determine the fraction of baryons which are missing. (6) Comparison of the primordial abundances of the light elements is a powerful test of the physics in big bang nucleosynthesis (BBN). (7) The changes in D/H with time help specify Galactic chemical evolution.

4.1   D is the Best Baryometer

D/H is very sensitive to η, but it is fairly insensitive to the expansion rate epoch, which depends on effective number of neutrino degrees of freedom. D is the third most abundant BBN product, much more abundant than Li. The chemical evolution of D is simple (unlike Li, and He^3), and we do not need ionization corrections: D I/D = H I/H (unlike He and Li).

4.2   Why QSOs?

QSO absorption systems have several advantages for the measurement of D, discussed in detail by Adams (1976) and Webb et al (1991). (1) We can see metal poor gas, which should have near primordial D/H. (2) Some are at high z and early times, before much destruction of D. (3) The QSO absorbers are widely distributed in the observable universe. (4) Each samples a large volume of space about 1 – 10 kpc across. But they also have several major disadvantages. (5) Absorbers which show D are very rare, about one in 30 QSOs at $z \simeq 3$. (6) Contamination by H is extremely likely. (7) Only very high quality spectra can distinguish D from contaminating H.

5   A Limit is not a Detection

A report that D/H is less than some high value does not imply that D/H is likely to be high. Rather it means that the measurement is poor, because of contamination or poor data. Over the past two years, there has been a flurry of papers reporting limits, detections and even some measurements of D/H (refs below). Pundits will note; a limit is not a detection, and it is certainly not a measurement of D/H.

Deuterium is very rarely seen in QSO absorption systems for two reasons: (1) the associated hydrogen is about 40,000 times more abundant and its absorption is spread over a large velocity range which readily covers the D line, (2) other hydrogen absorption from random Lyα clouds can also readily contaminate the D line. Each absorption system has certain characteristics that determine the likelihood that D can be seen and measured, and most such characteristics make a measurement of D/H impossible. A measurement can be made only for a select few systems, using spectra with both high signal-to-noise ratio (SNR) and high resolution (FWHM ≈ 10 km s^{-1}).

Songalia et al. (1994) and Carswell et al. (1994) reported the first data on D/H in a QSO absorption system towards QSO 0014+8118. Neither group declared this a measurement of D/H, and both suggested an upper limit of D/H $< 25 \times 10^{-5}$. This is a high upper limit on D/H, which does not rule out interesting values of D/H. But then a misconception arose as the studies were interpreted as possible measurements of D/H, suggesting a very high primordial D/H. Rugers & Hogan (1996a) re-analysed the Songalia et al. data and found a spike in the data which they claimed was real and could be fit with a more complex model. They claimed that the upper limit should be promoted to a measurement of high D/H. We do not believe that the spike is real, we do not believe the complex model, and we do not believe that this

QSO can yield a D/H measurement.

D/H has been reported in three other absorption systems (Rugers & Hogan 1996b; Carswell et al. 1996; Wampler et al. 1996). Even though Rugers & Hogan (1996b) report a measurement of D/H with large errors, none of these studies have measured D/H. None of the systems are suitable for measurements of D/H, and they allow only limits of $2 \times 10^{-5} < $ D/H $ < 25 \times 10^{-5}$.

5.1   There are only Two D/H Measurements

Deuterium has been seen in only two QSOs, and both give low D/H measurements (Tytler, Fan & Burles 1996; Burles & Tytler 1996) which are consistent with the high upper limits. The claimed measurements in Q0014+8118 are not credible.

6   The Deuterium Debate

The last two years have seen the rise of a new "political action committee" in cosmology: the High D/H Defence League. This group undertakes a variety of activities, reminiscent of the debates surrounding another interesting cosmological number: the Hubble constant. Members have published high D/H limits. They seek to promote limits into measurements, and they speculate that low D/H measurements are really high D/H which were subject to 10 sigma errors. They seek out circumstantial evidence for high D/H, and re-analyse and re-interpret data to strengthen the case for high D/H. They may promote exotic ways to destroy D, and cast doubt on a variety of evidence which happens to favor low D/H.

Debate is fun, healthy and interesting, but we can not give all arguments equal representation, or all data equal weight. Highly speculative ideas, like incomplete observations and inconclusive data analysis, should be reworked, completed and discussed in depth, before they are thrown onto the fire of public debate. We need to quantitatively distinguish the probable from the highly speculative, just as we must distinguish limits from detections. Without such filters, confusion will spread through astrophysics, leading to an atmosphere of distrust. Worst of all, the rush to publish undermines all quasar absorption work, which over the last 20 years has built up the highest reputation for novel and accurate analysis, demonstrated by an extremely high proportion of correct results.

7 The Atami Criteria

No single test determines whether a QSO spectrum gives a D/H limit or a measurement. We end with an update of the 19 questions which should be asked about any claimed upper limit on D [10].

1. How many QSOs were searched to find the D/H value? There is more chance of a false identification if only a few spectra were studied, because systems which give secure D/H are very rare.

2. Was there a systematic search for D in the spectra of many QSOs? What criteria were used to select QSOs for detailed observation?

3. The SNR should be high: $\simeq 100$ per 0.04Å is excellent. Lower SNR increases the chance that the D feature is contaminated with H.

4. The spectral resolution should be high: 8 km s^{-1} FWHM is excellent. Lower resolution increases the chance that the D feature is contaminated with H.

5. How large is the neutral H column density N(H I)? Count only the gas which shows D. High values are much more likely to give a real D detection. If log N(H I) $< 10^{17}$ we are unlikely to detect D/H $\simeq 2 \times 10^{-5}$ with the best Keck data. If D/H is low, every claim of D in a system with low N(H I) must be contaminated, and they will automatically give high D/H limits.

6. How well is N(H I) determined? If from a Lyman limit alone, is it know that there are no other systems which could account for the Lyman continuum absorption? If from Lyman lines, the more lines the better. One saturated line is inadequate in low SNR data, but may be acceptable with high SNR and high resolution. Favor N(H I) from many lines and the continuum.

7. How certain is it that there is a feature at the position of D?

8. How was the position of D determined? If from the D line itself, then there is a large chance of H contamination. Ideally from metal lines and narrow high order Lyman lines.

9. How many components were required to fit D? The fewer the better.

10. Are additional H components needed to fit around D? Bad idea. This greatly increases the chance of contamination.

11. How were the velocities of the components determined? If from D, then there is a high chance of contamination. Ideally from metal lines and narrow high order Lyman lines.

12. How good is the fit? It must be excellent in the critical regions which distinguish different values of D/H. Fits to high SNR spectra can look excellent even though they are not acceptable. Expand the vertical scale to clearly show 1 sigma errors on each pixel.

Table 2: RELIABILITY OF D/H LIMITS FROM QSO SPECTRA

Criterion	Deuterium (A)	(B)	(C)	System (D)	(E)	(F)	(G)
Many QSOs searched?	7	7	3	3	3	3	0
Systematic D/H search?	10	10	2	0	0	0	0
High SNR?	5	8	4	4	3	2	9
High spectral resolution?	10	10	10	10	5	5	10
High N(H I)?	6	8	2	8	2	10	0
Small σ for N(H I)?	6	7	8	2	7	0	4
Definite feature at D?	10	10	7	0	7	0	10
D v from other ions?	10	10	10	8	7	10	0
Few components to fit D?	5	5	3	0	10	0	5
Extra H to fit D?	0	0	-5	0	0	-5	0
Component v from where?	10	10	5	10	0	0	0
How good is fit to data?	10	10	5	2	5	8	10
Is fit unique?	0	0	0	-5	0	-5	0
Is b(D) known?	5	5	7	5	0	5	5
Was b(D) predicted?	7	7	0	7	0	5	0
Agreement on b(D)?	7	7	0	7	0	3	0
Metal abundance low?	10	10	10	7	0	5	0
Additional evidence?	5	5	0	0	0	5	0
H contamination corrected?	10	10	0	0	0	0	0
Total	133	139	71	58	49	51	53
(out of 170)							

D/H can only be detected when there is a high total score (QSOs A and B). In all other cases D is not detected, and we get upper limits on D/H. (A) Q1009+2956[2]; (B) Q1937-1009[11]; (C) 0014+8118 z=3.320[8, 3]; (D) 0014+8118 z=2.798[6]; (E) 1202-0752[12]; (F) 1442+2903[10].

13. Is the fit unique? Do different fits give different D/H?

14. Was the velocity dispersion $b = \sqrt{2}\sigma$ value measured for D? If not, why not? Lack of a measurement is a sign of a non-unique fit, poor data, or strong blending, all of which favor contamination by H.

15. Was the b value of D predicted? Whenever metal lines are seen, the temperature and turbulent velocity dispersion b_{tur} should be measured and used to give a prediction. If b is not measured, then b should lie between $b(\mathrm{H})$ and $b(\mathrm{H})/\sqrt{2}$. Gas which is suitable for the detection of D will be cool, with low b_{tur}, although thermal motions will probably dominate the b values.

16. How tight is the agreement between the predicted and measured $b(\mathrm{D})$?

17. Was the metal abundance measured? If not, why not? It could be high.

18. Is there additional evidence that the fit to the system is reasonable, such as standard element abundance ratios and reasonable temperatures?

19. Was there a Monte Carlo correction for Lyα forest lines at the position of D? If not, why not? Such corrections are difficult to calculate when fits are poor or ambiguous.

References

1. T.F. Adams, A&A, 50, 461, 1976.
2. S. Burles & D. Tytler, *Science*, submitted 1996; astro-ph 9603069.
3. R.F. Carswell et al. MNRAS, 268, L1, 1994.
4. D.H. Roberts, E.M. Burbiudge, G.R. Burbidge, A.H. Crowne, V.T. Junkkarinen and H. E. Smith, ApJ, 224, 344, 1978.
5. M. Rugers and C.J. Hogan, ApJ, 459, L1, 1996a.
6. M. Rugers and C. J. Hogan, AJ, 111, 2135, 1996b..
7. W.L.W. Sargent, P. Young, A. Boksenberg, R.F. Carswell and J.A.J. Whelan, ApJ, 230, 49, 1979.
8. A. Songaila, L.L. Cowie, C.J. Hogan, M. Rugers, Nature, 368, 599, 1994.
9. D. Tytler, in *New Light on Galaxy Evolution* ed. R. Bender and R.L. Davies (Rediel, Dordrecht, 1996).
10. D. Tytler D. & S. Burles in *Origin of Matter and Evolution of Galaxies* eds. T. Kajino, Y. Yoshii & S. Kubono (World Scientific Publ. Co., 1996) (astro-ph 9606110)
11. D. Tytler, X.M. Fan and S. Burles Nature, 381, 207, 1996.
12. E.J. Wampler et al. *AAp* in press 1996.
13. J.K. Webb et al. MNRAS, 250, 657, 1991.

A GLOBAL PERSPECTIVE ON STAR FORMATION

S. MICHAEL FALL

Space Telescope Science Institute
3700 San Martin Drive
Baltimore, MD 21218, USA

We outline a method to infer the global history of star formation in galaxies with input only from absorption-line observations of quasars. The application of the method to existing data leads to the conclusion that most stars formed at relatively low redshifts ($z \lesssim 2$). We combine the global rate of star formation with stellar population synthesis models to compute the mean comoving emissivity and mean intensity of background radiation from far-UV to far-IR wavelengths. These predictions are consistent with all the available measurements and observational limits, including recent results from *HST* and *COBE*.

1 Overview

This article concerns the evolution of, and relations between, various large-scale average properties of the population of galaxies as a whole. It is often convenient to express these "global" properties as mean comoving densities and to normalize them to the present closure density. We are particularly interested in the comoving densities of stars, gas, metals, and dust within galaxies, which we denote respectively by Ω_s, Ω_g, Ω_m, and Ω_d. The last three of these are meant to refer to the interstellar media (ISM) of galaxies, exclusive of the intergalactic medium (IGM), although in practice such a distinction may only be approximate. As defined here, Ω_m includes metals in both the gas and solid (i.e., dust) phases of the ISM. It is usually more informative to reexpress Ω_m and Ω_d in terms of the mean metallicity and mean dust-to-gas ratio, $Z \equiv \Omega_m/\Omega_g$ and $D/G \equiv \Omega_d/\Omega_g$. It is clear that all of these properties are related in the sense that, as new stars form, Ω_s will increase, while, in most cases, Ω_g will decrease, and Z and D/G will increase. One of our goals is to quantify such relations through the equations of "cosmic chemical evolution".

Until recently, there were no emission-based estimates of the global rate of star formation $\dot{\Omega}_s$ at $z \gtrsim 0.3$. The reason for this is that samples of galaxies selected by emission become progressively incomplete and include only brighter objects at higher redshifts. In contrast, samples of galaxies selected by absorption against background quasars do not suffer from this bias. Such observations are exquisitely sensitive to small column densities of absorbing or scattering particles. In principle at least, they enable us to estimate Ω_g, Ω_m, and even Ω_d as functions of redshift. From these and the equations of cosmic chemical evolution, we can then infer the global rate of star formation $\dot{\Omega}_s$. It is amusing to note that this idealistic program does not require the detection of a single stellar photon! Furthermore, if we are confident (or foolish) enough, we can combine our estimates of $\dot{\Omega}_s$ with stellar population synthesis models to compute the mean comoving emissivity $\mathcal{E}_\nu$ and the mean intensity of background radiation J_ν. One might then claim to have predicted the "emission history" of the universe from its "absorption history". This article describes a first attempt by Yichuan Pei, Stéphane Charlot, and the author to carry out such a program; a complete account of our work is given in references 1 and 2. Some related material can be found in references 3–6.

2   Absorption-Line Systems

Before proceeding, it is worth recalling some facts about the statistics of absorption-line systems. Let $f(N_x, z)$ be the column density distribution of particles of any type x that absorb or scatter light. These might, for example, be hydrogen atoms ($x = $ HI), metal ions ($x = m$), or dust grains ($x = d$). By definition, $H_0(1+z)^3|dt/dz|f(N_x,z)dN_x dz$ is the mean number of absorption-line systems with column densities of x between N_x and $N_x + dN_x$ and redshifts between z and $z + dz$ along the lines of sight to randomly selected background quasars. These lines of sight are very narrow (much less than a parsec across) and pierce the absorption-line systems at random angles and impact parameters. One can show that the mean comoving density of x is given by

$$\Omega_x(z) = \frac{8\pi G m_x}{3cH_0} \int_0^\infty dN_x f(N_x, z) N_x, \tag{1}$$

where m_x is the mass of a single particle (atom, ion, or grain). Equation (1) plays a central role in this subject. It enables us to estimate the mean comoving densities of many quantities of interest without knowing anything about the structure of the absorption-line systems. In particular, we do not need to know their sizes or shapes, whether they are smooth or clumpy, and so forth. A corollary of equation (1) is that the global metallicity, $Z \equiv \Omega_m/\Omega_g$, is given simply by an average over the metallicities of individual absorption-line systems weighted by their gas column densities.

The absorption-line systems of most interest in the present context are the damped Lyα (DLA) systems. It is widely believed that they trace the ISM of galaxies and protogalaxies and are the principal sites of star formation in the universe. There are excellent reasons to adopt this as a working hypothesis. First, the DLA systems have, by definition, $N_{\text{HI}} \gtrsim 10^{20}$ cm^{-2}, which is just below an apparent threshold for the onset of star formation[7]. Second, the DLA systems contain at least 80% of the HI in the universe and appear to be mostly neutral [4]. The other absorption-line systems, those with $N_{HI} \lesssim 10^{20}$ cm^{-2}, probably contain more gas in total than the DLA systems, but this must be diffuse and mostly ionized. In the following, we regard non-DLA systems as belonging to the IGM, even though some of them might actually be located in the outer halos of galaxies. This distinction – between the mostly-neutral ISM, where stars form, and the mostly-ionized IGM, where they do not – is clearly valid at the present epoch. Thus, the DLA systems are often referred to as DLA galaxies. It will be interesting to see exactly which types of galaxies they represent, but as we have already emphasized, this issue does not affect any of the global properties derived from equation (1).

The sample of known DLA galaxies now includes about 80 objects [8]. They are distributed over a wide range in redshift, $0 \lesssim z \lesssim 4$, although, as a consequence of selection effects, most of them are confined to the narrower range $2 \lesssim z \lesssim 3$. From observations of DLA galaxies in various subsets of this sample and comparisons with present-day galaxies, the following trends have emerged. The mean comoving density of HI decreases by almost an order of magnitude, from $\Omega_{HI} \approx (1-2) \times 10^{-3}h^{-1}$ at $z \approx 3$ to $\Omega_{HI} \approx 2 \times 10^{-4}h^{-1}$ at $z = 0$ [with $h \equiv H_0/(100$ km s^{-1} Mpc$^{-1})$] [4,8,9]. It is possible that Ω_{HI} increases between $z \approx 4$ and $z \approx 3$, but the evidence for this is weak [9]. The mean metallicity increases by about an order of magnitude, from $Z \approx 0.1Z_{\odot}$ or slightly less at $z \approx 2$ to $Z \approx Z_{\odot}$ at $z = 0$ [10,11]. The mean dust-to-gas ratio increases by a similar factor, while the mean dust-to-metals ratio remains roughly constant at about the present value in the local ISM [10,11,12]. These results are entirely consistent with the recent Keck observations by Lu et al [11,13,14]. The abundances of H_2 and CO appear to be much lower at $z \gtrsim 2$ than at $z = 0$ [15]. As a consequence of the relatively small samples involved, most of the numbers quoted here are uncertain by factors of 1.5 or more.

3   Cosmic Chemical Evolution

The global properties defined above are governed by a set of coupled equations, which are sometimes referred to as the equations of cosmic chemical evolution. In the approximation of instantaneous recycling (and $Z \ll 1$), they take the form

$$\frac{d}{dt}(\Omega_g + \Omega_s) = \dot{\Omega}_f, \tag{2}$$

$$\frac{d}{dt}(Z\Omega_g) + (Z-y)\frac{d}{dt}\Omega_s = Z_f\dot{\Omega}_f, \tag{3}$$

where y is the IMF-averaged yield. Equations (2) and (3) are strictly valid only when all galaxies evolve in the same way; otherwise, they should be regarded as approximations. The "source" terms on the right-hand sides of the equations allow for the exchange of material between the ISM of galaxies and the IGM; they represent the inflow or outflow of gas with metallicity Z_f at a rate $\dot{\Omega}_f$. To illustrate a range of possibilities, we consider three types of evolution: a closed-box model ($\dot{\Omega}_f = 0$), a model with inflow of metal-free gas ($\dot{\Omega}_f = +\nu\dot{\Omega}_s$, $Z_f = 0$), and a model with outflow of metal-enriched gas ($\dot{\Omega}_f = -\nu\dot{\Omega}_s$, $Z_f = Z$). Our inflow and outflow models are direct analogs of the standard models of chemical evolution in the disk and spheroid components of the Milky Way[16,17]. We fix the yield y in each model by requiring $Z = Z_\odot$ at $z = 0$. Then the only adjustable parameters are the "initial" comoving density of gas in galaxies $\Omega_{g\infty}$ (in practice, the value of Ω_g at $z \gtrsim 4$) and the relative inflow or outflow rate ν.

To complete the specification of the models, we make two other approximations, both motivated by the observations summarized in the previous section. (1) We neglect any ionized or molecular gas in the ISM of galaxies and set $\Omega_g = 1.3\Omega_{HI}$ (to account for He). (2) We assume that just over half of the metals in the ISM are depleted onto dust grains and set $D/G = 0.6Z$. The models are designed to reproduce (as input) the observed decrease in the mean comoving density of HI between $z \approx 3$ and $z = 0$. The only subtlety here is that the observed values of Ω_{HI} tend to underestimate the true values as a consequence of the obscuration of quasars by dust in foreground galaxies[18]. We make a self-consistent correction for this bias in the models by linking the obscuration of quasars to the chemical enrichment of galaxies. It is worth noting that, while this correction has a substantial effect on Ω_{HI}, especially at $z \sim 1$, it does not entail large numbers of "missing" quasars (only ~20% at $z = 2$ and ~40% at $z = 4$). The models reproduce (as output) the observed increase in the mean metallicity between $z \approx 2$ and $z = 0$ without any fine tuning of the parameters $\Omega_{g\infty}$ and ν. The reason for this is that most of the star formation and hence most of the metal production occur at $z \lesssim 2$.

Figure 1 shows the evolution of the comoving rate of metal production $\dot{\rho}_z$ in the models. This is given by $\dot{\rho}_z = y\psi$, with $\psi = (1 - R)^{-1}\dot{\rho}_s$ and $\dot{\rho}_s = (3H_0^2/8\pi G)\dot{\Omega}_s$, where $R \approx 0.3$ is the returned fraction. The predicted rates have maxima at $1 \lesssim z \lesssim 2$ and decline rapidly at lower redshifts. Figure 1 also shows estimates of, and lower limits on, $\dot{\rho}_z$ from recent ground-based surveys and the Hubble Deep Field[19,20,21,22]. These are proxies for global Hα and UV emissivities based on the close correspondence between UV emission and metal production in massive stars[3,22]. Evidently, the predicted and observed rates are in broad qualitative, and even some quantitative, agreement

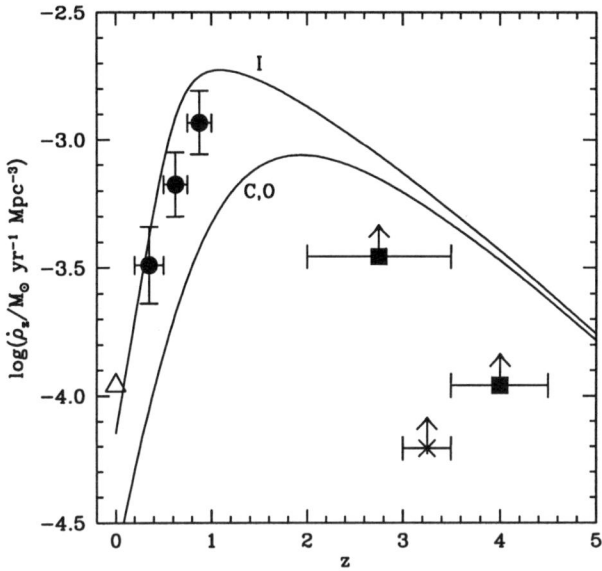

Figure 1: Comoving rate of metal production $\dot{\rho}_z$ as a function of redshift z (for $h = 0.5$, $q_0 = 0.5$, and $\Lambda = 0$). The curves are from the closed-box (C), inflow (I), and outflow (O) models with $\Omega_{g\infty} = 4 \times 10^{-3}h^{-1}$ and $\nu = 0.5$ (see Figure 1 of reference 1). The data points and lower limits represent global Hα and UV emissivities from ground-based and *HST* surveys (see Figure 9 of reference 22).

(given the uncertainties in both). This is remarkable because the models were constructed only with absorption-line systems in mind, not the emissivities represented in Figure 1. We have also combined our chemical evolution models with stellar population synthesis models to compute directly the mean comoving emissivity $\mathcal{E}_\nu$ at wavelengths from $10^{-1}\mu$m to $10^3\mu$m and, by an integration over redshift, the corresponding mean intensity of background radiation J_ν. These calculations include a self-consistent treatment of the absorption and reradiation of starlight by the dust within galaxies. The same models shown in Figure 1 also predict a far-IR/sub-mm background in nice agreement with a tentative detection based on *COBE* data [23].

References

1. Pei, Y.C., & Fall, S.M. 1995, ApJ, 454, 69
2. Fall, S.M., Charlot, S., & Pei, Y.C. 1996, ApJ, 464, L43
3. Cowie, L.L. 1988, in The Post-Recombination Universe, eds. N. Kaiser

& A.N. Lasenby (Dordrecht: Kluwer), 1

4. Lanzetta, K.M., Wolfe, A.M., & Turnshek, D.A. 1995, ApJ, 440, 435
5. Timmes, F.X., Lauroesch, J.T., & Truran, J.W. 1995, ApJ, 451, 468
6. Malaney, R.A., & Chaboyer, B. 1996, ApJ, 462, 57
7. Kennicutt, R.C. 1989, ApJ, 344, 685
8. Wolfe, A.M., Lanzetta, K.M., Foltz, C.B., & Chaffee, F.H. 1995, ApJ, 454, 698
9. Storrie-Lombardi, L.J., McMahon, R.G., & Irwin, M.J. 1996, MNRAS, in press (astro-ph/9608147)
10. Pettini, M., Smith, L.J., Hunstead, R.W., & King, D.L. 1994, ApJ, 426, 79
11. Pettini, M., King, D.L., Smith, L.J., & Hunstead, R.W. 1996, ApJ, in press (astro-ph/9607093)
12. Pei, Y.C., Fall, S.M., & Bechtold, J. 1991, ApJ, 378, 6
13. Lu, L., Sargent, W.L.W., & Barlow, T.A. 1996, ApJS, in press (astro-ph/9606044)
14. Kulkarni, V.P., Fall, S.M., & Truran, J.W. 1996, in preparation
15. Levshakov, S.A., Chaffee, F.H., Foltz, C.B., & Black, J.H. 1992, A&A, 262, 385
16. Larson, R.B. 1972, Nature Phys. Sci., 236, 7
17. Hartwick, F.D.A. 1976, ApJ, 209, 418
18. Fall, S.M., & Pei, Y.C. 1993, ApJ, 402, 479
19. Gallego, J., Zamorano, J., Aragón-Salamanca, A., & Rego, M. 1995, ApJ, 455, L1
20. Lilly, S.J., Le Fèvre, O., Hammer, F., & Crampton, D. 1996, ApJ, 460, L1
21. Steidel, C.C., Giavalisco, M., Pettini, M., Dickinson, M., & Adelberger, K.L. 1996, ApJ, 462, L17
22. Madau, P., Ferguson, H.C., Dickinson, M.E., Giavalisco, M., Steidel, C.C., & Fruchter, A. 1996, MNRAS, in press (astro-ph/9607172)
23. Puget, J.-L., Abergel, A., Bernard, J.-P., Boulanger, F., Burton, W.B., Désert, F.-X., & Hartmann, D. 1996, A&A, 308, L5

THE HST QUASAR ABSORPTION LINE SURVEY AND THE RELATIONSHIP OF ABSORBERS TO LARGE SCALE STRUCTURES

BUELL T. JANNUZI

National Optical Astronomy Observatories
P.O. Box 26732, Tucson, AZ 85726, USA

The *Hubble Space Telescope* quasar absorption line survey provides a large database for the study of the evolution of the gaseous content of the Universe. As our understanding improves of the relationship between the gaseous structures observed as absorption lines and the dark and luminous material observed as individual galaxies, groups, and clusters, we are able to make better use of the absorption line systems as probes of the formation and evolution of all structures. At the meeting I presented the latest results from our analysis of the survey database relevant to this topic. In this contribution I present two of these results: the possible identification of gas associated with a cluster or super-cluster of galaxies at z= 0.7 as traced by unusual absorption in the spectrum of PG 2302+029 and our current determination of the evolution of the number of Ly-α absorbers with redshift.

1 The HST Quasar Absorption Line Survey

The HST quasar absorption line survey was an HST key project for cycles 1-3[1]. The primary goal of the survey was to produce a large and homogeneous catalogue of absorbers suitable for the study of gaseous systems at low redshifts. Discussions of the design and execution of the survey can be found in papers by Schneider *et al.*[2] and Jannuzi[3]. In the next two sections I present our most recent results relevant to our efforts to understand the connections between the gaseous structures and other identified structures in the Universe.

2 Unusual Absorption in PG 2302+029: Ejected or Intervening?

We have discovered a high-ionization broad absorption line system at a redshift of $z_{abs} = 0.695$ in the spectrum of the $z_{em} = 1.052$ radio-quiet quasar PG 2302+039[4]. Broad absorption with FWHM from 3,000 to 5,000 km s^{-1} is detected from C IV, N V, and O VI in HST FOS spectra of the quasar (Figure 1). A narrow line system (barely resolved, FWHM $\sim$ 250 km s^{-1}) at $z_{abs} = 0.7016$ is resolved from the broad blend and includes absorption by Ly-α and the C IV, N V, and O VI doublets (Figure 2). The centroids of the broad system lines are displaced by $\sim$ 56,000 km s^{-1} to the blue of the quasar's broad emission lines. The reddest extent of the broad line absorption

309

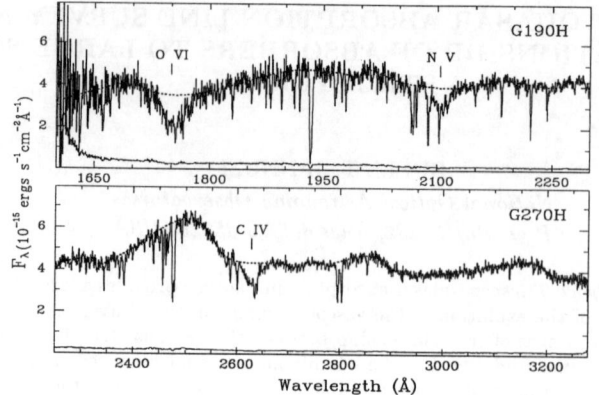

Figure 1: UV spectra of PG 2302+029 obtained with the FOS of the HST. The dotted line is the "continuum" fit and the lower line in each panel is the one σ uncertainty in the flux.

is more than 50,000 km s^{-1} from the rest frame of the quasar. This system's large velocity width and high ionization state are both characteristic of systems intrinsic to quasars (similar system in Q 2343+125 discussed by Hamann et al.[5]). However, the system's relatively symmetric profile, extremely large separation from the rest velocity of the quasar, and narrow absorption components are more characteristic of absorption by an intervening cluster or super-cluster of galaxies (see papers by Jakobsen[6,7], Sargent & Steidel[8], and Dinshaw & Impey[9] for discussions of such a system at $z = 2$ in the "Tololo Pair" quasars). We hope that future observations will allow a unique attribution of the cause of this broad system. However, whether the broad absorption complex of PG 2302+029 is caused by an intervening or an ejected system, its properties are unprecedented.

3   Evolution of Ly-α Absorbers

A main objective of the absorption line survey is to determine the nature of the evolution of low redshift Ly-α absorbers ($z < 1.6$). The ultimate goal is to relate the observed evolution to changes in other observed properties of the Universe. By the time this paper appears we will have completed our analysis using the entire database to determine the evolution of the number of Ly-α absorbers with redshift. As of July 1996 I can only present preliminary results determined from a sample of ≈ 550 Ly-α lines and using only those lines

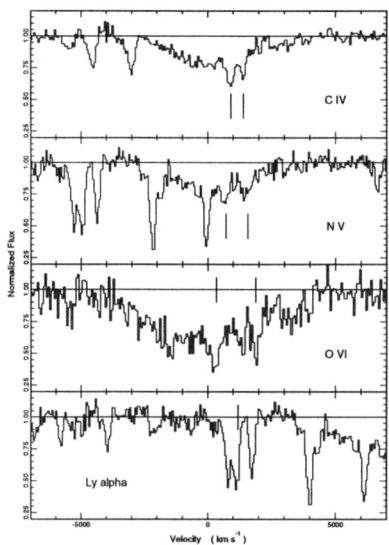

Figure 2: The detected broad absorption lines of PG 2302+029 are shown on a velocity scale. The locations of the lines in the narrow system are indicated with vertical tick marks.

that have redshifts larger than 3,000 km s^{-1} in the rest frame of the quasar. This subsample consists of approximately 75% of the complete database. For redshifts less than one, the density is remarkably flat (Figure 3), with $\gamma = 0.00 \pm 0.27$, $(dN/dz)_{z=0} = 33 \pm 3.8$, and $dN/dz = (dN/dz)_{z=0} \times (1+z)^{\gamma}$. This result is derived from a maximum likelihood estimation for the observed lines in those spectral regions where the 4.5 σ detection limit is less than a rest equivalent width of 0.24 Å. A fit to the absorbers with redshifts greater than 1.0 yields a more uncertain fit with $\gamma = 1.54 \pm 1.10$ and $(dN/dz)_{z=1} = 37 \pm 6.1$. The majority of the data not included in the current analysis is for quasars with redshifts between 1 and 1.8, so the uncertainty of this fit should be reduced when the entire data set is included. A fit to the entire redshift region from 0 to 1.6 yields a $\gamma = 0.46 \pm 0.187$ and $(dN/dz)_{z=0} = 28 \pm 2.8$. The slope of the observed low-redshift dN/dz relation differs significantly from the slope deduced from various ground-based samples that refer to redshifts $z > 1.6$.

Acknowledgments

I thank my collaborators, including the honoree of this meeting, Alec Boksenberg, for agreeing to let me share our results prior to joint publication. This research

312

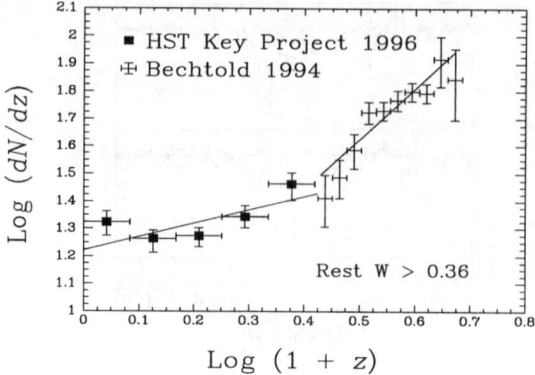

Figure 3: The number density of Ly-α absorbers as determined from our analysis of the key project database (as of July 1996) and the data of Bechtold (1994). For each data set a solid line shows the best fit power law. The fits were made using a maximum likelihood technique on the unbinned data.

is funded in part by the NSF's funding of NOAO and by grant GO-5343.01 from the STSCI, which is operated by the A.U.R.A., Inc., under NASA contract NAS5-26555. I thank F. Hamann, T. Barlow, and V. Junkkarinen for sharing their paper on Q 2343+125 prior to its publication.

References

1. J. N. Bahcall *et al.*, ApJS, 87, 1, 1993.
2. D. P. Schneider *et al.*, ApJS, 87, 45, 1993.
3. B. T. Jannuzi in *Science with the Hubble Space Telescope-II*, ed. P. Benvenuti, F. D. Macchetto, & E. J. Schreier (STSCI, 1996).
4. B. T. Jannuzi *et al.*, ApJ, Letters, in press, 1996.
5. F. Hamann, T. Barlow, & V. Junkkarinen, ApJ, submitted, 1997
6. P. Jakobsen & M. A. C. Perryman ApJ, 392, 432, 1992.
7. P. Jakobsen, M. A. C. Perryman, M. H. Ulrich, F. Machetto, & S. Di Serego Alighieri ApJ, 303, L27, 1986.
8. W. L. W. Sargent & C. C. Steidel ApJ, 322, 142, 1987.
9. N. Dinshaw & C. Impey ApJ, 458, 73, 1996.
10. J. Bechtold ApJS, 91, 1, 1994.

CHEMICAL EVOLUTION OF DAMPED LYMAN-ALPHA GALAXIES

LIMIN LU[a], WALLACE, L.W. SARGENT, & THOMAS A. BARLOW

Caltech, 105-24, Pasadena, CA 91125, USA

Studies of damped Lyα absorption systems found in the spectra of quasars provide valuable information on the chemical evolutionary history of high-redshift galaxies. Here we describe some key results from such a program.

1 Background

Damped Lyα absorption systems found in the spectra of high-redshift quasars refer to the class of intervening absorption systems that have the largest neutral hydrogen column density: $N(HI) \geq 2 \times 10^{20}$ cm^{-2}. The importance of damped Lyα systems for observational cosmology was not fully recognized until the first systematic survey by Wolfe and collaborators (Wolfe et al. 1986), who demonstrated that damped Lyα systems represent a significant population of high redshift objects with properties linking them to the progenitors of today's galaxies (cf. Wolfe 1988). The subject has since flourished, providing unique information on the properties and evolution of (especially high-redshift) galaxies and on structure formation in the early universe.

Many studies of elemental abundances in individual damped Lyα systems exist in the literature, which are too numerous to properly account for here. The most notable work was that of Pettini et al. (1994) who surveyed the abundances of Zn and Cr in 17 damped Lyα systems using medium resolution spectra. The results demonstrated that damped Lyα galaxies generally have low metallicities ($\sim 1/10$ solar), indicative of young galaxies in the early stages of chemical evolution.

2 Our Program

We have recently completed a new survey of elemental abundances in a sample of damped Lyα galaxies using spectra obtained with the 10m Keck telescope (Lu et al. 1996, hereafter LSBCV). In general, the Keck spectra are of much higher quality than those used in earlier studies, and the large format of the Keck echelle spectrograph allows the simultaneous analysis of many elements. The abundance data from our Keck observations are combined with similar

---

[a] Hubble Fellow

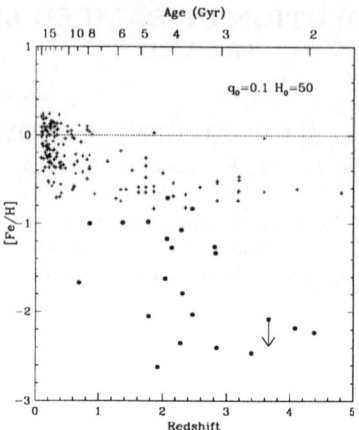

Figure 1: [Fe/H] vs z for damped Lyα systems (solid points) and for Galactic disk stars (pluses) from Edvardsson et al. (1993), where stellar ages have been converted into redshifts for $q_0 = 0.1$ and $H_0 = 50$. For reference, $[Fe/H] \equiv \lg(Fe/H)_{damp} - \lg(Fe/H)_\odot$.

data in the literature in order to study the chemical evolution of damped Lyα galaxies. Interested readers are referred to LSBCV for detailed analyses and discussion.

In figure 1, we show the distribution of [Fe/H] vs z for the sample of damped Lyα galaxies (solid points) and for the sample of solar-neighborhood Galactic disk stars (pluses) from the study of Edvardsson et al. (1993). The relative abundances of various elements, plotted against [Fe/H] of the systems, are shown in the six panels in figure 2. The main conclusions are summarized below.

1. In general, the damped Lyα galaxies have $-2.5 <$[Fe/H]< -1, corresponding to 1/300 to 1/10 solar metallicity. The N(H I)-weighted mean metallicity is 0.03 solar at $\langle z \rangle = 2.5$. The low metallicities are consistent with them being young galaxies in the early stages of chemical enrichment.

2. Comparisons of the distribution of [Fe/H] vs z for the sample of damped Lyα galaxies with the similar relation for the Milky Way disk as traced by stars in the solar neighborhood indicate that the damped Lyα galaxies are significantly less metal-enriched than the Galactic disk in its past. Taken at face value, this argues against the suggestion that damped Lyα galaxies are high-redshift (proto-) galactic disks (Wolfe 1988). Rather, the metallicities of the damped Lyα systems are consistent with those of globular clusters and nearby

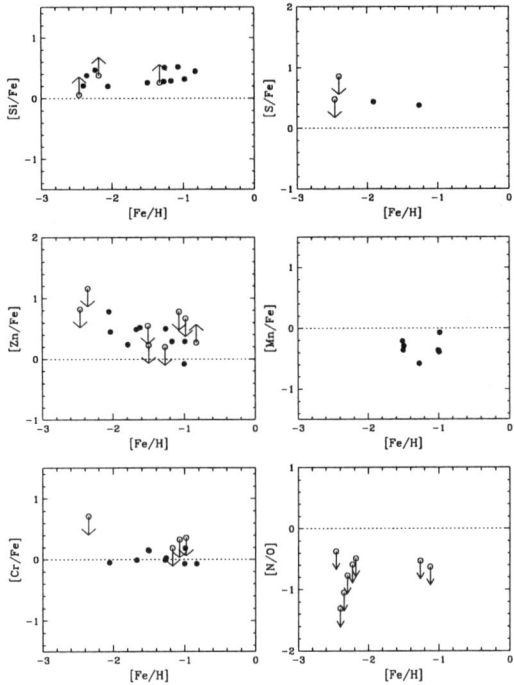

Figure 2: Relative abundance patterns for damped Lyα systems.

dwarf galaxies. This result may be in potential conflict with the evidence that damped Lyα systems appear to show kinematics characteristic of rotating disks (cf. Wolfe 1995).

3. There is evidence that the mean metallicity of damped Lyα galaxies increases abruptly at $z < 3$ compared to $z > 3$. If this trend is real, it may suggest that the time around $z = 3$ may be the epoch of galaxy formation in the sense that galaxies are beginning to form the bulk of their stars. Several other lines of evidence appear to point to the same general direction; these include a decline in the neutral gas contents of damped Lyα galaxies at $z > 3$ (Storrie-Lombardi et al. 1995), a rapid decline in the space density of quasars at $z > 3$ (cf. Schmidt, Schneider, & Gunn 1995), and the morphology of $z > 3$ galaxies, which Steidel et al. (1996) and Giavalisco et al. (1996) have argued to indicate the formation of stars in the spheroidal component of normal galaxies.

4. The relative abundance patterns shown in figure 2 are most consistent with the abundance pattern seen in Galactic halo stars (cf. Wheeler, Sneden, & Truran 1989), including the low N/O ratio, the overabundance of Si and S relative to Fe (i.e., enhancement of α-process elements over iron-peak elements), and the underabundance of Mn relative to Fe (i.e., the odd-even effect). The implication is that metal productions in the vast majority of damped Lyα systems in the sample were dominated by massive stars via Type II SN. There was little evidence for a significant contribution from Type Ia SN or from stellar mass loss of low-to-intermediate mass stars. However, the observed Zn/Fe ratio in damped Lyα systems is inconsistent with the abundance patterns seen in halo stars. This might suggest that there is a small degree of dust depletion of refractory elements (e.g., Fe) on top of the SN II nucleosynthesis pattern. However, the amount of depletion allowed by the data is not large enough to significantly affect any of the conclusions above. Detailed discussions on the effects of dust depletion are presented in LSBCV.

Acknowledgments

LL is supported by a Hubble Fellowship (HF1062.01-94A) from NASA. WWS acknowledges support from NSF grant AST92-21365.

References

1. Edvardsson, B. et al. 1993, A&A, 275, 101
2. Giavalisco, M., Steidel, C.C., & Macchetto, F.D. 1996, ApJ, in press
3. Lu, L., Sargent, W.L.W, Barlow, T.A., Churchill, C.W., & Vogt, S. 1996, ApJS, December issue
4. Pettini, M., Smith, L.J., Hunstead, R.W., & King, D.L. 1994, ApJ, 426, 79
5. Schmidt, M., Schneider, D.P., & Gunn, J.E. 1995, AJ, 100, 68
6. Steidel, C.C., Giavalisco, M., Pettini, M., Dickinson, M., & Adelberger, K.L. 1996, ApJ, 462, L17
7. Storrie-Lombardi, L.J., McMahon, R.G., Irwin, M.J., & Hazard, C. 1995, in QSO Absorption Lines, ed. G.Meylan (Springer-Verlag), 47
8. Wheeler, J.C., Sneden, C., & Truran, J.W. 1989, ARA&A, 27, 279
9. Wolfe, A.M. 1988, in QSO Absorption Lines: Probing the Universe, eds. Blades, Turnshek, and Norman (Cambridge Univ Press), p297
10. Wolfe, A.M. 1995, in QSO Absorption Lines, ed. G.Meylan (Springer-Verlag), p13
11. Wolfe, A.M., et al. 1986, ApJS, 61, 249

SEEING DOUBLE: PROBING THE UNIVERSE WITH QUASAR PAIRS

C.D. IMPEY

Steward Observatory, University of Arizona,
Tucson, AZ 85721, USA

Quasar absorption along paired and multiple lines of sight can be used to detect large scale structure at substantial lookback times. Superclustering of CIV absorbers has been seen towards a set of four quasars, including the pair Tol 1037−2704/Tol 1038−2712. Size measurements are also presented for Lyman-alpha absorbers using three quasar pairs which probe scales from tens to hundreds of kpc. The results show Lyman-alpha absorbers to be large and quiescent, with transverse sizes that may increase with decreasing redshift.

1 Introduction

Quasar absorption lines provide a powerful way of measuring the incidence of normal baryonic material from the local universe to redshift $z = 5$. Unlike the case with galaxies, quasar absorbers can be selected with an efficiency that is almost independent of redshift, given a bright enough illuminating source. The accurate velocity information gives kinematic information on the absorber, and different species of heavy elements give a clue to the chemical enrichment and ionization history of the universe (eg. Sargent [1]). With adjacent and multiple lines of sight, the three dimensional structure of the absorbers can be mapped out. The two types of probes discussed here are the CIV absorbers with $dN/dz \sim 1$, which are tracers for galaxy halos, and the copious low column density Lyman-alpha absorbers with $dN/dz \sim 100$, which until recently were thought to represent pristine intergalactic material.

2 Superclustering of CIV Absorbers

Heisler *et al* [2] were the first to detect a superclustering signal in the two-point correlation function of CIV absorbers, given mostly by the single line of sight toward the quasar PKS 0237−233. Subsequently, an overdensity of CIV absorbers was observed along the line of sight to the quasar pair Tol 1037−2704/Tol 1038−2712 (Jakobsen *et al* [3], Sargent and Steidel [4]). Recent work has added two nearby quasars to the original pair, with a total of 44 CIV systems distributed among the four lines of sight (Dinshaw and Impey [5]). In the search for spatial correlations, subcomponents with velocity splittings of less than 1000 kms^{-1} are treated as one absorber, so that the motions of

Table 1: Maximum liklihood size estimates of Lyman-alpha absorbers based on the numbers of coincident and anti-coincident lines.

Quasar Pair	Redshift Range	Most Probable Size (h_{100}^{-1} kpc)	95% Confidence Limits on Size (h_{100}^{-1} kpc)
Q 1343+264 A,B	$1.7 < z < 2.1$	$<R> = 125$	$80 < R < 580$
LB 9605, LB 9612	$1.1 < z < 1.7$	$<R> = 380$	$305 < R < 590$
Q 0107−025 A,B	$0.5 < z < 0.9$	$<R> = 505$	$360 < R < 1535$

clouds within a single galaxy halo are not aliased into a signal on larger scales. The result is a significant correlation signal with velocity separations of 4000-7000 kms^{-1}, corresponding to spatial scales of 30-40 h_{100}^{-1} Mpc. The velocity correlation is $\xi(v) = 2.35 \pm 0.55$, which is much larger than the expected two point correlation function of clusters. More lines of sight will be needed to decide whether this CIV overdensity is a rare fluctuation, or whether it is indicative of strong superclustering of galaxy halos at $z \sim 2$.

3 The Size of Lyman-Alpha Absorbers

The low column density hygrogen absorbers seen in quasar spectra represent an important baryonic component of the universe. No physical model of the absorbers can be constructed without information on their size and geometry. Three quasar pairs (which are not gravitationally lensed) have been observed to look for coincident and anti-coincident Lyman-alpha lines. If a larger number of coincidences are measured than would be expected by chance, then the absorbers must be at least the size of the transverse separation between the lines of sight. Ground-based observations of the pair Q1343+264 A,B yield 8 coincident lines within 150 kms^{-1} around $z \sim 2$, where only 0.2 would be expected by chance. FOS observations of the pair LB 9605 and LB 9612 with the Hubble Space Telescope can be used either to make a marginal measurement of characteristic size at $z \sim 1.5$, or to place a firm upper limit on absorber size at that redshift. FOS observations of a third pair, Q 0107−025 A,B, yield 4 coincident lines at $z \sim 0.7$, where only 0.2 would be expected by chance. The resulting maximum liklihood size estimates are shown in the table above.

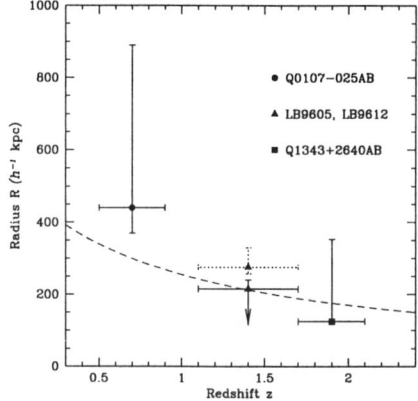

Figure 1: Estimates of absorber radius in three redshift intervals. The dotted error bars for the LB pair are corrected for the expected number of chance coincidences. The dashed line represents a freely expanding Hubble sphere with arbitrary normalization.

The maximum liklihood method was corroborated with Monte Carlo simulations, which yielded the same results. In the simulations, absorbers were assumed to be uniform spheres. However, it is clear that in general the sizes quoted in the table represent coherence lengths rather than diameters of spherical objects. Fang et al[6] found evidence that the characteristic absorber size increased with increasing line of sight separation, which might indicate clustered absorbers or a distribution of absorber sizes. The addition of the LB pair to their plot weakens this trend. However, the combination of the data on the three pairs in Figure 1 provides tantalizing evidence for an evolution in the size of the absorbers, increasing with time. The data span a redshift range during which the dN/dz evolution of the absorbers changes from rapid at $z = 2$ to essentially no evolution since $z = 1$. The pairs analysis is still limited by the small number of coincident and anti-coincident lines; more pairs or observations to lower equivalent width limits will be required to confirm this trend. It is consistent with absorbers in free expansion, or with the absorbers becoming more neutral as the UV photoionizing flux decreases at recent epochs.

4 Models and Simulations

A new statistical technique was used to test the relative liklihood of four geometric models: spherical absorbers with and without a size distribution, randomly inclined disks, and filaments with an axis ratio of 5:1. The absorbers
are given power law radial profiles of column density, such that a population of
the absorbers will match the column density distribution of the Lyman-alpha
"forest." Both coincidences and anti-coincidences figure into the liklihood calculation. Spherical absorbers with a fixed radius are ruled out. The two
favored geometries are filaments and randomly inclined disks. These results
are in striking accord with N-body simulations which incorporate gas dynamics, where Lyman-alpha absorption arises in diverse structures with coherence
lengths as large as 1 h_{100}^{-1} Mpc (Cen et al[7]). A direct comparison has been
made between the observations and SPH simulations at $z = 2$ (Hernquist et
al[8]), with the simulation spectra degraded to the resolution and SNR of the
observations, and lines are selected in the same way. On scales of 40 h_{100}^{-1} kpc,
the 95% confidence limits on the expected fraction of coincident pairs within
$\Delta v < 150$ kms^{-1} is 0.53-0.95 compared with a simulation prediction of 0.73
(Dinshaw et al [9]). On scales of 500 h_{100}^{-1} kpc, the observed fraction is 0.04-
0.29 compared with a simulation prediction of 0.39 (Crotts et al [10]). These
comparisons will be extended to a wider range of redshifts and pair separations.

Acknowledgments

I acknowledge Nadine Dinshaw, Craig Foltz, Ray Weymann, and Simon Morris,
a University of Arizona travel grant, and NSF grant AST 93-20715.

References

1. W.L.W. Sargent in *QSO Absorption Lines*, ed. C. Blades, D. Turnshek,
   and C. Norman (Cambridge University Press, Cambridge, 1988).
2. J. Heisler *et al* ApJ, 347, 52, 1989.
3. P. Jakobsen *et al*, ApJ, 303, L27, 1986.
4. W.L.W. Sargent and C. Steidel, ApJ, 322, 142, 1987.
5. N. Dinshaw and C.D. Impey, ApJ, 458, 73, 1996.
6. Y. Fang *et al*, ApJ, 462, 77, 1996.
7. R. Cen *et al* ApJ, 437, L9, 1994.
8. L. Hernquist *et al* ApJ, 457, L51, 1996.
9. N. Dinshaw *et al* ApJ, 437, L87, 1994.
10. A.P.S. Crotts, ApJ, 336, 550, 1989.

HIERARCHICAL CLUSTERING AND METAL ABSORPTION FROM GALAXIES IN THE PROCESS OF FORMATION

MARTIN G. HAEHNELT

Max-Planck-Institut für Astrophysik,
Karl-Schwarzschild-Straße 1, 85740 Garching, Germany

In a hierarchical cosmogony present-day galaxies build up by continuous merging of smaller structures. Hydrodynamical simulations were used to calculate the metal absorption properties of regions of ongoing galaxy formation. Predictions of the spatial distribution of H_I, C_{II}, C_{IV}, Si_{IV} and O_{VI} are presented. The observed column density ratios of the different ionic species can be well reproduced if a mean metallicity [Me/H]=−2.5, relative abundances as found in metal-poor stars and an UV background as proposed by Haardt & Madau [1] are assumed. The observed scatter in [C/H] is about a magnitude larger than that in the simulations suggesting an inhomogeneous metal distribution caused by *in situ* enrichment.

1 Introduction

While at low redshift metal absorption systems have been convincingly demonstrated to arise in the haloes of rather normal galaxies [2][3] as suggested by Bahcall & Spitzer [4], much less is known about the nature of metal absorption system at $z > 2$ [5][6][7]. Recently, numerical simulations performed with GRAPESPH [8] and CLOUDY [9] and assuming a cold dark matter cosmogony were used to show that the prominent complex C_{IV} absorption features observed at high redshift can be well reproduced by the absorping properties of regions in which galaxies form by hierarchical merging [10]. Here further results concerning the metal absorption properties of such regions are presented (see Navarro & Steinmetz [11] for more numerical details of the simulation and Rauch, Haehnelt & Steinmetz [12] for a detailed description of their metal absorption properties).

2 The spatial distribution of the ionic species

Figure 1 shows the projected column densities of the inner 700 kpc cube of one of the simulations at $z = 3.07$. The lowest H_I column densities (10^{12} to 10^{14} cm^{-2}) arise from large-scale sheet-like density enhancements in the IGM. In this diffuse gas, with densities around 10^{-5} cm^{-3}, high ionization species are prevalent and O_{VI} $\lambda 1031$ is often the strongest metal absorption line. Towards higher H_I column densities we start probing filaments embedded in the large-scale sheets. In these regions low column density $C_{IV}\lambda\lambda 1548, 1550$ lines from infalling gas with densities around 10^{-4} cm^{-2} (overdensities of about 10 to

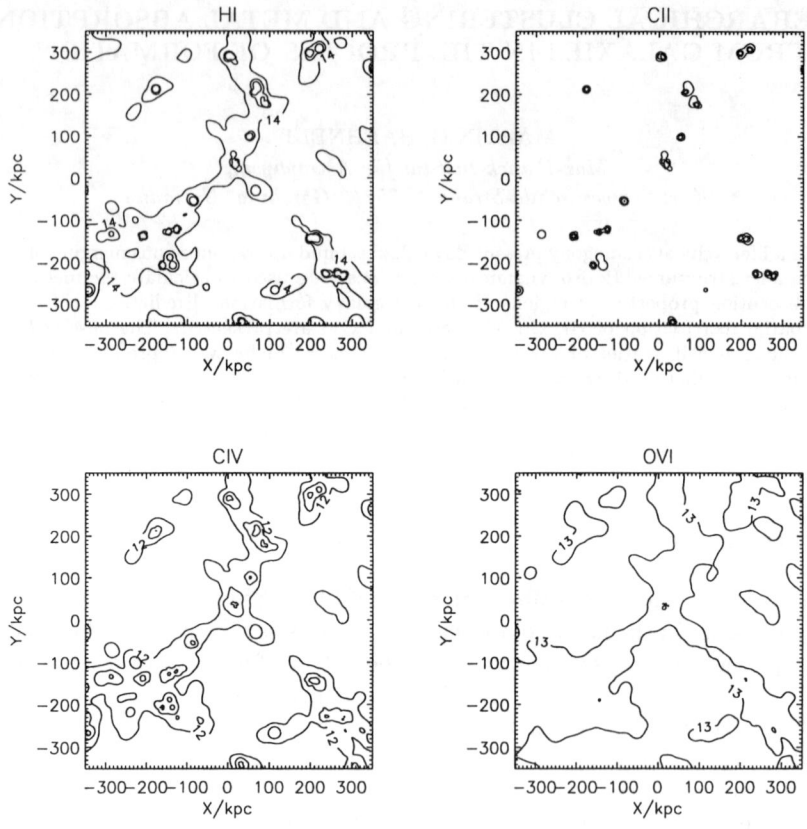

Figure 1: Projected column density of the inner 700 kpc of a simulation box at $z = 3.07$. Shown are logarithmic column density contours in steps of 1 dex for Hı, Cıı, Cıv and Ovı.

100) dominate the metal absorption features. Cıv remains the most easily visible metal ion in the as yet unvirialized regions around the protogalactic clumps which are later to merge into present-day galaxies. Still larger Hı column densities occur for lines-of-sight approaching the central regions of protogalactic clumps, giving rise to Lyman limit systems and eventually to damped Lyα absorbers. Total densities here exceed 10^{-4} cm^{-3}, and species like Cıı and Sııv become increasingly prominent.

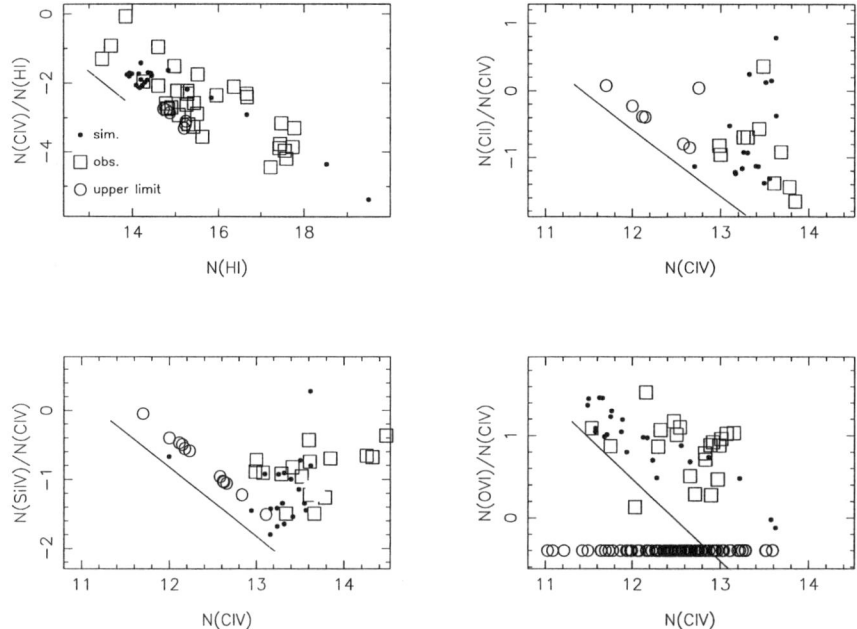

Figure 2: Filled circles denote column density ratios obtained from profile fitting the simulated data. An homogeneous metallicity [Me/H]=−2.5, relative abundances as in metal poor stars ([C]=0, [Si]=[O]=0.4) and the UV background proposed by Haardt & Madau have been assumed. The thin diagonal lines give upper limits for the simulated lines. Open squares mark observed values and open circles either observational upper limits (upper two and bottom left panels) or Ovi undetermined due to blending (bottom right panel).

3 Column density ratios

Detailed information on the accompanying metal absorption has recently become available for H_I column densities as small as a few times $10^{14} \, cm^{-2}$. As demonstrated in figure 2 good overall agreement between the results from the artificial spectra and the observed properties [13] [14] is obtained. This implies either that much of the as yet unvirialized gas had been subject to a widespread phase of stellar nucleosynthesis well before redshift three, or, alternatively, that metal transport outward from fully collapsed regions has been efficient. The scatter of the column density ratios for Si_{IV}/C_{IV}, C_{II}/C_{IV}, and O_{VI}/C_{IV} versus C_{IV} is consistent with the observational results, so the range of ionization conditions appears to be well captured. Nevertheless, the observed scatter in [C/H] is larger by a factor three to ten than predicted by the numerical simulations where all metals were distributed homogeneously. This may indicate that some of the metal enrichment took place *in situ* (with incomplete mixing prior to observation). Alternatively there may be a wider spread in physical conditions (e.g. spatial variations of the UVfield) than assumed by the simulations.

4 Conclusions

High-resolution hydrodynamical simulations of galaxy forming regions can substantially aid the interpretation of metal absorption systems. Even so substantial simplifications are present in the current work the good agreement of simulated and observed column density ratios gives reason to believe that the underlying mechanism producing many of the metal absorption systems at high redshift has been correctly identified.

Acknowledgement

I thank Michael Rauch and Matthias Steinmetz for allowing me to present joint results of our colloboration.

References

1. Haardt, F., Madau, P., 1996, ApJ, 461, 20
2. Bergeron J., Boissé P., 1991, A&A, 243, 344
3. Steidel C.C., 1995, in *QSO Absorption Lines*, Proc. ESO Workshop, ed. G.Meylan (Heidelberg: Springer), p. 139.
4. Bahcall J.N., Spitzer L., 1969, ApJ, 156, L63
5. Sargent W.L.W., Boksenberg A., Steidel C.C., 1988, ApJS, 68, 539
6. Petitjean P., Bergeron J.A., 1994, A&A, 283, 759
7. Aragon-Salamanca A., Ellis R.S., Schwartzenberg J.-M., Bergeron J.A., 1994, ApJ, 421, 27
8. Steinmetz M., 1996, MNRAS, 278, 1005
9. Ferland G.J., 1993, University of Kentucky Department of Physics and Astronomy Internal Report
10. Haehnelt M.G., Steinmetz M., Rauch M., 1996, ApJ, 465, L95
11. Navarro J., Steinmetz M., 1996, ApJ, submitted
12. Rauch M., Haehnelt M.G., Steinmetz M., 1996, ApJ, submitted
13. Songaila A., Cowie L.L., 1996, AJ, 122, 335
14. Rauch M., private communication

A SEARCH FOR QSO ABSORPTION LINES FROM NEARBY STARBURST GALAXIES

DAVID V. BOWEN

Royal Observatory, Blackford Hill, Edinburgh, EH9 3HJ, U.K.

COLIN A. NORMAN & TIM HECKMAN

Bloomberg Center, Johns Hopkins University, Baltimore, MD 21218, USA

J. CHRIS BLADES & LAURA DANLY

STScI, 3700 San Martin Drive, Baltimore, MD 21218, USA

1   Should QSO absorption lines be related to starburst galaxies?

Starburst galaxies are known to pump prodigious amounts of mass, energy, and momentum into their circumgalactic halos and the surrounding intergalactic medium. Gaseous outflows can be seen in warm gas (via optical line emission), hot gas (via X-ray emission), and relativistic plasma (via radio synchrotron emission). In particular, these flows represent one of the few *direct* observations which show that interstellar gas can be distributed many kpc outside of a galaxy. In the search for the origin of QSO absorption line systems, it is therefore important to test whether starburst outflows could be responsible for the absorption lines seen in QSO spectra at high-redshift.

We are therefore conducting a programme to search for absorption lines toward nearby starburst galaxies, using background QSOs as probes. Not only do we wish to examine whether absorption from outflows can mimic the high-z absorption systems, but the observations should provide important information on the physical state of these complex galactic processes, information which cannot be found in any other way. We have therefore observed two QSOs behind NGC 253 and two QSOs behind NGC 520, using the FOS aboard the *Hubble Space Telescope (HST)*, to search for absorption from these outflows.

2   Results

NGC 520: Strong Mg II absorption is seen towards Q0121+0327, 26 h^{-1} kpc from the center of the galaxy[a], at a velocity of 2285 km s^{-1}, close to the systemic velocity of the galaxy. The sightline toward Q0122+0338 passes twice as far away (52 h^{-1} kpc), but there is evidence for weak Mg II absorption at

---

[a]$h = H_0/100$, where H_0 is the Hubble constant, $q_0 = 0$

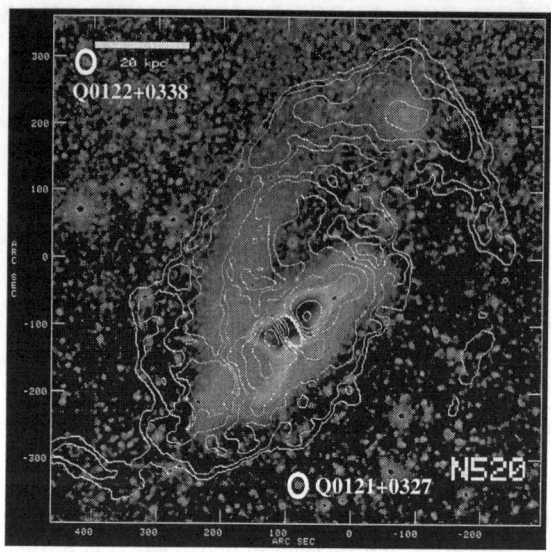

21cm HI contour map of NGC 520 superimposed on optical image. Lowest HI contour shown is 2×10^{19} cm^{-2} . Positions of QSOs are indicated by white circles.   Courtesy of John Hibbard (http://galileo.ifa.hawaii.edu/~hibbard/home.html)

2490 km s^{-1}, higher than the galaxy's systemic velocity. The Mg II absorbing gas detected could, in principle, be caused by outflows from the center of the galaxy. However, as Fig. 1 shows, NGC 520 is embedded in an extended H I halo arising from interactions with a neighbouring dwarf galaxy. There are several cases now of QSO absorption lines arising in tidally disrupted gas, and it seems more likely that this is yet another example.

NGC 253: the two sightlines toward Q0042$-$2450 & Q0048$-$2608 both pass 55 h^{-1} kpc, each in opposite quadrants, along the extension of the galaxy's minor axis. We detect no Mg II along either sightline to 3σ equivalent width limits of 0.23 Å. The lack of absorption is surprising, considering that the existence of an outflow from the galaxy is well established. However, the non-detection does not preclude the existence of more diffuse, more highly ionized gas at large distances. ROSAT observations of NGC 253 show that gas at $\sim$ 10 kpc is at a temperature of $T \sim 2 \times 10^6$ K; for adiabatic cooling in a free wind, T declines with radius as $r^{-4/3}$, so at a distance of 55 h^{-1} kpc, $T \sim 2 \times 10^5$ K. Hence future searches for highly ionized gas, such as Si IV, C IV, N V & O VI with *HST* are particularly important.

Lyα ABSORBERS IN THE NEARBY UNIVERSE

DAVID V. BOWEN

Royal Observatory, Blackford Hill, Edinburgh, EH9 3HJ

MAX PETTINI

Royal Greenwich Observatory, Madingley Rd., Cambridge CB3 0EZ

We present initial results of a galaxy redshift survey designed to determine the association between Lyα forest lines and galaxies in the nearby universe.

Until the launch of the *Hubble Space Telescope* (*HST*), the origin of the dense Lyα-forest absorption lines seen in high redshift QSO spectra remained a mystery. Since Lyα lines can only be detected from the ground above redshifts of $z \approx 1.8$, finding the galaxies responsible for the absorption systems was impossible. The discovery of Lyα absorption lines in UV *HST* spectra has, however, now made it possible to search for galaxies which might be associated with these absorption systems. In particular, identifying the association of Lyα clouds with galaxies at the lowest redshifts ($z < 0.1$) is vitally important for understanding the evolution of the Lyα-forest.

To investigate the origin of the nearby Lyα-forest, we require complete knowledge of the distribution of galaxies along a QSO line of sight. We have thus begun a programme to measure galaxy redshifts using the multi-fibre spectrographs WYFFOS on the *WHT* and HYDRA on the *WIYN* telescopes. These instruments are ideally suited to a low-z survey of galaxies since their wide field of view samples group- to cluster-like scale lengths. We illustrate our results using the field of Q1821+643, a QSO which has been observed with both FOS [1] and GHRS [2], and for which Lyα-forest lines have been identified. We have measured the redshifts of all non-stellar objects with magnitudes $R \leq 18.5$ within 30' of the QSO sightline. This corresponds, for example, to an absolute magnitude of $M_R = -17.4$ and separation of $1.2h^{-1}$ Mpc[a] at $z = 0.05$. We have obtained spectra of 43 galaxies with 3 WYFFOS configurations, and 18 galaxies with one configuration of HYDRA. The distribution of the galaxies is shown in Figure 1.

We find that towards Q1821+643, the majority (7 of 8) of Lyα-forest clouds in the low redshift universe are not associated with bright galaxies on scales of < 1 h^{-1} Mpc. Two galaxies are detected within 300 h^{-1} kpc of the QSO sightline; one bright ($M_R = -20.3$) galaxy, 108 h^{-1} kpc away, is

---

[a]$h = H_0/100$, where H_0 is the Hubble constant, $q_0 = 0$

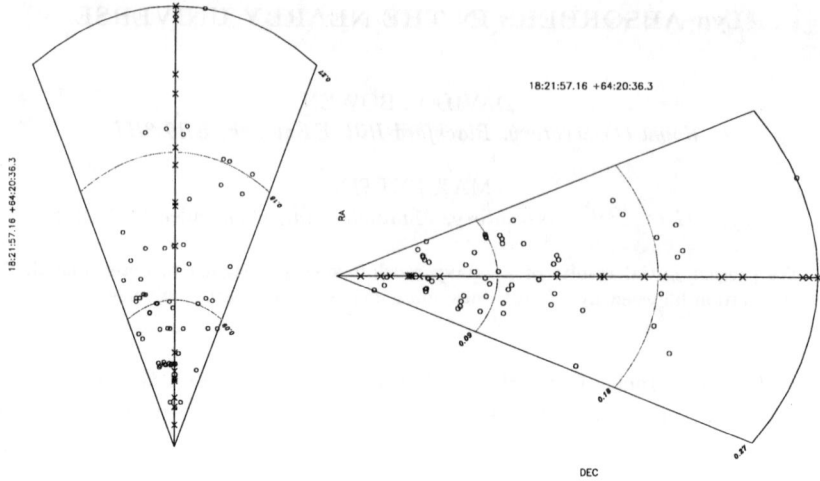

Figure 1: Wedge diagrams showing the spatial distribution of galaxies observed with WYF-FOS and HYDRA as a function of z, plotted in RA (left) and DEC (right). Redshifts of Lyα lines are plotted along the QSO line of sight as crosses, while galaxies are shown as circles.

associated with strong Lyα absorption, but no C IV absorption, while a fainter ($M_R = -16.3$) galaxy, 293 h^{-1} kpc away, shows no absorption, even to the sensitive limits obtainable with the GHRS. The distribution of galaxies shown above suggests that a cluster of Lyα clouds at $z = 0.040 - 0.045$ resides in a void foreground to a small shell of galaxies at $z = 0.05$.

These data provide important information on the associations between galaxies and Lyα clouds in the nearby universe. Comparing these and future results with data obtained at higher redshift will enable us to trace the evolution of the Lyα-forest over the last few billion years.

References

1. Bahcall, J. N., Jannuzi, B. T., Schneider, D. P., Hartig, G.F. & Green, R.F. 1992, ApJ, 397, 68
2. Savage, B. D., Sembach, K. R., & Lu, L. 1995, ApJ, 449, 145

EVOLUTION OF THE QUASAR METALLIC ABSORPTION SYSTEMS

K. DENDA

*Institute of Astronomy, University of Tokyo, 2-21-1 Osawa, Mitaka,
Tokyo 181, JAPAN*

I examine a consistent picture of evolution of the quasar metallic absorption systems based on observation and photoionization model. In order to explain the observed number per unit redshift (dn/dz) of Lyman limit system (LLS) and Mg II system at low redshifts, gas consumption in gaseous halo is required. Since the effect of chemical evolution cancels out the effect of evolution of the UV background, dn/dz of such systems show apparently "no evolution" feature. On the other hand, dn/dz of C IV system is sensitive to the ionization states of absorber, and may give constraints both on the amplitude of a break at the He II ionization edge in the ionizing spectrum and on gas density of absorbers.

1 Observational Data and Model

I compare the numerical results with the observational data obtained by Steidel[1] and Bahcall et al.[2] for C IV system, for LLS Sargent et al.[3] and Stengler-Larrea et al.[4], and Steidel & Sargent[5] for Mg II system.

The integrated UV background radiation is calculated as same as Madau[6], except that He I absorption is included, and that new H I column-density distribution is adopted (Bechtold[7]; Stengler-Larrea et al.[4]),

$$f(N_{\text{H I}}, z) \propto \begin{cases} N_{\text{H I}}^{-1.5}(1+z)^{1.32}; & (10^{13} < N_{\text{H I}} < 1.58 \times 10^{17} \text{cm}^{-2}), \\ N_{\text{H I}}^{-1.5}(1+z)^{1.50}; & (1.58 \times 10^{17} < N_{\text{H I}} < 10^{22} \text{cm}^{-2}). \end{cases} \quad (1)$$

Absorbing Halo is assumed to be a gaseous sphere with a density distribution, $n_{\text{H}} \propto r^{-2}$. The total gas mass is $M_{gas} = 10^{10}$–$10^{11} M_{\odot}$. I apply the chemical evolution model of DLA (Lanzetta et al.[8]), except that I set $Z(z=0)/Z_{\odot} = 0.1$. Ionization states are determined under the photoionization and thermal equilibria. Observed number of absorbers per unit redshift dn/dz is calculated

$$\frac{dn}{dz} \propto n_a \int \left| \frac{pdp}{dN_X} \right| dN_X (1+z)^{1-q_0}; \quad q_0 = 0.5. \quad (2)$$

Here I assume that number density of absorbers n_a keeps comoving constant.

2 Results

Gas consumption is crucial in order to explain the dn/dz of LLS and Mg II system at low redshifts. If carbon is much highly ionized, the dn/dz of C IV

330

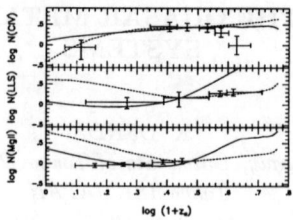

Figure 1: Evolution of dn/dz of quasar metallic absorption systems, derived by the UV background (dashed line), and by the UV background + chemical evolution (solid line).

system cannot be reproduced. When the ionizing spectrum is given, the lower limit on gas density is determined. When the gas density is fixed, the He II break in the UV spectrum may determine the dn/dz of the C IV system. The C IV column-density distribution provides the upper limit on gas density. If gas density is high, the resultant slope of $N_{C\ IV}$ is flatter than that observed (Petitjean & Bergeron [9]). If the filling factor is unity, the typical halo has therefore the total gas mass of $M_g \sim 5 \times 10^{10} M_\odot$ at $z \simeq 2$. When I adopt the chemical evolution model considering the dust obscuration (Pei & Fall [10]), my models can hardly explain the observation of the C IV systems.

Acknowledgment

My travel expense was partly supported by a Grant-in-Aid of the Foundation for Promotion of Astronomy in Japan.

References

1. C.C. Steidel, ApJS, 72, 1, 1990.
2. J.N. Bahcall *et al.*, ApJS, 87, 1, 1993.
3. W.L.W. Sargent *et al.*, ApJS, 69, 703, 1989.
4. E.A. Stengler-Larrea *et al.*, ApJ, 444, 64, 1995.
5. C.C. Steidel and W.L.W. Sargent, ApJS, 80, 1, 1992.
6. P. Madau, ApJ, 389, L1, 1992.
7. J. Bechtold, ApJS, 91, 1, 1994.
8. K.M. Lanzetta *et al.*, ApJ, 440, 435, 1995.
9. P. Petitjean and J. Bergeron, A&A, 283, 759, 1994.
10. Y.C. Pei and S.M. Fall, ApJ, 454, 69, 1995.

PROPERTIES OF THE Lyα ABSORBING SYSTEMS AT HIGH-z

A. DE LA FUENTE
INSA/VILSPA, Villafranca del Castillo, Apartado 50727,28080-Madrid, Spain

P. M. RODRIGUEZ–PASCUAL
ESA IUE Observatory, Apartado 50727, 28080-Madrid, Spain.
Affiliated to the Astrophysics Division, Space Science Department

J. L. SANZ
Instituto de Física de Cantabria (CSIC-UC), Santander (Cantabria), Spain

M. C. RECONDO
ESA IUE Observatory, Apartado 50727, 28080-Madrid, Spain

Summary

A global sample of Lyα lines from seven QSOs ranging from redshift 2.14 to 4.51, observed at a similar S/N ($\geq$ 10) ratio and high-resolution has been selected from the literature.

The study [Ref. 1] of this sample confirms a plateau and a strong fall-off (cut-off point at log $N_{HI} \simeq 14$) in the N_{HI} distribution. The b distributions above and below such a cut-off show similar asymmetric bell-shape form but a significant different mean and median values ($< b >= 28, 36$ km/s and $b_{med} = 32, 37$ km/s, for weak and strong lines respectively), arguing against the independence of b and N_{HI}. Supporting such a relation, the diagram b-N_{HI} suffers a deficit of strong lines with low b values (K-S probability of the data coming from uncorrelated distributions $p_{K-S} = 0.047$).

We have found that the number density of systems per unit interval of redshift can be fitted to a power-law with index $\gamma = 2.35 \pm 0.22$ when all the QSOs are included but a higher slope $\gamma = 3.22 \pm 0.41$ is obtained for redshift range $2.1 < z < 3.7$.

The characteristic Doppler parameters do not show evolution with z, for weak (log N_{HI} <14) absorbing systems, while for strong (log N_{HI} >14) ones, typical b values decrease with z.

References

1. de la Fuente, A., Rodríguez-Pascual, P. M. ,Sanz, J. L., Recondo, M. C. MNRAS, 281, 463, 1996.

GHRS SPECTROSCOPY OF Q0302-003

SARA R. HEAP

Goddard Space Flight Center, Greenbelt MD 20771 U.S.A

The Goddard High Resolution Spectrograph (GHRS) has obtained spectra of QSO 0302-003 in an effort to detect and measure HeII Lyα absorption from the diffuse intergalactic medium and/or discrete clouds. The strong observed absorption – even in the "Dobrzycki-Bechtold void" – suggests that at least some of it originates in the diffuse gas.

The flux of a quasar shortward of the rest-frame wavelength of HeII Ly α (304Å) should be depressed due to absorption by He^+ ions in a diffuse intergalactic medium (the Gunn-Peterson effect) and/or absorption by He^+ ions in clumps or clouds that make up the Ly α forest[1,2]. Such absorption was recently detected in a spectrum of the $z = 3.286$ quasar, Q0302-003, obtained with the Faint Object Camera[3]. The spectral resolution and signal-to-noise of the FOC data, however, were not sufficient to show whether the absorption is produced by a uniformly distributed medium or by a forest of Lyα lines. Last year, Songaila, Hu & Cowie[4] obtained a high-resolution spectrum of Q0302-003 using the HIRES echelle spectrograph on Keck. The spectrum shows a multitude of Lyα–forest clouds[5], some with column densities as low as 2×10^{12} cm^{-2}. Songaila et al. argue that absorption in these clouds may be sufficient to account for the observed UV absorption. Q0302-003 has now been observed[6] by the Goddard High Resolution Spectrograph (GHRS) at 0.6 Å resolution, and it is a planned target of observation by the Space Telescope Imaging Spectrograph (STIS), which is scheduled for installation on the Hubble Space Telescope in February 1997. In this paper, we describe the findings from the GHRS observations.

Figure 1 shows the reduced GHRS spectra of Q0302: a 100-min exposure obtained in a standard integration (top), and a 400−min exposure obtained using a special noise-censoring technique called the "FLYLIM" mode (bottom). The GHRS spectra are similar to the FOC spectrum in showing emission by HeII Ly α redshifted to 1302.1 Å, and in showing a depressed flux level shortward of Ly α. Between 1280-1295 Å, there is a ledge in the flux level which may be the signature of the proximity effect. The standard ACCUM-mode spectrum indicates an opacity in the HeII trough (1240-1280 Å) of $\tau = 1.7 \pm 0.3$, where the uncertainties take into account errors in the extrapolated continuous flux (dashed line) and the measured flux in the trough, but do not include possible systematic errors due to sky emission partially filling in the trough.

The FLYLIM spectrum yields a much higher opacity, nominally $\tau = 4$. The spectrum, however is liable to an unknown, but potentially serious, systematic error inherent in FLYLIM data, so a safer estimate is a lower limit, $\tau > 1.7$.

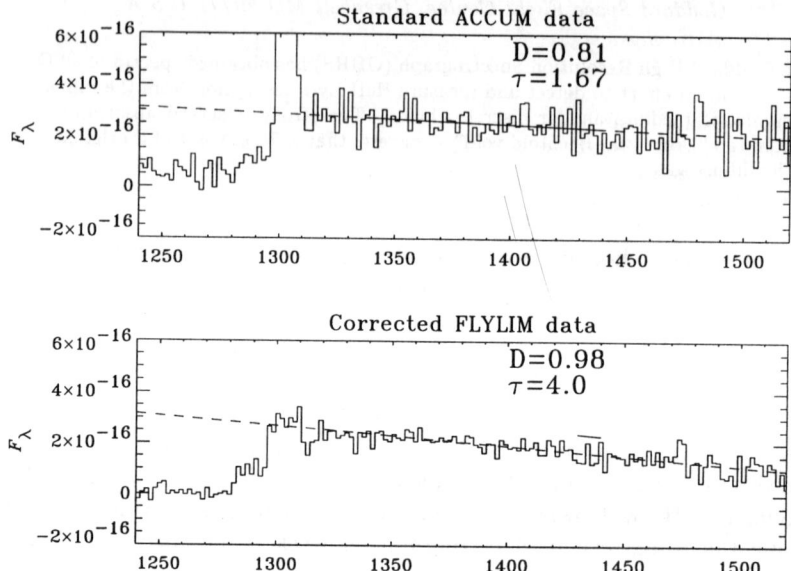

If there were no diffuse IGM or population of weakly absorbing clouds, then there should be a residual flux in the HeII trough, particularly in a spectral window between 1260 and 1275 Å (the "Dobrzycki - Bechtold void"[7]). That no such flux is observed suggests the presence of a diffuse IGM.

References
1. Gunn, J. & Peterson, B. 1965, ApJ 142, 1633
2. Miralda-Escudé, J. 1993, MNRAS 262, 273
3. Jakobsen, P. et al. 1994, Nature, 370, 35
4. Songaila, A., Hu, E. & Cowie, L. 1995, Nature 375, 124
5. Hu E., Kim T.-S., Cowie L. et al. 1995, AJ 110, 1526
6. Hogan C., Anderson S. & Rugers M. 1996, Nature, in press
7. Dobrzycki, A. & Bechtold, J. 1991, ApJ 377, L69

INTERGALACTIC HELIUM ABSORPTION AT HIGH REDSHIFT

P. JAKOBSEN

Astrophysics Division, Space Science Department of ESA,
ESTEC, 2200 AG Noordwijk, The Netherlands

Strong redshift-smeared HeII λ304 absorption from singly ionized intergalactic helium has to date been detected in the far-UV spectra of three quasars, twice at $z > 3$ with the HST and once at $z < 2.7$ with the Hopkins Ultraviolet Telescope flown on the Astro-2 mission. The three measurements are mutually consistent, although none of the observations can distinguish directly between HeII absorption in unresolved low column Lyman forest clouds or a truly diffuse intergalactic medium.

1 HeII Detections

1.1 Q0302−003

Q0302−003 ($z = 3.29$, $V = 18.4$) was first detected in the far-UV in the HST/FOC prism survey by Jakobsen et al. [1], and then re-observed with the COSTAR-corrected FOC [2]. The inferred limit on the HeII optical depth is $\tau_{\mathrm{HeII}} > 1.7$ (90% confidence). Q0302−003 has recently also been observed with the GHRS by Hogan, Anderson & Rugers [3]. The (flylim) GHRS spectrum is in good agreement with the re-calibrated FOC spectrum and may show evidence for a proximity effect in the HeII line.

1.2 PKS1935−692

PKS1935−692 ($z = 3.18$, $V = 18.8$) was first detected in the UV in the HST/FOC imaging snapshot survey of Tytler and coworkers, and re-observed spectroscopically with the FOS [4]. The FOS spectrum of PKS1935−692 suggested less intense HeII absorption ($\tau_{\mathrm{HeII}} \sim 1$) than seen toward Q0302−003. However, this has not been confirmed by follow-up observations taken with the FOC objective prism (Jakobsen & Tytler, Program DD 6156). The limit on the HeII optical depth measured from the FOC spectrum is $\tau_{\mathrm{HeII}} > 1.5$ (90% confidence), in good agreement with that seen toward Q0302−003.

1.3 HS1700+6416

HS1700+6416 ($z = 2.72$, $V = 16$) was first detected in the UV by Reimers et al. [5] using IUE. Although its redshift is too small for HeII to be reached with

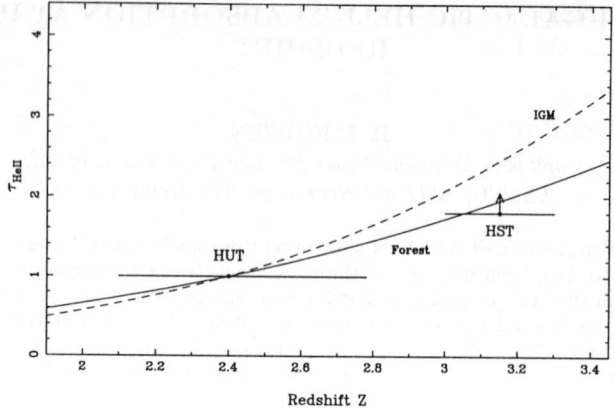

Figure 1: Comparison of the combined HST and HUT measurements of τ_{HeII} and the redshift ranges probed by the data. The two continuous curves show the expected evolution in redshift assuming a non-evolving ionizing background in the two extreme cases where the absorption is completely dominated by unresolved line absorption in the HeII forest and true Gunn-Peterson absorption in a diffuse IGM.

HST, HeII absorption toward HS1700+6416 has been successfully detected by Davidsen, Kriss & Zheng with the Hopkins Ultraviolet Telescope flown on the Astro-2 mission [6]. The HeII absorption seen toward HS1700+6416 is of strength $\tau_{\text{HeII}} = 1.00 \pm 0.07$.

References

1. Jakobsen, P. et al. ApJ 417,528 1993
2. Jakobsen, P., Boksenberg, A., Deharveng, J. M., Greenfield, P., Jedrze-jewski, R. and Paresce, F. Nature 370,35 1994
3. Hogan, C. J., Anderson, S. F. and Rugers, M. H. Nature (submitted)
4. Tytler, D., Fan, X. M., Burles, S., Cottrell, L., Davis, C., Kirkman, D. and Zuo, L. in *QSO Absorption Lines*, ed. G. Meylan (Springer, Berlin, 1995)
5. Reimers, D., Clavel, J., Groote, D., Engels, D., Hagen, H. J., Naylor, T., Wamsteker, W. and Hopp, U. A&A 218,71 1989
6. Davidsen, A. F., Kriss, G. A. and Zheng, W. Nature 380,47 1995

CONSTRAINTS ON THE SPATIAL UNIFORMITY OF LYMAN ALPHA FOREST CLOUDS AT $z = 1$

VESA JUNKKARINEN, E.A. BEAVER, E. MARGARET BURBIDGE,
ROSS D. COHEN, R.W. LYONS, LIN ZUO

Center for Astrophysics and Space Sciences,
University of California, San Diego,
La Jolla, CA 92093-0111, USA

We have analyzed HST/FOS spectra of components A and B of the gravitationally lensed QSO 0957+561 [4] ($z_e = 1.4136$, observed angular separation 6.2 arcsec). In the redshift range 0.83 to 1.41, we find 37 Lyα forest absorption lines in both components with no significant anticoincidences. However, three of the lines have equivalent width (W_λ) differences between components A and B probably produced by column density differences between the two lines of sight through a single cloud. The A and B lines of sight sample linear separations in the Lyα forest from near zero to 5.5 h^{-1} kpc (h = H$_o$ / 100 km s^{-1} Mpc^{-1}, q$_o$ = 0.5). We have modeled these observations with a Monte–Carlo simulation using spherical clouds with radial density gradients sampled by the observed linear separations. We find that the W_λ differences and lack of anticoincidences could be produced by clouds with monotonic radial density gradients and radii of order 30 h^{-1} kpc to 80 h^{-1} kpc. If the cloud radii or size scales are very large, R > 100 h^{-1} kpc, as shown by observations of other more widely separated lines of sight [3 1 2], then the simulations show that the Lyα forest has more structure than a monotonic radial density gradient in a spherical cloud. These observations suggest that the gas producing the Lyα forest is organized in large structures (we don't observe any edges) but these structures are not particularly smooth on scales of a few kpc.

1   HST/FOS UV Spectra of Q0957+561 A and B

Moderate resolution (2.5Å FWHM) HST/FOS spectra (2220Å – 3280Å) of 0957+561 A and B were obtained on 26 January 1995. The integration times were 6500 s for component A and 6560 s for B. The spectra show the usual broad emission lines observed in luminous QSOs and a large number of absorption lines. Most of the strong absorption lines belong to an absorption system at $z_a = 1.3911$ [4] which has a large H I column density [5]. Weaker C IV systems are present at $z_a = 1.1249$ and probably $z_a = 1.0717$. There are also 37 (mostly weaker) lines that have no identifications in known systems and hence are probably Lyα. The lines all have a detection of $\geq 4\sigma$ in at least one spectrum. There are no convincing anticoincidences. Equivalent widths were measured for all of the absorption lines in the Q0957+561 A and B spectra. Three of the 37 Lyα forest lines have W_λ's that are different at more than 4σ (formal error) and these differences appear reasonable on the spectra.

338

In particular these lines are away from emission lines where the continuum placement sometimes causes problems. The three lines with apparent W_λ differences occur at redshifts of 0.925, 1.226, and 1.294 which correspond to linear separations of 4.78, 1.43 and 0.86 h^{-1} kpc. The W_λ differences are relatively large; 30-50% for the three lines which have rest W_λ's of roughly 0.4 Å.

2 Monte Carlo Simulations and Discussion

The lack of absorption line anticoincidences and the observed W_λ differences can be modeled using Monte Carlo methods[3]. The observed 37 linear separations are used along with a spherical cloud model with a monotonic dependence of H I density vs. radius. The total column densities and velocity widths (b values) are chosen to match the typical observed W_λ's. Clouds with constant H I density tend to have a steep W_λ gradient near the edge. Clouds with steep W_λ gradients near the edge can fit the observations over some range of cloud radii (the range depends on the choice of parameters, the maximum range, for 95% confidence in the Monte Carlo simulations, is around 30 h^{-1} $\leq$ R $\leq$ 80 h^{-1} kpc). Clouds with very smooth W_λ gradients are ruled out completely because small clouds produce too many anticoincidences while large clouds can never produce W_λ differences similar to those observed over small linear separations. These simulations rule out very large spherical clouds (R $>$ 100 h^{-1} kpc) with monotonic density gradients in general. Since observations of more widely separated lines of sight[3][1][2] indicate large clouds, the W_λ differences in 0957+561 suggest that these clouds are not smooth on a scale of a few kpc.

Acknowledgments

This research has been supported in part by NASA grant NAG5–1630.

References

1. N. Dinshaw, C.D. Impey, C.B. Foltz, R.J. Weymann, and F.H. Chaffee, ApJ, 437, L87, 1994.
2. Y. Fang, R.C. Duncan, A.P.S. Crotts, and J. Bechtold, ApJ, 462, 77, 1996.
3. A. Smette, J. Surdej, P.A. Shaver, C.B. Foltz, F.H. Chaffee, R.J. Weymann, R.E. Williams, and P. Magain, ApJ, 389, 39, 1992.
4. D. Walsh, R.F. Carswell, and R.J. Weymann, Nature, 279, 381, 1979.
5. L. Zuo, E.A. Beaver, E.M. Burbidge, R.D. Cohen, V.T. Junkkarinen and R.W. Lyons, ApJ submitted 1996.

THE UV ABSORPTION LINE SPECTRUM OF HS 1307+4617

S. KÖHLER AND D. REIMERS

Hamburger Sternwarte, Gojenbergsweg 112, 21029 Hamburg, Germany

Ultraviolet spectra of the $z = 2.13$ QSO HS 1307+4617 have been obtained with the FOS and the GHRS onboard the HST covering the wavelength range from 1150 to 3280 Å. We report on He I and Neon absorption in Lyman Limit systems. We briefly describe the detection of Ne VIII 774 and O V 629 emission as well as the spectral energy distribution of HS 1307+4617 in the UV.

1 He I and Ne absorption in LLS

So far 8 heavy-element absorbing systems have been identified, among them 4 weak Lyman Limit Systems (LLS). All LLS with an adequate redshift show strong He I absorption. Resonance series lines at 584, 537 and 522 Å are simultaneously observed allowing an estimate of the neutral helium column densities. Column densities and b-values for neutral hydrogen are well defined for LLS by modelling the Lyman edge and the Lyman series lines. We find H I/He I column density ratios of 42, 43 and 35 adopting turbulent line broadening ($b_{HeI} = b_{HI}$) for the systems at $z =1.3052$, 1.435, 1.7732, respectively. Adopting thermal broadening ($b_{HeI}=0.5\ b_{HI}$) we find H I/He I column density ratios of 4 and 10 for the systems at $z =1.3052$ and 1.435. Absorption lines by He I were first discovered in HST spectra of HS 1700+6416 with ratios H I/He I=20, 22, 38 adopting $b_{HeI}= b_{HI}$ (Reimers & Vogel [1]). But due to the higher redshift of HS 1700+6416 the analysis of the UV spectra is more affected by line blending and the H I/He I ratios are more uncertain.

In several high-redshift absorber systems towards HS 1700+6416 we detected absorption lines of neon with column densities in the range $\log N(\text{Ne III})= 14.7\text{-}15.5$, $\log N(\text{Ne IV})=14.7\text{-}15.5$, $\log N(\text{Ne V}) \leq 15.5$, $\log N(\text{Ne VI})=14.6\text{-}14.85$ and $\log N(\text{Ne VII})=14.2\text{-}15.0$ adopting both b derived from H I and b=100 km s^{-1} (Köhler, Reimers & Wamsteker [2]). Considering the absorber at $z = 1.7732$ towards HS 1307+4617 with the largest H I column density ($\log N(\text{H I})= 16.78$) we find evidence for absorption by Ne IV, Ne V, Ne VII, S III, O III, O IV, O V and C III. We derive similar neon column densities as for absorbers towards HS 1700+6416. Considering a low-ionization phase which accounts for the observed H I, He I, O III, O IV, O V, Ne III, Ne IV and Ne V column densities we obtain again quite large neon abundances ([Ne/O]$\approx 0.5\text{-}0.7$) compared to oxygen.

340

2 Spectral energy distribution

For the normalization of the spectra a continuum was calculated adopting a power law ($f_\nu \propto \nu^\alpha$) for the intrinsic quasar spectrum and taking into account the neutral hydrogen continuum absorption by the four LLS (method as described in Vogel & Reimers[3]). Three different values for α were needed to describe the continuum shape of the reddening corrected data: $\alpha = -1.1$ in the range 2200 to 3280 Å, $\alpha = -0.5$ in the range 1710 to 2220 Å and $\alpha = -1.6$ for $\lambda \leq 1710$ Å. The excess emission at wavelengths $\lambda_{obs} \geq 2850$ Å is probably due to several QSO emission lines (S VI 937, C III 977, N III 990, O VI 1033, Lyβ). The broad emission at ≈ 2400 Å is a further indication of Ne VIII 774 emission by QSOs. Furthermore, we claim the detection of O V 629 emission.

In order to study the spectral energy distribution the continuum derived for the dereddened spectra was corrected for the H I continuum absorption of the four LLS and transferred to luminosities. An additional correction for the cumulative H I continuum absorption of Lyα clouds with $\log N(\text{H I}) \leq 16$ cm^{-2} (transmission curve computed by Jakobsen) leads to higher luminosities and – even more important – to changes in the continuum slopes. The maximum in the $\log(\nu L_\nu)$ distribution is obviously located in the range $\log \nu = 15.5$–15.7 Hz in the restframe of the QSO. The flat shape of the distribution in the UV suggests a power law index $\alpha = -1$ for rest wavelengths $\lambda = 500$–1050 Å. A similar behaviour of the UV continuum is observed for the $z = 2.72$ QSO HS 1700+6416 (Vogel & Reimers[3]; Köhler, Reimers & Wamsteker[2]).

Results of model calculations for extended accretion disks by Scherdin (1996, PhD thesis Hamburg) (for a non-rotating black hole of mass 10^9 M$_\odot$ and an accretion rate of 40 M$_\odot$/year) can at least qualitatively explain the observed spectral shape of the energy distribution. The models by Scherdin allow the simultaneous calculation of both the continuum flux and the emission line fluxes, thus in future model calculations we will try to reproduce both the observed emission lines and the spectral shape of the energy distribution.

Acknowledgments

This work has been supported by the BMBF (DARA) under No. 50 OR 96 016.

References

1. D. Reimers and S. Vogel, A&A, 276, L13, 1993.
2. S. Köhler, D. Reimers and W. Wamsteker, A&A, 312, 33, 1996.
3. S. Vogel and D. Reimers, A&A, 294, 377, 1995.

WEAK LINE ANALYSIS FROM LYMAN-α CLOUD SIMULATIONS - PROFILE DISTORTIONS FROM INTERNAL VELOCITY FIELDS

A.P.B. LEONARD

Department of Physics, Nuclear & Astrophysics Laboratory, Keble Road,
Oxford OX1 3NP, England

We investigate the nature of weak line systems produced by low column density gas in a CDM universe, using high resolution SPH simulations. On small scales the cloud internal velocities are shown to transform the gas distribution in velocity space, often into a series of thin spikes caused by velocity caustics. Though this effect is generally masked by thermal broadening for HI, this may not be true for CIV lines, possibly introducing a fake clustering signature into their distribution.

1 Internal properties of Lyman α clouds.

We have simulated the collapse of cosmologically distributed gas in the CDM universe over 5 small (500 kpc radius) regions of a 10 Mpc box to very high resolution, using the TreeSPH code (e.g. Katz et al. 1996). The simulations primarily aimed to study the formation of cooled clumps of gas forming in low-v_c halos in different conditions (Quinn Katz & Efstathiou 1996, QKE), however they also trace the low-density gas surrounding these clumps. Our gas mass resolution was $2 \times 10^5 M_\odot$ whilst the gravitational softening length was fixed to 0.5 kpc. In velocity units this gives us the possibility of resolving velocity components of absorption features on scales less than 1 kms^{-1} at $z = 2.4$.

We find that at low column densities where weak Ly α absorption would result, i.e. $10^{13} < N(HI) < 10^{14}$ cm^{-2}, the gas is in extended, still collapsing, structures; often filaments $\sim$ 50 kpc thick and as long as our high resolution region extends at $z = 2.4$ (see HI map in Figure 3b, QKE).

2 Effect of internal velocities on absorption features.

To ascertain the effect of the dynamics on weak Ly α profiles, we calculated mock absorption for many lines of sight at each redshift, the generic features of which are well illustrated in the example shown in the left hand figure below[a]. The thin solid line shows what the Ly α profile would look like without thermal broadening, and where the peculiar velocities of the particles were set to zero.

---

[a]In our analysis we can set $J_{21} = 1$ *a posteriori*, less than that used by QKE but in line with proximity effect predictions, since IGM temperatures are insensitive to J_{21} at these levels. $(J_\nu(z) = 10^{21} \times J_{21}/(1 + 5/(1 + z))^4 (\nu/\nu_{LL})^{-1}$ erg cm^{-1} Hz^{-1} sr$^{-1})$.

The thick solid line shows the same but with the peculiar velocities included as simulated. We see that including the velocities of the gas does not broaden the distribution at all in the conventional sense, but instead transforms it into a series of very thin spikes overlaid on a shallow absorption trough. It is because the velocity field of the gas is both ordered, and discontinuous (due to shocks) that we see this effect occurring (see Miralda-Escudé & Rees 1993).

Despite the underlying distribution differences, the thermally 'smoothed' profiles (dotted) that one would observe are almost identical for the two cases in this example. To illustrate the non-trivial effects on observed line parameters, we show column densities *vs*. Doppler parameters derived using the Voigt profile fitting code VPFIT (R. F. Carswell et al.), for 33 lines of sight through a single cloud at z = 4 (SNR= 50, $\Delta v_{res} = 8 \text{kms}^{-1}$, $\Delta v_{samp} = 3 \text{kms}^{-1}$), again with and without peculiar velocities. We see that the velocity structure both broadens and thins the observed lines, generally to a small extent. We also see that the measured column densities appear to be unaffected in all cases.

The effect of velocities would be much more visible in CIV absorption, where thermal broadening is reduced by a factor $\sqrt{12}$. Preliminary calculations show that, assuming $N(\text{CIV})/N(\text{HI}) = 10^{-2}$ throughout and looking through higher density gas ($N(\text{HI}) \sim 10^{15} \text{cm}^{-2}$, for $J_{21} = 0.1$), the CIV observed profiles are significantly changed, enough to require many components for adequate voigt profile fitting. It is possible then that this effect could be injecting a strong small scale clustering signal into CIV profile data.

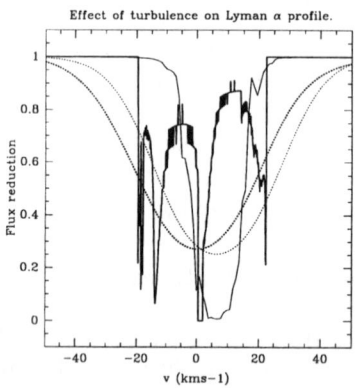

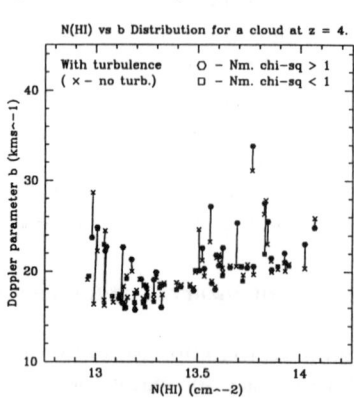

1. Katz, N., Weinberg, D.H., Hernquist, L., Miralda-Escudé, J, 1996. ApJ, **457**, L57.
2. Quinn, T.R., Katz, N., Efstathiou, G., 1996. MNRAS, *in press*.
3. Miralda-Escudé, J., Rees, M.J., 1993. MNRAS, **260**, 617.

THE DAMPED Lyα SYSTEM AT z=1.66 IN FRONT OF THE DOUBLE QSO HE 1104–1805 AB.

S. LOPEZ AND D. REIMERS

Hamburger Sternwarte, Gojenbergsweg 112, 21029 Hamburg, Germany

We present a preliminary study of the z=1.66 absorption system in HST and ESO–NTT spectra of the two components of the gravitationally lensed QSO candidate HE 1104–1805 AB (z=2.31, m_B(A)=16.7, m_B(B)=18.6, 3″ separation [1]), which appears as a damped Lyα system (DLA) only in the spectrum of the A image, as found by Smette et al [2]. We argue that the galaxy giving rise to this absorption system alone is not likely to be the lensing agent, if this object is indeed lensed.

1 Metallicity at z=1.66

The HST+FOS spectra (S/N≃20(A) and 17(B)) show a damped Lyα absorption line only in the spectrum of the A component. Corresponding Lyman limit edges are seen in A (optically thick) and B (τ_B≃2 or logN_{HI}≃17.5). ¿From the ESO–NTT+EMMI spectra (FWHM=0.6–0.9 Å, S/N≃50(A) and 15(B)), values of [Zn II/H I]=–1.03±0.18 and [Cr II/H I]=–1.48±0.03 are derived for A, assuming that the Zn II and Cr II lines lie on the linear part of the curve of growth and that logN_{HI}=20.8 as determined by Smette et al.[2] This metallicity is above the value Z ≃ 1/15 $Z_\odot$ found by Pettini et al [3] in a sample of DLAs at redshifts between z ≃ 2 and 3. Sightline A reflects therefore a high metal content at z=1.66 compared to values found at higher redshifts. Cr depletion [Cr II/Zn II]=–0.45 is low compared to that found in typical DLAs by these authors (where the scatter in the [Cr II/Zn II] values is however large.). These Cr and Zn lines are not detected in B at the 3σ level.

2 Comparing the Two Sightlines

All low and high ionization lines commonly found in DLAs are detected at z=1.66144 in the optical spectrum of component A. In B, only the Mg II and C IV doublets are detected at the 3σ level. At least three C IV absorbing sub-systems spanning velocity intervals of 360 (A) and 180 (B) km s^{-1} and three Mg II sub–systems spanning a velocity interval of 360 km s^{-1} along both lines of sight are observed. The C IV absorbing components show a rather different velocity structure in A and B, in contrast to the Mg II lines, which follow a similar pattern along the two lines of sight (Δv_{max} ~ 20 km s^{-1}). Additionally, while the equivalent widths of the C IV lines are comparable in A and B, the Mg II absorption is clearly stronger in the former spectrum.

In the HST spectra, one would expect a splitting of the O VI 1032, 1038 lines similar to that seen for the C IV lines, under the assumption that the absorption of both ions occurs in the same regions. Multicomponent fits were thus performed, but only the fits for the lines in the A spectrum succeeded. Total rest frame equivalent widths of W=1.18 and W=1.15 Å are respectively derived within $\sigma_W \simeq 0.1$ Å errors, in agreement with the strongest O VI absorption lines found by Burles and Tytler[4].

3   A Lensing Agent at z=1.66 ?

Assuming the absorption lines to arise in a rotating disk-type galaxy, with sightline A passing through halo+disk gas and B through halo gas only, the observed correlation between Mg II lines in A and B makes it difficult to firmly support a model in which the galaxy center lies between the two light beams. In fact, a lens at lower redshift would allow to better explain the similar line patterns as due to sightlines being much closer than 25 h_{50}^{-1}kpc (the projected distance between both lines of sight at z=1.66). Furthermore, the relatively high mass required by a deflector giving rise to this separation is not compatible with that of a typical spiral galaxy. Therefore, if HE 1104–1805 AB is a lensed QSO[2,5], we discard a spiral galaxy at z=1.66 as lens candidate.

Concerning the line strengths, the optical data suggest that C IV absorption takes place in larger volumes than those of Mg II. Additionally, the stronger Mg II lines in the spectrum of the bright component (A) indicate that the absorber center is closer to the light beam A. However, this cannot be given as an *a priori* argument against a z=1.66 lens hypothesis, since in a gravitational lensing geometry a sufficiently extended deflector mass would lead to this configuration. Indeed, if the absorption arises in three or more metal line systems covering a proper distance of a few Mpc, one of which gives rise to the DLA, then this is still consistent with galaxy clustering and should not be excluded, even as lensing model.

This work has been supported by the BMBF (DARA) under No.50OR96016. We gratefully acknowledge discussions with S. Köhler and A. Smette.

References

1. L. Wisotzki *et al*, A&A, 278, L15, 1993.
2. A. Smette *et al*, A&AS, 113, 199, 1995.
3. M. Pettini *et al*, ApJ, 1996, submitted
4. S. Burles and D. Tytler, ApJ, 460, 584, 1996.
5. L. Wisotzki *et al*, A&A, 297, L59, 1995.

MODELING THE EVOLUTION OF THE LYMAN-ALPHA FOREST

J.P. MÜCKET, R. RIEDIGER

Astrophysikalisches Institut Potsdam, An der Sternwarte 16,
14482 Potsdam, Germany

We study the time evolution of the number density of Lyα clouds in a CDM model using numerical simulations. We demonstrate that two Lyα cloud populations may be distinguished: a 'void' component and a component associated with dense structures.

We have performed simulations described in detail by Mücket et al. (1996) modeling the cloud distribution along a line of sight up to a fictitious QSO at redshift $z = 5$. In particular, the UV flux has been obtained consistently by the same simulation. We have assumed that the flux generated at given redshift z is proportional to the rate at which the baryonic material cools below $T_4 = 0.5$ ($T = 10^4 T_4$) in the simulation. Besides heating due to the photoionizing UV flux we have considered in rough approximation heating due to star formation processes. In Fig. 1 the obtained time dependent flux intensity is shown. The simulations used 128^3 particles on a 256^3 grid and a box size of 12.8 Mpc corresponding to a co-moving cell size of 50 kpc ($h = 0.5$, $\Omega_b = 0.05$). During the simulation we distinguished between two different populations of clouds: (1) Clouds being involved in shell-crossing processes (shocks) and therefore associated with structures of enhanced density like filaments and sheets. (2) Gas clouds being located in the surroundings of the structures and in underdense regions (voids) of the matter distribution.

Results are shown in Fig. 2. The square symbols give the data points one has to compare with (Lu et al. (1991); Petitjean et al. (1993); Bahcall et al. (1993)). The left hand figure presents the time-dependence of dn/dz for all clouds ($\gamma \approx 2.6$ for $1.5 < z < 5.$ and $\gamma \approx 0.6$ for $0 < z < 1.5$, where a power law $dn/dz \propto (1 + z)^\gamma$ is assumed); cp. also with observations presented by Januzzi (this volume). The right hand figure shows the contribution of the two populations separately. The dashed curve gives the time evolution of the "void" population showing a very strong evolution ($\gamma \geq 5$) of the number density according to $N_{HI} \propto (1+z)^5$ if the optically thin clouds are co-expanding. The curve for the time dependence of the "filament" population leads to $\gamma \approx 2.5$ till $z \approx 1.5$ and shows then a clear flattening ($\gamma \approx 0.6$ for $z < 1.5$). That would lead to the conclusion that Lyα clouds with $N_{HI} > 10^{14}$ cm^{-2} observed at low redshifts are all associated with dense structures. At high redshifts the

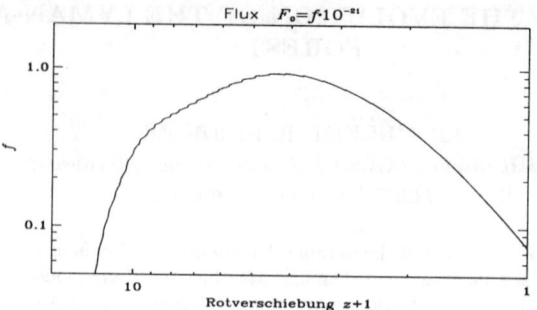

Figure 1: The computed flux as a function of the redshift $1 + z$

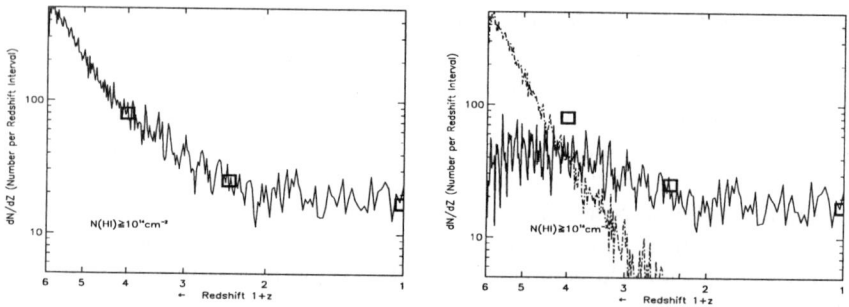

Figure 2: The evolution of the number density per redshift for $N_{HI} \geq 10^{14}$: total contribution (left) and separate contributions of the two Lyα cloud populations

main contribution to the number density comes from the "void" population. The investigations show that a considerable part of clouds with lower column densities ($N_{HI} \leq 10^{13}$ cm^{-2} survives till $z = 0$ (cp. with observation of Stocke et al. (1995)). The obtained HI distribution has been used to generate complete spectra along the l.o.s. and to determine the N_{HI} and Doppler parameter distribution. Taking in mind the general assumptions of our model the results lead to the conclusion that the Lyα absorption line observations can provide direct information about the distribution of the dark matter structures in the universe.

References

1. J.P. Mücket, P. Petitjean, R. Kates, R. Riediger, A&A, 308, 17, 1996.

MEASURING Ω_{baryon} AND THE IONIZING RADIATION BACKGROUND FROM THE OPACITY OF THE LYMAN ALPHA FOREST

M. RAUCH, J. MIRALDA-ESCUDÉ, T.A. BARLOW, R. CEN, L. HERNQUIST, N. KATZ, J.P. OSTRIKER, W.L.W. SARGENT, D. WEINBERG

Astronomy Department, 105-24 California Institute of Technology, Pasadena, CA 91125, USA, etc.

Recent advances in high resolution spectroscopy together with the increasingly realistic hydrodynamical simulations of the universe now available allow us to make detailed comparisons between the predicted and observed Lyman α forest, with the aim of measuring cosmological parameters.

One of the most easily measurable quantities is the *distribution of pixel intensities* (or flux decrements, or optical depths), i.e. the amount of light per unit velocity absorbed by intervening HI clouds from the beam of QSO. The optical depth τ is a measure of the distribution of the neutral hydrogen in real and velocity space, $\tau \propto dN_{HI}/dv$, where dN_{HI} is the neutral hydrogen column density spread out over velocity interval dv. Simulations of the fate of baryonic matter in the high redshift universe predict density, temperature, macroscopic motion, and thus the ionization state of the gas as a function of the cosmological model, the ionizing radiation background, and the total Ω_b in the universe. Over a wide range of realistic gas densities the gas is highly ionized, and photoionization dominates the ionization equilibrium. Then the optical depth scales with the ratio of the total density squared, divided by the ionization rate due to background radiation, Γ, i.e., $\tau \propto \Omega_b^2 h^3/\Gamma$. Comparing observed and predicted τ distributions we can measure this quantity, as a function of redshift. If we are able to obtain an independent estimate of one of the parameters e.g., the ionizing flux, we can determine $\Omega_b h^{3/2}$.

We have observed a sample of 7 QSOs with the Keck telescope, and compared the observed distribution of flux decrements $D = e^{-\tau}$ with the predicted distribution from the hydro-simulation of a CDM+Λ universe by Cen et al. (1994). With a suitable global scaling of the simulated optical depths for each of three redshift bins excellent agreement with the observed distribution can be attained (Fig. 1).

To derive a lower limit on Ω_b we adopt a temporally constant UV background with $\Gamma > 7 \times 10^{-13}\,\mathrm{s}^{-1}$. This value corresponds to $J_\nu \approx 1.6 - 2.5 \times 10^{-22}$ (depending on the spectral shape). It is a *lower limit* in that it includes only the radiation background from QSOs alone, as estimated on the basis of the QSO luminosity function). The corresponding lower limits for $\Omega_b h^{3/2}$ at $< z >$

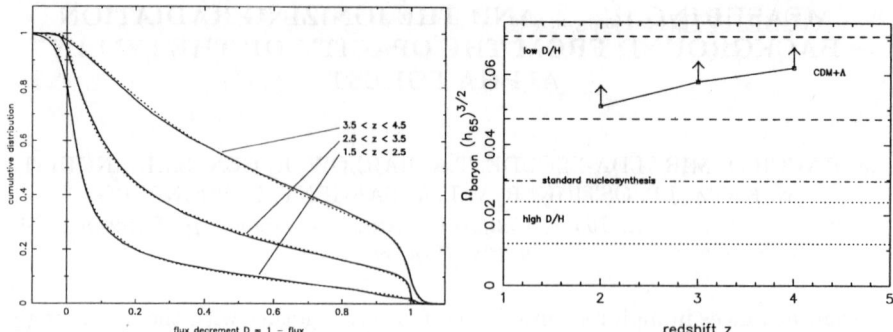

Figure 1: left: Comparison between observed (solid line) and simulated (best fit: dotted line) cumulative flux decrement distribution for the CDM+Λ model of Cen et al. 1994. Right: lower limits on $\Omega_b h^{3/2}$, assuming a minimum ionization rate of $\Gamma > 7 \times 10^{-13}$.

$= 2$, 3, and 4 are shown in the figure on the right. The region between the dashed lines is the 95% confidence area for the "low D/H" value (Tytler, Fan & Burles 1996) as derived by Hata et al. (1996). The dotted lines show the corresponding limits for the "high D/H" value (e.g Rugers & Hogan 1996). The dash-dotted line denotes the earlier "canonical" value from primordial nucleosynthesis calculations (Walker et al. 1991). Our measurement supports a high Ω_{baryon} (low D/H) universe, at least for the cosmological model applied here. It seems to indicate that the "high D/H" measurements (D/H $\sim 10^{-4}$) may be upper limits, rather than actual detections of deuterium. Alternatively, D/H may not be a good measure of the global baryon density.

The method presented here is currently the most direct way for measuring the universal baryonic Ω, being independent of primordial nucleosynthesis. Baryons are counted in the reservoir where most of them actually reside. The main uncertainty stems from the assumed intensity of the ionizing radiation background. However, from the firm lower limit on Γ due to QSOs alone we can determine a *lower limit for the amount of baryons in the universe*:

$$\Omega_{baryon} h_{65}^{3/2} > 0.051$$

Further (probably smaller) uncertainties may enter through the dependence of the ionization state of the gas on the cosmological model, an effect currently under investigation for various other scenarios. Although the precise cosmological model may be difficult to ascertain with this method, the degree of agreement between observed and simulated flux distribution functions achieved by manipulating only one free parameter may be considered as strong evidence in favour of a hierarchical structure formation scenario.

OBSERVING GALAXY FORMATION IN DAMPED Lyα ABSORBERS

L. J. STORRIE-LOMBARDI

Center for Astrophysics & Space Sciences, U.C. San Diego
MS 0111, 9500 Gilman Drive, La Jolla, CA 92093-0111 USA

How and when galaxies formed are questions at the forefront of work in observational cosmology. Damped Lyα systems are a population of high column density HI absorbers (log $N_{HI} \geq 20.3$ cm^{-2}) that dominate the baryonic mass contributed by HI and have properties indicating they are galaxy progenitors [1]. We present preliminary results from our recent Keck data for $z > 3$ damped systems, combined with previous results, to construct the evolution of the comoving neutral gas density, $\Omega_g(z)$, in the redshift interval $z=[0.008,4.7]$. Our observations confirm the turnover in Ω_g at z~3 previously indicated by the data from the APM Damped Lyα Survey [2] The shape of the Ω_g curve has profound implications for galaxy formation theories. A fall-off in the curve with increasing redshift points to an epoch where the damped Lyα systems are still collapsing.

We have undertaken a spectroscopic programme using LRIS on the Keck telescope to confirm candidate damped Lyα systems at high redshift, observing 27 $z > 3.5$ QSOs (Storrie-Lombardi & Wolfe, in preparation). It includes the majority of the candidate absorbers discovered in the APM damped Lyα survey[2,3] plus ~ 20 high redshift QSOs not previously surveyed from the literature. This has been combined with the large $z < 3$ data set compiled in Wolfe et al. [4] resulting in 76 damped systems with log $N_{HI} \geq 20.3$ atoms cm^{-2} from 618 QSO spectra covering the redshift range $0.008 \leq z \leq 4.7$.

The evolution of the mean cosmological mass density in neutral gas, $\Omega_g(z)$, contributed by damped Lyα absorbers is shown in figure 1 ($q_0=0.5$). The region Ω_{star} is the $\pm 1\sigma$ range for the mass density in stars in nearby galaxies[5]. The point at z=0 is the value inferred from 21 cm emission from local galaxies [6,7]. These results point to an epoch at z $\gtrsim$ 3 prior to which the highest column density damped systems are still forming. Using the high redshift damped candidate features discovered in the APM Damped Lyα Survey[2,3], a flattening in $\Omega_g(z)$ at z~2 and a possible turnover at z~3 was first indicated[11]. The results from our Keck spectroscopy, providing accurate column density measurements for the previous candidate systems, in addition to more redshift path surveyed, suggest that the turnover is real. This is consistent with recent results from Lu et al. [8] where they conclude that $z \sim 3$ could be the epoch of galaxy formation, or more precisely the epoch at which the mean metallicity of galaxies increased significantly. It is also consistent with the declining space density of QSOs at $z > 3$ [9] and the high rate of star formation detected in

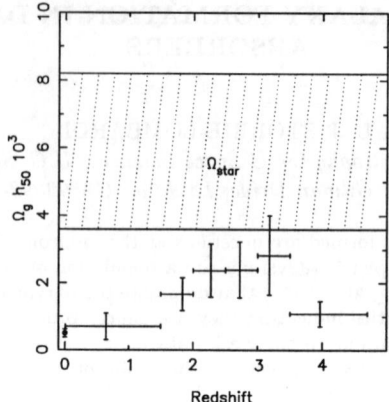

Figure 1: The mean cosmological mass density in neutral gas, $\Omega_g(z)$, contributed by damped Lyα absorbers (q$_0$=0.5). The region Ω_{star} is the $\pm1\sigma$ range for the mass density in stars in nearby galaxies[5]. The point at z=0 is the value inferred from 21 cm emission from local galaxies [6,7].

galactic bulges at $z \approx 3.5$ [10].

1. Wolfe, A.M. 1995, in QSO Absorption Lines, ed. G. Meylan, (Springer: New York), 13
2. Storrie-Lombardi, L.J., McMahon, R.G., Irwin, M.J., & Hazard, C. 1996, ApJ, 468, 128 [1996a]
3. Storrie-Lombardi, L.J., Irwin, M.J., & McMahon, R.G. 1996, MNRAS, in press [1996b]
4. Wolfe, A.M., Lanzetta, K.M., Foltz, C.B. & Chaffee, F. 1995, ApJ, 454, 698
5. Gnedin, N.Y. & Ostriker, J.P. 1992, ApJ, 400,1
6. Fall, S.M. & Pei, Y.C. 1993, ApJ, 402, 479
7. Rao, S. & Briggs, F. 1993, ApJ, 419, 515
8. Lu, L., Sargent, W.L.W., & Barlow, T.A. 1996, ApJ, in press
9. Schmidt, M., Schneider, D.P., Gunn, J.E. 1995 AJ, 110, 68
10. Steidel, C.C., Giavalisco, M., Pettini, M., Dickinson, M., & Adelberger, K.L. 1996 ApJ, 462, L17
11. Storrie-Lombardi, L.J., McMahon, R.G., & Irwin, M.J. 1996, MNRAS, in press [1996c]

IONIZATION OF METAL ABSORPTION SYSTEMS

A. DE LA VARGA

Hamburger Sternwarte, Gojenbergsweg 112, 21029 Hamburg, Germany

E. STENGLER-LARREA

Instituto de Física de Cantabria, Facultad de Ciencias, Santander, Spain

We present the initial stages of a study focused to analyse the properties and evolution of the heavy-element absorbing systems in quasar spectra. Metal abundances can be derived from the observed ionic column densities adopting a certain ionization model for the absorbers. The model dependence of such estimates will be discussed. Studying absorbers showing several ionization stages and considering as many lines of sight as possible can help to constrain the shape and intensity of the ionizing background radiation and disentangle its origin.

Metal absorption systems observed at high redshift could arise in gas condensations which are evolving into galaxies. At lower redshift they are shown to be related to galactic halos. We are completing a sample of metal absorbers at $z > 2$ and a further one at low and intermediate redshift, in order to probe the physical properties and the evolution of the metal-rich gas.

To estimate the ionization state of the clouds models have been computed using the code CLOUDY (Ferland[1]). The clouds are assumed to be photoionized by the background of quasars, young galaxies and the local ionization from the galaxy related to the metal absorber. The intensity of the ionizing flux at the Lyman limit, J_{LL}, was varied within limits in agreement with studies of the proximity effect [2]. Several shapes for the ionizing conitinuum [3,4,5,6,7] were tested and both slope and size of the break at the HeII edge were systematically varied. Allowed values for the ionization parameter U are inferred by comparing the column density ratios predicted by our models with the observed ones, since those ratios do not depend on the assumed metallicity (at least for $[Z/H] \leq -1$). One of the main aims of the new observations (optical and UV range) is to cover many ionic species for most systems so that components of different ionization structure and inhomogeneities in the gas can be detected and modelled.

To illustrate the work in progress we have applied our models to some metal absorbers from the literature. Uncertainties in the obseved N(HI) make it difficult to determine a reliable metallicity for the systems. Assuming logN(HI) $\simeq 16.0$ for the two component system at $z = 3.53$ studied by Savaglio *et al.* [8], the abundances would be $-2.3 \leq [C/H] \leq -0.9$ and Si/C $\simeq 0.04$-3.7 times solar, depending on the adopted shape for the ionizing background. Hard backgrounds (e.g. a single power law $f_\nu \propto \nu^{-\alpha}$ with $\alpha < 1.2$) can only fit the observations for cloud dimensions along the line of sight much smaller than the transversal correlation lengths constrained by quasar pairs observations. Adopting the ionizing background from HM[3] we obtain [C/H] ~ -1.5 and Si/C ~ 2.3, with $\log U \sim -2.5$ and -1.8 for the low and high redshift component, respectively.

The partial Lyman Limit sytems (PLLS) detected by Songaila & Cowie[9] at $z \simeq 2.5\text{-}3.4$ can be modelled for $-3.0 \lesssim \log U \lesssim -1.8$ and for a large spread of carbon abundances ($-3.0 \lesssim$ [C/H] $\lesssim -0.6$), depending on the assumed shape of the background. They found an overabundance of Si/C about 3 times solar for the Lyman α forest clouds. Adopting different photoinization models for their PLLS we find values for Si/C consistent with 1 to 3 times solar (we obtain similar results for the absorbers at $z \simeq 3\text{-}3.8$[8]) but the sample is still too small to conclude possible different enhacements for both types types of absorber.

In order to reproduce simultaneously the OVI and CIV column densities measured by Burles & Tytler[10] for absorbers at $z \sim 1$ a rather hard ionizing radiation field is required, such as a single power law of index $\alpha < 1.5$. Otherwise, longitudinal absorber dimensions much larger than 100 kpc h_{50}^{-1} or very large values for J_{LL} would be implied. High resolution observations of the OVI doublet are required to constrain the ionization state and the temperature of the gas, and determine which are the main ionization mechanisms.

The observed parameters for the previous examples were insufficient to test more sophisticated models for the absorbers. Our sample will allow more detailed studies of systems showing different ionization phases. Moreover, constraints on the ionizing flux and, therefore, on the effect of a local ionization will be discussed.

E.A.S. acknowledges support from the "Soc. de Ciencias Aranzadi, Gobierno Vasco". A.V. acknowledges support from the Euroconference-Programme.

References

1. G.F. Ferland, Internal Rep., Univ. of Kentucky,1993
2. e.g. E. Giallongo *et al.*,ApJ, 466, 46, 1996.
3. F. Haard and P. Madau, ApJ, 461, 20, 1996.(HM)
4. A. Meiksin and P. Madau, ApJ, 412, 34, 1993.
5. J. Bechtold *et al.*, ApJ, 315, 180, 1987.
6. J. Miralda-Escudé and J.P. Ostriker, ApJ, 350, 1, 1990.
7. M. Donahue and J.M. Shull, ApJ, 383, 511, 1991.
8. S. Savaglio *et al.* (astro-ph 9606063) ApJ in press, 1996
9. A. Songaila and L. Cowie AJ, 335, 112, 1996.
10. S. Burles & D. Tytler ApJ, 460, 584, 1996.

MASSIVE STARS AND THE IONISING BACKGROUND

J. VERNET[1,2] and D. VALLS–GABAUD[1,3]

[1] *Observatoire de Strasbourg, 11 Rue de l'Université, 67000 Strasbourg, France*
[2] *ESO, Karl-Schwarzschild-Straße 2, 85748 Garching bei München, Germany*
[3] *Royal Greenwich Observatory, Madingley Road, Cambridge CB3 0EZ, UK*

Using the observed cosmic evolution of HI, and the latest model atmospheres for massive stars, we predict the FUV emissivity of galaxies. The resulting ionising background from galaxies is smaller than the one produced by quasars.

Recent observations of the HeII GP effect and of ionised species at high z indicate that the ionising background is softer than previously thought. The spectra from quasars being too hard, a background dominated by galaxies may be a solution. Using the observed evolution of HI, as traced by damped Ly-α systems, we derive the density of star formation as a function of z. With the latest model atmospheres for massive stars from the Geneva group, we predict the FUV emissivity of galaxies (see Figure 1). At 2800 Å, the prediction agrees with the observed evolution of the luminosity function in the CFRS survey.

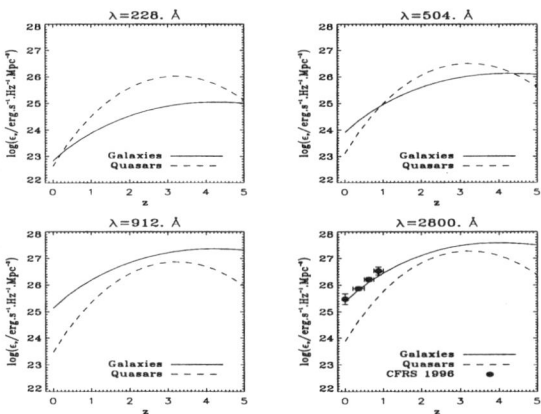

Figure 1: Evolution of the emissivity of galaxies and quasars as a function of redshift.

Even though at shorter wavelengths the contribution from galaxies may dominate the total emissivity, when the absorption produced by the intervening absorbers is taken into account the ionising background from galaxies is about a factor of 3 smaller than the one produced by quasars.

NITROGEN ABUNDANCES IN DLA SYSTEMS AND CHEMICAL EVOLUTION AT HIGH REDSHIFT

G. VLADILO, P. MOLARO

Osservatorio Astronomico di Trieste, Italy

F. MATTEUCCI

Department of Astronomy, University of Trieste, Italy

Spectroscopic determinations of chemical abundances of Damped Ly α (DLA) systems are, together with deep field imaging and with spectroscopy of Lyman break galaxies, a powerful tool for understanding the nature of "normal" high redshift galaxies. In particular, abundance ratios of elements with different nucleosynthetic origins can be used to constrain the first stages of galactic chemical evolution. Here we briefly discuss advantages and problems related to the measurement of the N/O ratio in DLA absorbers and the implications of this kind of measurements for understanding chemical evolution at high redshift.

Damped Lyman α (DLA) systems are the class of QSO absorbers with highest hydrogen column density and are believed to be the progenitors of present-day disk galaxies[1]. The study of elemental abundances in DLA systems by means of high resolution spectroscopy is a unique tool for probing the chemical history of the early universe. Element-to-element abundances are especially important as a diagnostic of the first elemental buildup and the N/O ratio is particularly useful for constraining chemical evolution models owing to the different origins of nitrogen and oxygen[2].

Nitrogen and oxygen present several advantages with respect to other elements as far as abundance determinations in DLA systems are concerned. *No need for ionization corrections:* because DLA systems are optically thick to photons with hν > 13.6 eV, nitrogen (I.P.=14.53 eV) and oxygen (I.P.=13.61 eV) are expected to be mainly in neutral form and this is confirmed by detailed ionization models[3,4]. *Negligible contributions from intervening HII regions:* warm ($T \approx 10^4$) ionized gas along the sight line may contain significant amounts of singly ionized species like FeII and SiII, but not of neutrals with I.P. $\approx$ 13.6 eV like NI and OI. *Negligible gas-to-dust depletion:* even though DLA systems show a dust-to-gas ratio roughly 10% of the Galactic one[5,6] refractory elements like Si and Fe may be still underabundant in the gas since they are highly depleted (1-2 dex) in the ISM; this is not the case of nitrogen and oxygen which are weakly depleted ($\leq$ 0.4 dex) in the ISM[7].

The difficulty in measuring nitrogen in DLA systems is the *contamination by Lyα forest* of the two resonance triplets of NI (λ 1134Å and 1200Å).

However, it is unlikely that each of the 6 NI transitions is blended with a Ly α interloper and clear detections have been obtained in a few systems [8,9]. The problem in measuring oxygen is the *saturation of the OI line at* λ 1302 Å. This leaves the broadening parameter $b(OI)$ poorly constrained, with a large error in the OI column density. However, when NI is clearly identified, it is possible to measure $b(NI)$ and to constrain $b(OI)$ [9]. If NI and OI trace the same regions of the DLA one can in fact show that $0.94\, b(NI) \leq b(OI) \leq b(NI)$, the lower value corresponding to thermal broadening and the higher to pure turbulence.

The comparison of available measurements of N/O ratios in DLA systems [8,9,10,11,12] with evolutionary tracks predicted by galactic chemical evolution models [2] indicate that the class of DLA absorbers is not homogeneous. The N/O ratios found in the QSO 0347-383 $z_{abs} = 3.025$ system [9], the marginal detection in the QSO 2344+124 $z_{abs} = 2.538$ system [10], and the upper limits in the QSO 1946+7658 $z_{abs} = 2.884$ and QSO 2348-147 $z_{abs} = 2.279$ systems [10,12] are consistent with conventional models of galactic chemical evolution. On the other hand, the N/O ratios measured in the QSO 0000-26 $z_{abs} = 3.390$ and QSO 1331+1700 $z_{abs} = 1.776$ systems [8,11] can only be reproduced by models with a primary production of nitrogen, one intense starburst and differential galactic winds. The abundance pattern in these latter systems is similar to what is observed in dwarf irregular galaxies such as IZw18 and suggests that some of the DLA absorbers may be associated to dwarf galaxies.

References

1. Wolfe A.M., Turnshek D.A., Smith H.E., Cohen R.D., ApJS 61, 249, 1986.
2. Matteucci F., Molaro P., Vladilo G., A&A in press, 1996.
3. Fan X.-M., Tytler D., ApJ Suppl. 94, 17, 1994.
4. Lu L., Savage B.D., Tripp T.M., Meyer D.M., ApJ 447, 597, 1995.
5. Fall S.M., Pei Y.C., ApJ 337, 7, 1989.
6. Pettini M., Smith L.J., Hunstead R.W., King D.L., ApJ 426, 79, 1994.
7. Cardelli J.A., Savage B.D., Ebbets D.C., ApJ 383, L23, 1991.
8. Molaro P., D'Odorico S., Fontana A., Savaglio S., Vladilo G., A&A 308, 1, 1996.
9. Vladilo G., Molaro P., Matteucci F., Centurión M., in *The Early Universe with the VLT*, Garching 1-4 April 1996, in press, 1996.
10. Lipman K., PhD thesis, Cambridge University, 1995.
11. Green R.F. *et al.*, in *ESO Workshop on QSO Absorption Lines*, ed. G. Meylan, Springer Verlag, 85, 1995.
12. Pettini M., Lipman K., Hunstead R.W., ApJ 451, 100, 1995.

QUASARS & ACTIVE GALACTIC NUCLEI

THE REDSHIFT ONE 3CR RADIO GALAXIES

PHILIP N. BEST and MALCOLM S. LONGAIR

Cavendish Laboratory, Madingley Road, Cambridge CB3 0HE

The 3CR radio galaxies exhibit the same strong cosmological evolutionary effects which are found in the population of active galaxies in general. HST, VLA and UKIRT observations have been made of an almost complete sample of 28 narrow-line 3CR radio galaxies in the redshift interval $0.6 < z < 1.8$. The HST images show clearly the alignment effect between the optical morphologies of the galaxies and the radio axes of the double radio sources and these can be placed in an evolutionary sequence with increasing separation of the hot-spots of the radio structures. In the case of 3C34, there is evidence for jet-induced star-formation in a galaxy which lies in the path of the radio jet. It is suggested that similar processes play a role in creating the aligned structures. The implications of these studies for understanding the K-z relationship for 3CR radio galaxies are discussed.

1 Introduction

There is now overwhelming evidence for the evolution of the population of active galaxies with cosmological epoch. To a good approximation, the radio galaxies and radio quasars, the optically selected quasars and the faint X-ray sources display the same form of evolution, which involves a very large increase in the comoving number density of sources of a given luminosity out to redshifts of about 2 (see, for example, Dunlop [1]). At larger redshifts, the change of the comoving space density with redshift is less well defined, although there is growing evidence that it decreases beyond $z \sim 2$. The brightest extragalactic radio sources in the sky exhibit this strong evolutionary behaviour and so they should provide clues to the physical nature of the strong cosmological evolution displayed by the active galaxies. It is intriguing that there is growing evidence that star and metal formation proceeded vigorously over the same redshift interval, $1 < z < 3$ — it seems likely that the evolution of the active galaxies is intimately related to the process by which the galaxies in general formed.

2 The 3CR Radio Galaxies

The 3CR radio sources are of particular importance in these studies. From the complete sample of Laing *et al* [5] we have concentrated upon a complete subsample of 127 sources which satisfy the selection criteria $S_{178\,\mathrm{MHz}} \geq 9.7$ Jy, $|b| \geq 10°$, and $10° < \delta < 55°$. For all the sources in this sample, identifications and redshifts are now available. The sample consists of 24 radio quasars and 103 radio galaxies. The radio galaxies can be further divided into 79 FR2 and 24 FR1 radio galaxies, according to the scheme described by Fanaroff and Riley [2]. The FR2 sources are the classic double radio sources with 'hot-spots' towards the leading edges of the double source components and these represent all the most luminous radio sources in the sample. We will concentrate entirely upon the FR2 sources in this analysis.

We have performed a number of statistical tests of the properties of the FR2 radio sources in the sample. First, the FR2 radio galaxies and radio quasars have similar redshift distributions. Second, the V/V_{max} tests give similar results for both samples; specifically, for the radio quasars and broad-line radio galaxies, $\langle V/V_{max} \rangle = 0.686 \pm 0.042$, whereas for the narrow-line radio galaxies it has a value of 0.697 ± 0.031. In both these cases, the values of $\langle V/V_{max} \rangle$ are greater than the value 0.5 expected for a uniform distribution. These values of the $\langle V/V_{max} \rangle$ are direct evidence for the strong evolutionary effects inferred to take place between redshifts 0 and 2. Thus, the cosmological evolutionary effects present in the active galaxies are present in the sample of narrow-line radio galaxies.

This all makes a great deal of sense if one is a believer in orientation-based unified scenarios for luminous radio galaxies and quasars. The properties of the radio sources in the 3CR samples seem to be in remarkable agreement with a unification scenario in which the radio galaxies and the radio quasars belong to the same family of objects observed at different angles to the line of sight — the axis of the obscuring torus lies within 45° of the line of sight for the radio quasars, whereas for the radio galaxies it lies within 45° of the plane of the sky [3,4]. If this picture is correct, the 3CR narrow-line radio galaxies present a splendid opportunity for studying the host galaxies of the 3CR radio quasars — because the quasar nucleus is obscured, the narrow-line radio galaxies show directly the nature of the parent galaxies of the 3CR radio quasars.

3   HST Observations of Large Redshift 3CR Radio Galaxies

Our HST programme consisted of imaging observations with the WFPC2 of almost all the narrow line 3CR radio galaxies with redshifts in the interval $0.6 < z < 1.8$. The values of V/V_{max} for these galaxies all lie in the range 0.5 to 1. The 28 radio galaxies were all observed by the HST in wavebands corresponding approximately to rest-frame U and B wavebands. In addition, all the radio sources were mapped at 8.4 GHz with the VLA in A configuration, providing an angular resolution of 0.18 arcsec. Additional observations in B and C configurations were made for those sources with large angular sizes. All the galaxies were imaged at 2.2 μm with IRCAM3 of UKIRT, providing an angular resolution of about 1 arcsec.

In interpreting the HST images, it is a useful rule of thumb that a standard L^* elliptical galaxy at a redshift of one is a diffuse low-surface brightness object and only objects of high surface brightness and/or flat spectra in the optical waveband stand out prominently [6,7]. As a result, if the radio galaxies were standard elliptical galaxies, little structure would be observed in the HST images — an example of this is observed in the case of 3C34 (see below). The most remarkable results of our observations are illustrated in Figures 1 and 2 which show a montage of all 8 3CR radio galaxies in the redshift interval $1 < z < 1.3$. The advantage of restricting attention to this narrow redshift range is that all the radio sources have more or less the same intrinsic radio luminosity. The HST images bear little resemblance to the expected appearance of giant elliptical galaxies. They all illustrate dramatically the alignment effect discovered by Chambers, McCarthy and their colleagues in 1987 [8,9]. The images in Figures 1 and 2 are shown on the same linear scale, so that in 3C266 there is the smallest projected separation between the hot-spots of the double radio sources and in 3C 356 the largest. According to the theory of double radio sources, the hot-spots and diffuse radio structures are supplied by beams or jets from the nucleus with more or less the same intrinsic power and the greater the separation of the hot-spots, the older the source. According to the unification scenario, the difference between the actual and projected separations between the hot-spots should not be very great because the radio axes should all be observed within about 45° of the plane of the sky. A further piece of empirical evidence for the structures being more or less in the plane of the sky is the frequency with which there appear to be dark lanes running across the central regions of the radio galaxies as can be seen from inspection of Figures 1 and 2. We presume that these dark lanes are associated with large-scale obscuring material which may be related in some way to the obscuring torus about the quasar-like nucleus.

It can be seen that the remarkable structures are most prominent in those sources with the smaller physical separations between the hot-spots, and are well aligned with the beams which are presumed to supply energy to the hot-spots. The aligned high-brightness structures are more or less confined to the vicinity the giant elliptical galaxies, as may be seen by comparing the optical and infrared images in Figure 1. In contrast to the optical images, there is little evidence for the aligned components in the infrared images. Although they are of lower angular resolution than the HST observations, convolution of the optical images produces images significantly more extended and aligned with the radio axes than the infrared images are observed to be. The implication is that the high-brightness optical components seen in the HST images must have flat spectra.

362

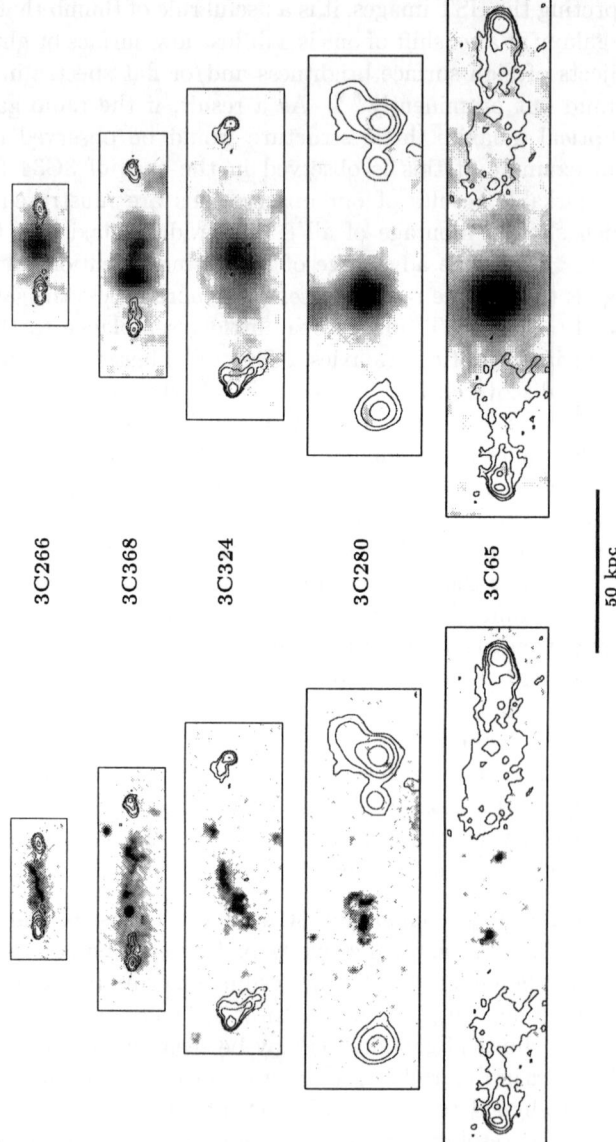

Figure 1: HST (left panels) and UKIRT (right panels) images of the radio galaxies 3C 266, 368, 324, 280 and 65 with the VLA radio contours superimposed. The images are drawn to the same physical scale (Best, Longair & Röttgering 1996a).

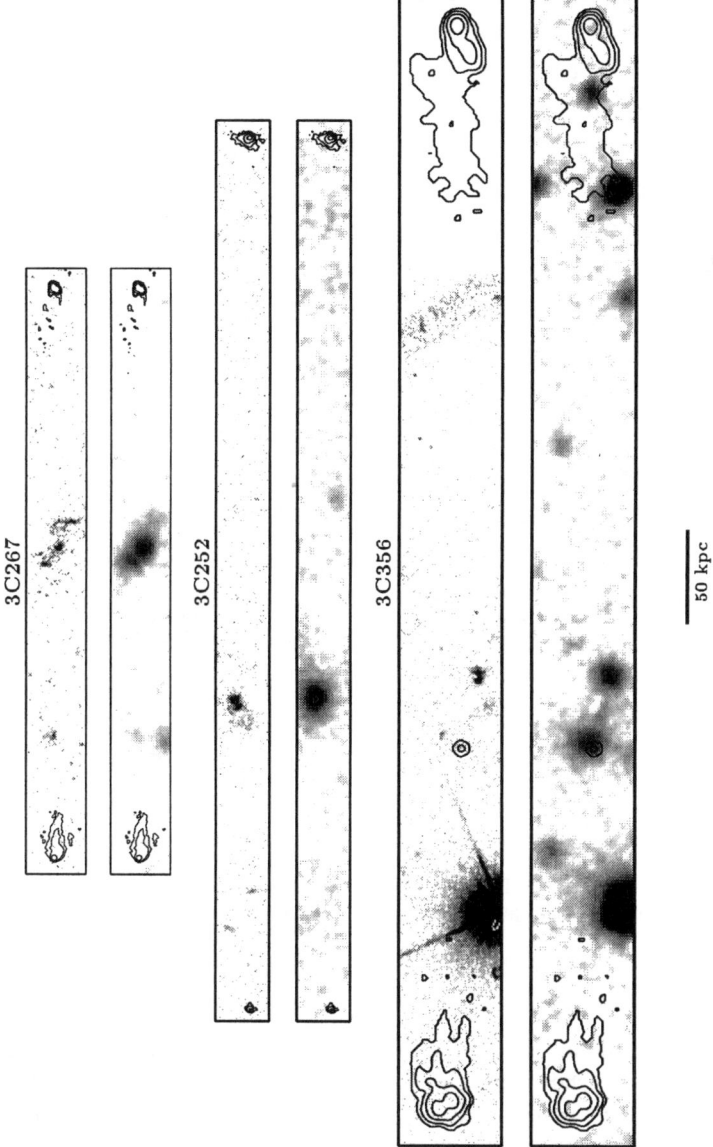

Figure 2: HST (upper panels) and UKIRT (lower panels) images of the radio galaxies 3C 267, 252 and 356 with the VLA radio contours superimposed. The images are drawn to the same physical scale (Best, Longair and Röttgering 1996a).

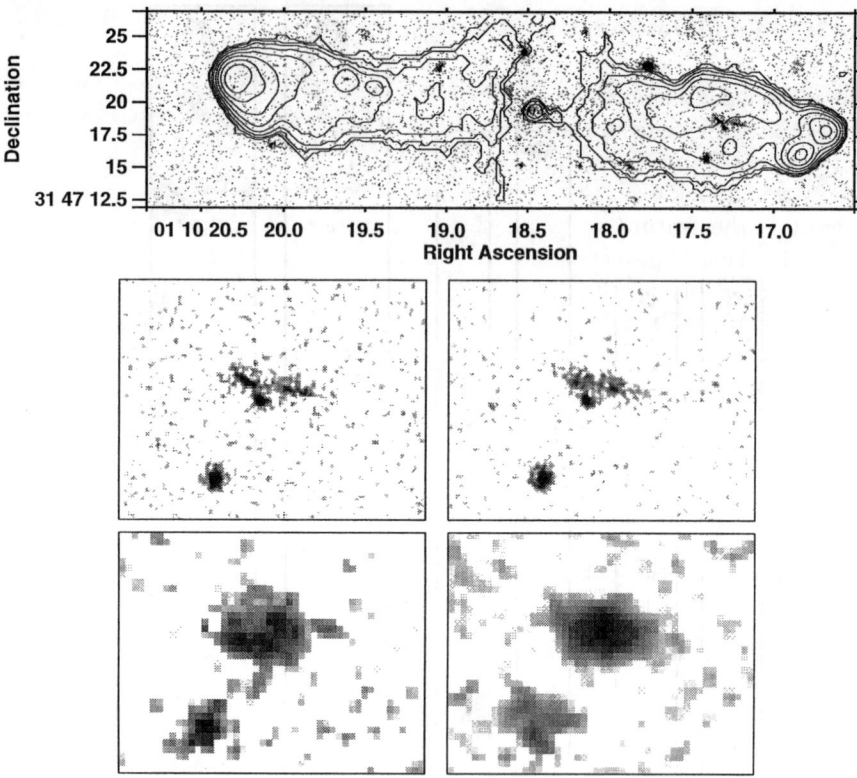

Figure 3: (a) The rest–frame U HST image of the radio galaxy 3C 34 with contours of the radio source structure superimposed. (b) and (c) Images of the structure of the optical 'jet' which lies along the line from the nucleus of the galaxy to the western hot-spot as observed through the f555W and f785LP filters respectively. (d) J and (e) K images of the galaxy associated with the optical jet observed with UKIRT (Best, Longair & Röttgering 1996b).

One great bonus of studying these radio sources is that age estimates for the sequence shown in Figure 1 can be found from synchrotron ageing arguments. In the cases of 3C 266 and 280, these ages are found to be about $(1-2) \times 10^6$ years respectively, corresponding to advance speeds of the hot-spots through the intergalactic gas of about $0.15c$. The inference is that the structures are relatively short-lived, all the activity being completed in about 10^7 years. It is evident that the radio jets have a major perturbing effect upon the ambient gas.

4 3C 34 and Jet-induced Star Formation

The radio galaxy associated with 3C 34 is of particular interest. In this case, there is little evidence for high-brightness features within the galaxy itself. In fact, the rest-frame U image gives a good impression of what a giant elliptical galaxy would look like at a redshift $z = 0.69$ (Figure 3a). What is of particular interest is the feature within the eastern radio lobe, about halfway to the hot-spot. This is a very blue object, elongated along the axis of the beam which presumably powers the hot-spots. The HST images are shown in more detail in Figures 3b and c. The infrared images of this object (Figures 3d and e) suggest that there is a galaxy located at this point and this is supported by the discovery of a strong depolarisation patch in the deep radio maps of Johnson *et al*[11]. We have argued that this feature of the HST images is the result of the interaction of the radio beam with the interstellar gas of a galaxy which is a member of the cluster about the host galaxy of 3C 34[12]. We suggest that the features seen in the HST images are due to star formation produced by a jet-cloud interaction. Again, we are able to determine a time-scale and age for the starburst since a synchrotron age is available for the radio source of about $(1 - 3) \times 10^7$ years. Therefore, the starburst is now observed about 5×10^6 years after it was excited by the radio jet. We have been able to model the spectral energy distribution of these strange elongated structures as a decaying starburst. To produce the observed intensity of radiation would require a star-formation rate of about 100 $M_{\odot}$ yr^{-1} for about 10^6 years. The resulting spectrum of the galaxy plus decaying starburst is shown in Figure 4 and compared with our photometry in the optical and infrared wavebands.

It is interesting to compare this example of jet-induced star formation with that associated with the Cartwheel galaxy. In that case, star formation in a ring has been induced by a strong shock due to the passage of a 'missile' close to the galaxy nucleus, and a star formation rate of 67 $M_{\odot}$ yr^{-1} is derived[13].

We suggest that induced star formation may well play a role in understanding the high surface brightness structures seen in Figures 1 and 2. The evolution suggested by these images may be understood in terms of decaying starbursts, similar to that found in 3C34, induced by the passage of the radio jet. The strong optical polarisation observed in many of these sources is interpreted as scattering of the obscured optical emission from the nucleus by the ionised gas and dust in the vicinity of the star forming regions. According to unification schemes it is inevitable that there should be a polarised scattered component since the beams are ejected within 45° of the plane of the sky. It also seems inevitable that there should be strong shocks associated with the interaction of the beam with the interstellar and ambient intergalactic gas, res-

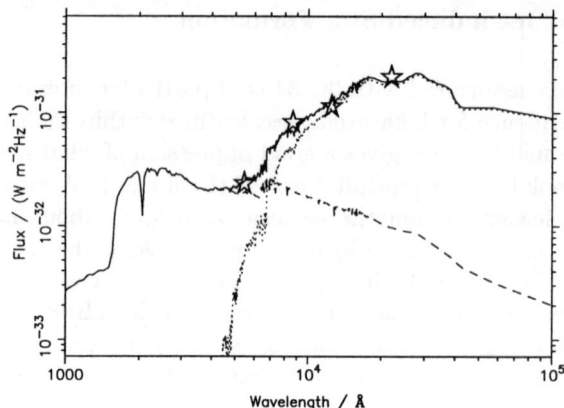

Figure 4: The spectrum of the galaxy shown in Figures 3(b)–(e). The stars indicate the broad-band intensities of the galaxy and the solid line shows a fit to these data consisting of a standard galaxy spectrum (dotted line) plus a decaying starburst (dashed line) observed 5×10^6 years after it was excited by the radio jet (Best, Longair & Röttgering 1996b).

ulting in the formation of copious amounts of ionised gas. In our interpretation of the evolutionary sequence seen in Figures 1 and 2, the sequence is associated with the decay of the starburst and the relaxation of the star clusters within the potential of the host galaxy. In whatever way we interpret the observations, however, it seems inevitable that there must be large quantities of cold gas within the host galaxy.

5 The K-z Relation for 3CR Radio Galaxies

In the light of these new observations, we have re-examined the infrared apparent magnitude-redshift relation at a wavelength of 2.2 μm for the 3CR narrow-line radio galaxies. Lilly and Longair [14] found a splendid correlation with a remarkably small dispersion about the mean relation and, if the effects of evolution were ignored, a best-fit value of $q_0 \approx 3.5$ was obtained, which was not a popular conclusion. When the simplest correction for the passive evolution of the K luminosities of these galaxies over cosmological time-scales was included, a best-fit value of $q_0 \approx 0.5 \pm 0.5$ was found. One of the basic problems is that there is no understanding of why the narrow-line radio galaxies should possess such a tight redshift-K magnitude relation. There must be some astrophysical reason why it is only luminous elliptical galaxies which produce the characteristic FRII types of luminous radio source. Whatever is going on, there must be some systematic changes taking place with cosmic epoch.

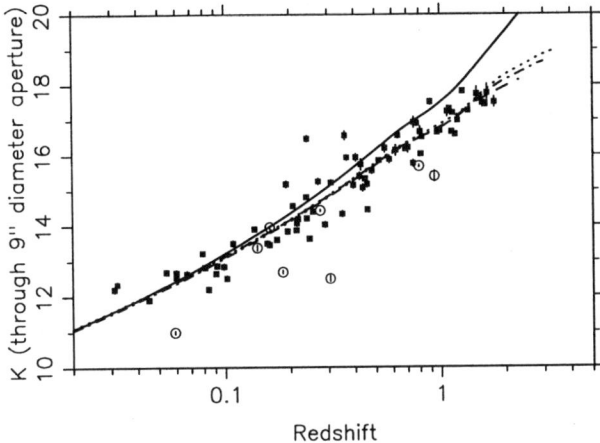

Figure 5: The redshift–apparent magnitude relation in the infrared K waveband for the radio galaxies associated with powerful radio sources. The solid line shows the expectation of a standard world model with $q_0 = 0.5$ on the assumption that their properties have remained unchanged with cosmic epoch. The dashed lines shows the expected relations when account is taken of passive evolution of the stellar populations of these galaxies with cosmic epoch, with different assumptions about their epoch of formation. It can be seen that it makes little difference to the expected relation if the stellar populations formed at redshifts 5, 10 or 20.

The discovery of the alignment effect in 1987 indicated that the optical luminosity of the galaxy is influenced by its radio properties and this is vividly illustrated by our HST images. As is clear from Figures 1 and 2, the elongated optical emission must have a flat spectrum since there is no evidence for these elongated structures in the corresponding infrared images. Analyse of the light distributions of the infrared images shows that they follow those expected of giant elliptical galaxies. Thus, although the optical images are indeed strongly influenced by the presence of the radio source, the infrared images show the old stellar population of the underlying giant elliptical galaxy. Therefore, the K-z relation should be little affected by the extraordinary structures observed in the optical waveband.

There have been suggestions that the infrared magnitudes could be contaminated by emission from a compact nucleus which might be obscured in the optical waveband. We have analysed the infrared light distributions to search for nuclear components and find, in all but two cases, that they are similar to those expected of a giant elliptical galaxy. In the two cases in which there is evidence for a weak nuclear component, the radio galaxies 3C 22 and 41, there was also evidence from the radio maps for the presence of bright nuclear components and radio jets. Finally, we checked the photometry of the old infrared observations which were made using single element photometry with blind offsets and chopping. Our new infrared observations show that the 1984 infrared magnitudes were remarkably good within the expected uncertainties.

The upshot of all of this is that we have a new determination of the K-z relation (Figure 5). The solid points show the infrared magnitudes of the old stellar components of the radio galaxies. The open circles indicate the broad-line radio galaxies as well as 3C22 and 3C41 at $z \sim 0.8$. Also shown in Figure 5 are various lines indicating the expected $K-z$ relations for different assumptions about the evolution of the stellar populations of the radio galaxies. The solid line shows the expected relation if the spectrum of a giant elliptical galaxy is redshifted in a $q_0 = 0.5$ world model. The three dashed lines show the expected relation if the galaxies were formed at a large redshift and the stellar populations allowed to evolve passively to the present epoch. The three lines correspond to different assumed redshifts of formation of 5, 10 and 20. It can be seen that passive evolution models can account for the observed K-z relation. Notice that this means that these galaxies must have formed their stellar populations at a large redshift, probably at a redshift greater than 3.

References

1. J.S. Dunlop, in *Frontiers of Space & Ground-based Astronomy*, ed. W. Wamsteker, M.S. Longair & Y. Kondo (p. 395. Dordrecht: Kluwer Academic Publishers., 1994).
2. B.L. Fanaroff & J.M. Riley, MNRAS, 167, 31P, 1974.
3. P.D. Barthel, in *First Stromlo Symposium: Physics of Active Galactic Nuclei*, ed. G.V. Bicknell, M.A. Dopita & P.J. Quinn (p. 175. San Francisco: ASP Conf. Series., 1994).
4. P.N. Best *et al*, MNRAS, 275, 1171, 1995.
5. R.A. Laing, J.M. Riley & M.S. Longair, MNRAS, 204, 151, 1983.
6. R.G. Abraham, *et al*, MNRAS, 1996, in press.
7. M. Giavalisco *et al*, 1996, (preprint).
8. K.C. Chambers, G.K. Miley & W. van Breugel, Nature, 329, 604, 1987.
9. P.J. McCarthy *et al*, ApJ, 321, L29, 1987.
10. P.N. Best, M.S. Longair & H.J.A. Röttgering, MNRAS, 280, L9, 1996a.
11. R.A. Johnson, J.P. Leahy & S.T. Garrington, MNRAS, 273, 877, 1995.
12. P.N. Best, M.S. Longair & H.J.A. Röttgering, MNRAS, 1996b, submitted.
13. J.L. Higdon, ApJ, 455, 524, 1995.
14. S.J. Lilly & M.S. Longair, MNRAS, 211, 833, 1984.

UV AND LY-α IMAGING OF 4C41.17 AT $Z = 3.800$

W.J.M. VAN BREUGEL

Institute of Geophysics & Planetary Physics, LLNL, L-413, PO Box 808, Livermore, CA 94550, USA

G.K. MILEY

Leiden Observatory, PO Box 9513, Leiden, 2300 RA, The Netherlands

P.J. MCCARTHY

The Observatories of the Carnegie Institution of Washington, 813 Santa Barbara St., Pasadena, CA 91101, USA

H. SPINRAD

Astronomy Department, University of California, 601 Campbell Hall, Berkeley, CA 94720, USA

We present rest-frame UV and narrow-band Ly-α HST observations of the $z = 3.800$ radio galaxy 4C41.17. The observations suggest that we may be witnessing, for the first time, the formation of a galaxy due to the merging of (star forming?) subclumps, as predicted in dissipative galaxy formation scenarios, and of the effects of a young (quasar like ?) AGN on this environment.

1 Introduction

4C41.17 at $z = 3.800$ ($m_R = 23.8$) was the first HzRG discovered using a steep radio spectrum selection technique. With $F_{Ly-\alpha} = 3.5 \pm 2 \times 10^{-15}$ erg s^{-1} cm^{-2} ($6'' \times 4''$ aperture) it is also the brightest $z > 3$ $Ly - \alpha$ galaxy and thus a prime target for a detailed emission-line study. Previous observations with the aberrated HST showed that the radio-aligned optical continuum of 4C41.17 is very clumpy, with scale-sizes of a kpc. Here we present much improved observations with the refurbished HST, allowing a more detailed examination of of 4C41.17 and its environment. The images were taken through the F702W filter (line free; 6.0 hrs exposure), the F569W filter (including the Ly-α line; 2.0 hrs), and a LRF filter centered at 5830 Å (Ly-α ; 2.0 hrs). One of the field objects in the HST images was also seen at radio wavelengths, and was used to align the optical and radio frames with a relative accuracy better than 0.1$''$.

2   Rest-frame UV

The aligned rest-frame emission from 4C41.17 shows a complex, elongated morphology with numerous compact (kpc- sized), bright ($\sim$ 0.03 - 0.3 μJy) components (Fig. 1). The radio core (AGN) is located slightly East from a bright UV knot in a gap between two main regions. Together with previously discovered evidence for dust and cold gas in 4C41.17 from sub-mm (Dunlop *et al.* 1994) and Fabry-Perot Ly-α imaging (Hippelein and Meisenheimer 1993) this suggests that the AGN may be obscured at UV wavelengths.

While the UV, radio and Ly-α emission are all aligned, there is no one-to-one correspondence between these features, except at the location of the brightest Ly-α knot. The UV knots are all too bright to be caused by optical synchrotron emission. Instead, by analogy to $z \sim 1$ radio galaxies, a large fraction of the UV emission might be due to nebular continuum or scattered light from a hidden quasar. Indeed, the total rest-frame UV luminosity of the aligned clumps in 4C41.17 is $log[\nu L_{\nu,aligned}] = 46.9$ ($erg\,s^{-1}$), which is comparable to that of radio loud steep spectrum quasars ($46.4 < log[\nu L_{\nu}] = 47.6$ at 1500 Å for $1 < z < 2$ quasars).

3   Ly-α

The brightest Ly-α emission is associated with the second brightest radio knot and appears to have a V-shaped structure with small extensions pointing backwards toward the nucleus. Considering that radio knots probably pinpoint locations with strong shocks in jets, this Ly-α feature might be associated with shocked gas. Further East, or 'downstream' from this region, the Ly-α emission becomes diffuse and appears to spread in a cone with its apex in the vicinity of the radio core, and a cone angle of Ly-α is 20°- 25°. No bright Ly-α emission was detected from 4C41.17-South.

4   4C41.17-South

Perhaps the most important result from our HST observations is the discovery of an amorphous, 'non-aligned', clumpy group of objects south of 4C41.17 which is embedded in a halo of diffuse continuum emission and which appears connected with the 4C41.17 system (Fig. 2). Keck near-IR images have shown that this companion system is very blue ($\alpha_{1500Å} \sim 0$).

It is of interest to compare this with the recent discovery of a population of normal star-forming galaxies at $z = 3.25$ (Steidel *et al.* 1996). Following the Steidel *et al.* modelling assumptions and scaling for the difference in redshift

F702W

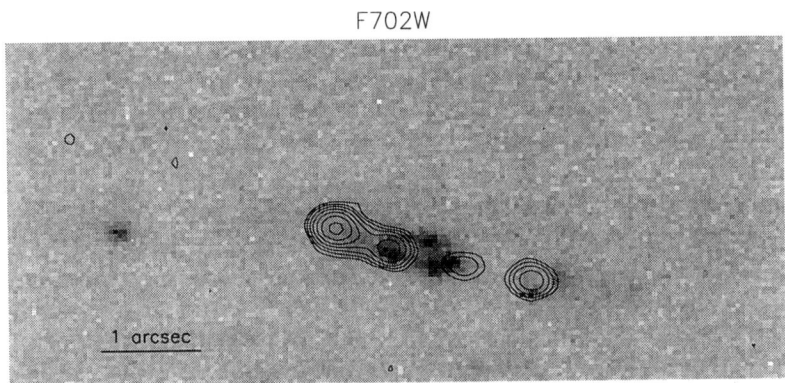

Figure 1: HST WFPC2 PC/F702W image of 4C41.17 with radio contours superimposed.

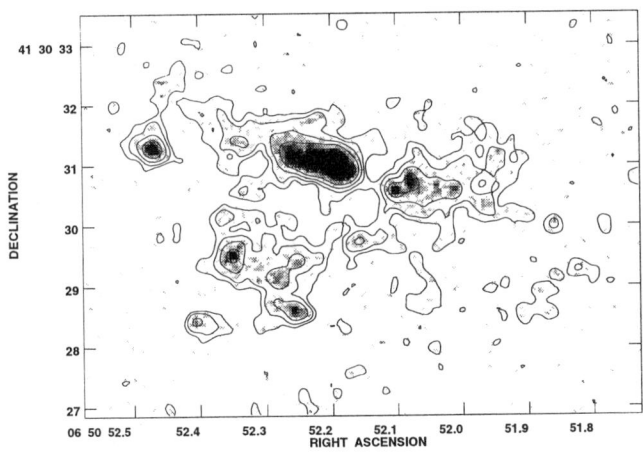

Figure 2: Smoothed version of Fig 1 showing the clumpy companion system.

Table 1: Starformation in 4C41.17 South?

F702W	Ap. diam.	kpc	F_ν [μJy]	log L_λ [erg/s/Å]	S.F.R [$M_\odot$/yr]
knots	0.3″	5	0.06	39.8	0.5
total	2.0″	34	0.8	40.9	6.3

we find that the compact, kpc-sized clumps in 4C41.17-South exhibit a modest star formation rate (S.F.R) of 0.5 $M_\odot$/yr on average (see Table 1; $H_0 = 50$, $q_o = 0.05$). However, 60% of the UV emission comes in 4C41.17-South comes from diffuse, low surface brightness emission and most of the star formation might be distributed over a much wider area ($\sim$ 30 kpc diameter) compared to the normal starforming galaxies at these redshifts. We estimate that the total S.F.R for 4C41.17-South might be $\sim$ 6.3 $M_\odot$/yr. It is likely therefore that in 4C41.17-South we are indeed witnessing the large scale collapse and merging of star-forming regions. This system may ultimately evolve into a massive elliptical, the usual galaxy type hosting powerful radio galaxies when observed at much lower redshifts.

5   Conclusion

The HST observations of 4C41.17 suggest that we may be viewing, for the very first time, the formation of a galaxy through merging of kiloparsec sized, very active starforming regions ('building blocks'), as predicted in dissipative galaxy formation scenarios. These star formation clumps could be sources for dust and dense gas and thus, for example, provide scattering mirrors or ionization material as they are intercepted by the collimated UV-continuum radiation from the 4C41.17 AGN. The AGN itself, one might hypothesize, could have formed through rapid stellar evolution and the formation of a black hole in the center one of these clumps, possibly triggered by the merging process. For further details and appropriate references see van Breugel et al. 1996 (in prep.).

Acknowledgments

The work by WvB described here was performed at IGPP/LLNL under the auspices of the U.S. Dept. of Energy under contract W-7405-ENG-48.

THE HOST GALAXIES OF DISTANT RADIO SOURCES

ARJUN DEY

National Optical Astronomy Observatories,
950 N. Cherry Ave., Tucson, AZ 85726, USA

A detailed study of the host galaxies of distant ($z > 1$) radio sources using HST and the Keck Telescope reveals a coherent view of these spectacular objects that explains both the alignment effect and the uniform $K - z$ Hubble diagram. Much of the spatially extended rest-frame UV continuum emission in the most powerful radio galaxies is not starlight, but a combination of scattered light from a hidden active nucleus and nebular emission. In contrast, the UV light in lower power radio sources is mostly uncontaminated by AGN-related emission, and in two galaxies at $z \sim 1.5$ is found to be dominated by starlight from an old (> 3.5 Gyr) stellar population. Since the stellar components of radio galaxies appear to be dynamically relaxed and old, even at $z > 1$, these massive objects were likely formed at very high redshifts ($z > 10$). Age-dating the stellar components in the host galaxies of distant radio sources can therefore provide strong constraints on the earliest epoch of massive galaxy formation and, perhaps, the cosmological parameters.

1 Introduction

The host galaxies of radio sources are among the most enigmatic objects in the Universe. The association of nearby radio sources with gE and cD host galaxies[1] has led to the speculation that the hosts of high-redshift radio sources may be evolutionary progenitors of the present-day massive ellipticals. Indeed, near-IR observations have demonstrated that the rest-frame optical morphologies of these objects are fairly symmetric structures, resembling elliptical galaxies [2,3]. Moreover, the host galaxies of luminous radio sources tend to define a fairly narrow sequence in the $K - z$ Hubble diagram [4]. Since this sequence is well-represented by the predicted 'passive' evolution of a massive ($5 - 10\ L^*$) galaxy with high formation redshift [5], it is believed that the rest-frame optical emission is dominated by starlight. A comparison of the $K - z$ diagram of radio galaxy hosts [5] with that of field galaxies [6] clearly demonstrates the importance of this population to our understanding of the evolution of the most massive galaxies.

However, the rest-frame UV morphologies of powerful radio galaxies are in stark contrast to their relatively normal optical morphologies. At $z > 1$, radio galaxies tend to be very luminous objects with spectacular UV morphologies [2], extending over >50 kpc. These UV structures tend to be aligned with the radio axes [7,8], suggesting a causal relationship between the active nucleus and the extended continuum emission.

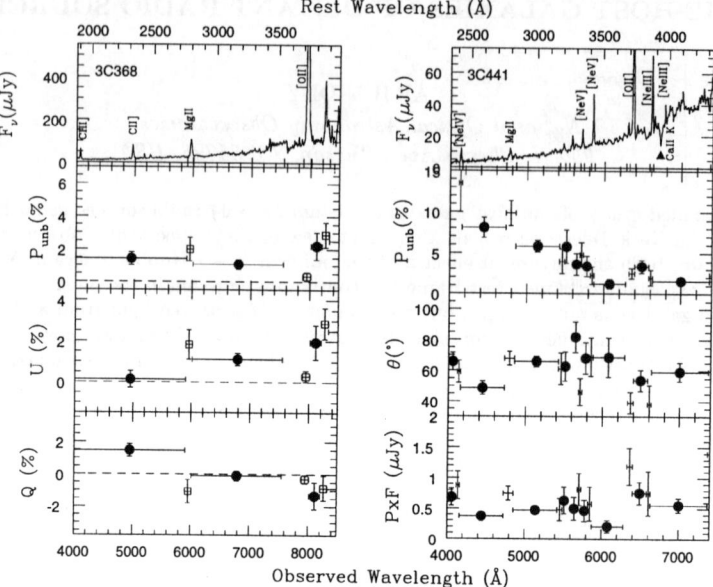

Figure 1: Spectropolarimetry of 3C368 (z=1.132) and 3C441 (z=0.707) obtained with the W. M. Keck Telescope. 3C368 is unpolarized, with $P < 3\%$; hence only the Stokes parameters (Q and U) and the unbiased percentage polarization (P_{unb}) are shown. In contrast, 3C441 is strongly polarized and shows a monotonically decreasing P and wavelength-independent polarized flux. Note the CaII K stellar absorption in 3C441.

2 The Origin of the UV Emission

A significant fraction (perhaps all) of the UV continuum emission in high-redshift, powerful radio galaxies is not starlight, but a combination of scattered AGN light and thermal nebular continuum [9,10]. Polarimetric observations by several groups [9,11,12,13,14] have now firmly demonstrated that the rest-frame UV emission in powerful, high-redshift radio galaxies is strongly polarized; of 15 $z > 0.7$ radio galaxies studied thus far, only 3 appear to be unpolarized with $P < 5\%$ (*e.g.*, Fig 1). In nearly all known cases the position angle of the electric vector is perpendicular to the major axis of UV emission, suggesting that the polarization is due to scattering (off ambient electrons and dust) of light from a hidden central source. In a few cases, the broad emission feature of MgIIλ2800 is detected in the polarized flux; the properties of this line are consistent with the central source being a quasar, hidden from our direct view and radiating anisotropically. Therefore, extracting any information about the stellar content of powerful radio galaxies requires understanding, accounting

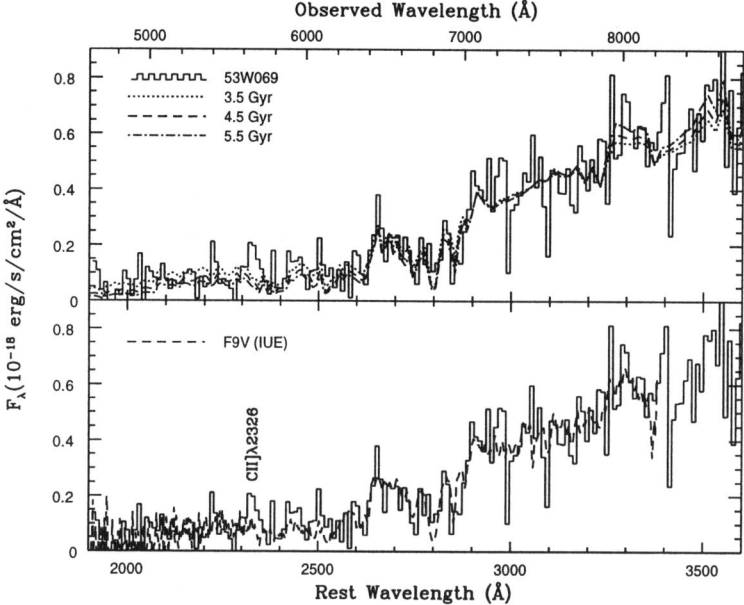

Figure 2: The histogram in both panels shows the Keck spectrum of LBDS 53W069 at a redshift $z=1.432$. The dotted, dashed, and dot-dash line in the upper panel are respectively 3.5 Gyr, 4.5 Gyr and 5.5 Gyr-old population synthesis models of Jimenez et al. 1996 [19]. The lower panel shows the spectrum of 53W069 compared with the mean IUE spectrum of an F9 main sequence star. These comparisons suggest that the age of 53W069 is at least 4.5 Gyr.

for and subtracting the scattered AGN and nebular components.

3 Starlight in Radio Galaxies

Although the UV emission is largely a spectre of the AGN, starlight begins to dominate at longer wavelengths. Deep spectroscopy of radio galaxies with the W. M. Keck Telescope [15,16] has revealed CaII K stellar absorption in several cases (*e.g.*, 3C441 in Fig 1). Moreover, the relative dominance of the AGN light at UV wavelengths appears to be a function of radio source luminosity; less powerful radio sources have less spectacular UV morphologies, and even their UV emission tends to be dominated by starlight. Hence, lower power radio sources may be better probes of cosmology and galaxy evolution.

A deep spectroscopic study using the Keck Telescope of weak radio sources associated with red host galaxies has indeed revealed that the UV emission in these objects is dominated by starlight. For example, the $z=1.43$ radio source

LBDS 53W069 ($S_{1.4GHz} \approx 3.8$ mJy) is associated with a faint, red ($I-K \approx 4.1$) host galaxy, and has a spectrum extremely similar to that of a late-type F star (Fig 2). Since the UV light in populations with ages >1 Gyr is dominated by the turn-off population, a spectral type of F9 implies a large age for this object. Indeed, population synthesis of this spectrum results in an age ~ 4.5 Gyr. The existence of such an old object at this early epoch places strong constraints on the earliest epoch of massive galaxy formation (for $H_0=50$, $\Omega_0=0.2$, 53W069 must have been formed at $z_f > 10$), and on the cosmological parameters [17,18].

Acknowledgments

I am very grateful to my collaborators Wil van Breugel, Andrea Cimatti and Ski Antonucci, Hyron Spinrad, James Dunlop, John Peacock, Raul Jimenez, Daniel Stern and Rogier Windhorst for their permission to present data prior to publication. I gratefully acknowledge the support of an NOAO Postdoctoral Fellowship.

References

1. T. A. Matthews, W. W. Morgan & M. Schmidt, ApJ, 140, 35, 1964.
2. M. S. Longair, 1997, in this book
3. M. Dickinson, A. Dey & H. Spinrad, 1996, in *Galaxies in the Young Universe*, ed. H. Hippelein & K. Meisenheimer
4. S. J. Lilly & M. S. Longair, MNRAS, 211, 833, 1984.
5. P. J. McCarthy, ARA&A, 31, 693, 1993.
6. L. Cowie, 1996, in this book.
7. P. J. McCarthy *et al.*, ApJ, 321, L29, 1987.
8. K. C. Chambers, G. K. Miley, & W. van Breugel, Nature, 329, 604, 1987.
9. S. di Serego Alighieri, R. A. E. Fosbury, P. J. Quinn & C. N. Tadhunter, Nature, 341, 307, 1989.
10. R. Dickson *et al.*, MNRAS, 273, L29, 1995.
11. S. di Serego Alighieri, A. Cimatti, & R. Fosbury, ApJ, 431, 123, 1994.
12. A. Dey *et al.*, ApJ, 465, 157, 1996.
13. A. Cimatti *et al.*, ApJ, 465, 145, 1996.
14. B. T. Jannuzi *et al.*, ApJ, 454, L111, 1995.
15. A. Dey & H. Spinrad, ApJ, 459, 133, 1996.
16. A. Stockton, M. Kellogg & S. E. Ridgway, ApJ, 443, L69, 1995.
17. J. S. Dunlop *et al.*, Nature, 381, 581, 1996.
18. J. S. Dunlop, 1997, in this book
19. R. Jimenez *et al.*, 1996, in preparation

HST IMAGING OF HIGH-REDSHIFT QUASARS

TIMOTHY HECKMAN

Dept. of Physics and Astronomy, Johns Hopkins University,
Baltimore, MD 21218, USA

MATTHEW LEHNERT

Sterrewacht Leiden, Postbus 9513, 2300RA Leiden, The Netherlands

Elucidation of the underlying causes of the rapid cosmological evolution of the quasar population would have important implications for understanding the early heating of the IGM, the formation and early evolution of galaxies, and of course the quasar phenomenon itself. We report here the preliminary results of several HST programs to image the 'fuzz' around small samples of high-z quasars. We find that both the UV continuum and Lyα emission is routinely resolved around radio-loud quasars, but only in a minority of the radio-quiet quasars. Radio-loud quasars appear to inhabit extraordinary galaxies (similar to radio galaxies). Radio-quiet quasars (the dominant quasar population) appear to be hosted by smaller, less massive galaxies (whose properties could be similar to the newly discovered class of high-z field galaxies).

1 Introduction and Motivation

The study of the environments of high-redshift AGN is a vital enterprise that connects to the 'big picture' issues in contemporary cosmology that have been the subject of this meeting. The focus in this paper is on quasars, since others (Dey, Dunlop, Longair, and van Breugel) have ably discussed high-z radio galaxies.

First of all, quasars would be astrophysically significant objects in their own right, even if they had absolutely nothing to do with broader cosmogonical issues. They are the most luminous objects in the Universe and provide unique laboratories for the investigation of extreme physical processes and conditions. They are also an energetically significant component of the Universe: integrated over cosmic time, they have produced of-order 10^{-1} as much radiant energy as stars. They most likely dominated the metagalactic ionizing background at high redshift (z $\sim$ 1 to 5), and in so-doing regulated the physical state of the Lyα forest, of pre-galactic gas clouds, and of the gaseous halos of galaxies. This 'feedback' from quasars therefore had a profound impact on the early evolution of the baryonic component of the Universe.

Today, we see only the dying embers of the quasar conflagration that blazed in the early Universe: the amount of emission provided by quasars has declined by a factor of about 10^2 from its peak during the 'quasar epoch' at z = 2 to

3 to the present-day (cf. Boyle[1]). This strong and rapid cosmic evolution of the quasar population is certainly one of the most remarkable and mysterious phenomena in all of astronomy (e.g. Turner[2]). It is hard to resist speculating that it is connected in some way to galaxy formation and early evolution (e.g. Rees[3]).

Indeed, the ubiquity of supermassive black hole candidates in the nuclei of the nearest galaxies implies that many (most?) galaxies went through through an early quasar phase. This statement has some quantitative basis. The amount of 'burnt matter' associated with 'dead quasars' amounts to about 10^{-3} of the mass contained in galactic bulges in the present universe. Kormendy & Richstone [4] suggest that the mass of a supermassive black hole is related to the mass of the stellar bulge in which it resides. Their Figure 14 then implies that the candidate supermassive black holes in eight nearby galaxies indeed have masses of-order 10^{-3} to 10^{-2} of the bulge (as required if 'dead quasars' are common in bright present-day galaxies).

Thus, the underlying causes of the rapid cosmological evolution of the quasar population are a mystery around which are woven many different strands of contemporary cosmology. One key part of the quest to solve this mystery is the determination of the nature of the host galaxies of high-z quasars, and a comparison of these galaxies to both the host galaxies of quasars at low-redshift and to the field population of galaxies at high-redshift. Although ground-based images can provide important complimentary information, such a program is really only feasible with the high angular resolution provided by the HST (e.g. 1 arcsec $\sim$ 8 kpc at z = 2 for h = 0.75 and q_0 = 0.1, which we adopt throughout).

Accordingly, we will describe preliminary results from a modest-scale program of HST imaging observations of the environments of both radio-loud and radio-quiet quasars at high-redshift (z>1).

2 The Difficulties

While conceptually straightforward, using the HST to determine the redshift dependence of the properties of quasar host galaxies has been hampered by a number of problems (even after the refurbishment of the HST optics):

1) Investigations at high-z have focused largely on the rare class of extremely radio-loud AGN (primarily radio galaxies, but including radio-loud quasars). One might then well worry about how generally applicable the spectacular data on high-z radio-loud objects are to the AGN phenomenon as a whole (especially in view of the 'alignment effect').

2) The samples studied have effectively been flux-limited (e.g. the 3CR cata-

log). This results in a very strong correlation between redshift and quasar luminosity. Thus, even if z-dependent properties are found, it is not straightforward to determine whether these are physically related to quasar luminosity or to cosmic epoch.

3) The quasars have not typically been imaged at the same *rest* wavelength as a function of redshift. That is, with HST we are usually in a position of comparing rest-frame optical images of low-z quasar hosts to rest-frame UV images of high-z hosts. This is particularly worrisome because the UV emission is very sensitive to young stars, nebular continuum, and dust scattering of quasar light (and thus can be dominated by material that contains relatively little mass and may be quite transient).

4) The broad-band filters for the images used often transmit strong line emission. Not only does this make the interpretation of the images uncertain, the degree to which line emission contaminates the image will be a strong function of redshift as lines migrate in and out of the bandpass.

5) The high-z quasars imaged to date are so powerful that only quite extraordinary host galaxies could be detected. Thus, null results are not very illuminating.

6) The HST is - after all - a modest-sized telescope, whose primary imager has a relatively large read-noise per small pixel (especially with the Planetary Camera, which must be used to adequately sample the point-spread function). It takes a lot of orbits to go deep!

3 Our Programs

We are involved in several different programs with various collaborators to image the host galaxies of high-redshift quasars with HST:

1. In collaboration with George Miley and Wil van Breugel, we have imaged 5 radio-loud quasars.

2. In collaboration with James Lowenthal, we have imaged 5 radio-quiet quasars.

Both samples have a mean redshift of about 2.2. The quasars have typical B magnitudes between 19 and 20.5, with $< M_B > \sim -25.5$. Since the comoving density of quasars reaches its peak in this redshift range, and since M_B = -25.5 is near the 'knee' in the quasar luminosity function at this redshift, the radio-quiet sample is typical of the population that is responsible for much of the radiant energy ever produced by quasars over the history of the Universe.

The data consist of F555W PC images ($\lambda_{rest} \sim 1800$Å) with total integration times of about 2000 sec per quasar and images of redshifted Lyα obtained with the WFC and the appropriate [OII] quad filter with integration times of about 5000 sec per quasar. Both sets of images have been CR-split to allow rejection of cosmic rays. The limiting sensitivities (3 rms) are about 27.5 V magnitudes for a point source and 25.5 V magnitudes arcsec^{-2} for the F555W images and 7×10^{-17} erg cm^{-2} s^{-1} for a point source and 7×10^{-16} erg cm^{-2} s^{-1} arcsec^{-2} for the Lyα images. Thus, the images are of modest depth, and in the case of diffuse trans-arcsec-scale structure do not go as deep as ground-based images we have obtained of some of these same quasars - cf. Heckman et al[5]).

3. With Wil van Breugel, Pat McCarthy, and Vijay Kapahi, we have imaged 10 radio-loud quasars from the Molonglo survey over the redshift range z = 0.3 to 3. We have been awarded time to observe 10 more in Cycle 6.

In this case, we have used a suite of filters on the WFPC2, so that we have imaged every quasar at approximately the same *rest* wavelength ($\sim$2000Å). The images have a total integration time of about 2000 sec per quasar (e.g. they reach a depth comparable to the F555W images described above).

In addition to these programs, some other notable HST programs for imaging high-z quasars include the 3CR snapshot survey (PI - Bill Sparks), the 3CR '$z \sim 1$' program of Ridgway and Stockton, and the images of BR 1202-0725 (a radio-quiet quasar at z =4.7!) described at this conference by E. Hu.

4   Methodology

We have used several different strategies to determine whether the HST images of the high-z quasars are spatially-resolved, and if so, to determine the basic structural properties of the 'fuzz'.

First, we have collected an atlas of archival or publically-available images of HST stars taken with the appropriate filter (as near in time as possible to our observations). Next, we have measured the encircled energy diagrams (EED's - by which we mean the fraction of the flux from a point source interior to a radius r, as a function of r). We have then intercompared all the EED's taken through a given filter to determine the reproducibility of the EED. This intercomparison implies that we can detect fuzz that contributes more than about 5% as much light as the quasar itself (within a radius of about 1 to 1.5 arcsec). This is consistent with the known effects of PSF variations in time due to the so-called 'breathing' of the telescope (cf. Biretta et al[6]).

We have also compared these empirical EED's and the two-dimensional PSF's to those generated by Tiny Tim. We find good agreement, provided that we use only azimuthal averages (Tiny Tim does not reproduce the detailed 2-dimensional structure of the PSF) and restrict the radial range under consideration to r < 2 arcsec (beyond which incompletely-understood large angle scattering in the WFPC2 becomes important - Biretta et al[6]).

We have then used the library of stellar images to construct an average EED (in one dimension) and PSF (in 2-D) for the F555W filter. In the case of the [OII] quad filter, there are too few archival stellar images to do this, and we have had to rely on Tiny Tim. Comparing these EED's to our the EED for each quasar, we have been able to estimate the amount of excess light around the quasar. We have used the PSF's to subtract the quasar light, scaling the PSF so that the the total flux in the difference image is defined to be zero inside a radius of 2 PC pixels (0.09 arcsec). This is a conservative choice, since the actual underlying fuzz will certainly have a finite central surface brightness.

5 Results to Date

5.1 Radio-Loud Quasars

In nearly every case we have examined carefully to date, the rest- frame UV continuum associated with the radio-loud quasars is spatially-resolved by the HST (see top two panels in Figure 1). This fuzz constitutes 5% to 30% of the total continuum arising from within a radius of about 2 arcsec ($\sim$ 15 kpc). The integrated magnitude of the fuzz is typically V $\sim$ 22 $\pm$ 0.5.

Comparing these images to ground-based images of the same quasars (Heckman et al[5]; Lehnert et al[7]) shows that these two kinds of data are complementary: the HST images show the lumpy and asymmetric morphology of the relatively compact regions of high-surface-brightness, while the ground-based images reveal lower-surface-brightness structures having larger angular scales. The typical UV luminosity of the fuzz is roughly 10^{11} $L_\odot$. This is a factor of 3 to 10 more luminous than the typical 'Lyman drop-out' field galaxies studied by Steidel et al[8] and Giavalisco et al[9], and a factor of 2 to 10 more luminous in the UV than the UV-brightest local (z $\sim$ 0) starburst galaxies.

¿From the first set of images obtained of the set of Molonglo quasars (program 3 above), we find that - despite strong cosmological dimming - it is actually *easier* to detect UV fuzz around the quasars as the redshift increases! As we have emphasized in section 2 above, it is not clear whether the physical connection here is between fuzz properties and redshift or fuzz properties and quasar luminosity.

As in the radio galaxies, there are some examples in which the continuum fuzz is aligned with the radio source. However, clear counterexamples also occur, and we can not yet rigorously demonstrate the existence of a strong alignment effect (based on the small sample analyzed to date). In at least one case (PKS 0445+097), much of the misaligned 'fuzz' turns out to be an intervening galaxy responsible for MgII absorption at z=0.84 (Lehnert & Becker[10]).

The narrow-band images of Lyα are also spatially-resolved, with the fuzz being detected over scales ranging from 0.1 to several arcsec. Again, comparing these images to the ground-based images and spectra discussed in Heckman et al[5] and Heckman et al[11] respectively, shows that HST misses the diffuse, low-surface brightness structures, while highlighting the compact, high-surface-brightness features. The typical luminosity of the spatially-resolved Lyα emission is about 10^{44} erg s^{-1}. To produce this emission via photoionization and recombination, gas around the quasar need intercept only a few % of the ionizing radiation from the quasar.

As shown in Figure 1, there is clear evidence that the gas "knows" about the radio source. Not only is the emission aligned with the axis of the radio plasma (as would be expected if the ionizing radiation from the quasar escapes from the nucleus preferentially along the jet axis), but there are also some cases of detailed morphological correspondence which suggest 'jet-cloud' interactions that light-up the gas. However, there are counterexamples in which the Lyα emission is oriented roughly perpendicular to the radio source axis.

In general, there is no detailed one-to-one correspondence between the structures seen in the UV continuum images and the Lyα images. This suggests that nebular emission is not the major contributor to the UV continuum fuzz.

5.2 Radio-Quiet Quasars

In striking contrast to the radio-loud quasars described above, we we have only been able to convincingly resolve the UV continuum and Lyα fuzz in 2 of the 5 radio-quiet quasars imaged to date. This striking difference confirms and extends the results of Lowenthal et al[12], who imaged several samples of high-z radio-quiet quasars from the ground in the near-IR and optical (rest-frame optical and UV).

The UV continuum fuzz has V ~ 22.5 in the brightest resolved case and V > 24 in the unresolved cases. The fuzz properties could then overlap with those of the field population of 'Lyman drop-out' galaxies. The implied UV luminosities range from $< few \times 10^{10} L_\odot$ to $\sim 10^{11} L_\odot$, consistent with typical UV luminosities of local starburst and star-forming galaxies (cf. Wang & Heckman[13]).

The lack of Lyα fuzz implies that typically less than about 1% of the ionizing photons produced by the quasar are intercepted by gas clouds in the host galaxy over radial scales of about 1 to 20 kpc. This is a somewhat surprising result in the context of models in which quasars are hosted by dynamically

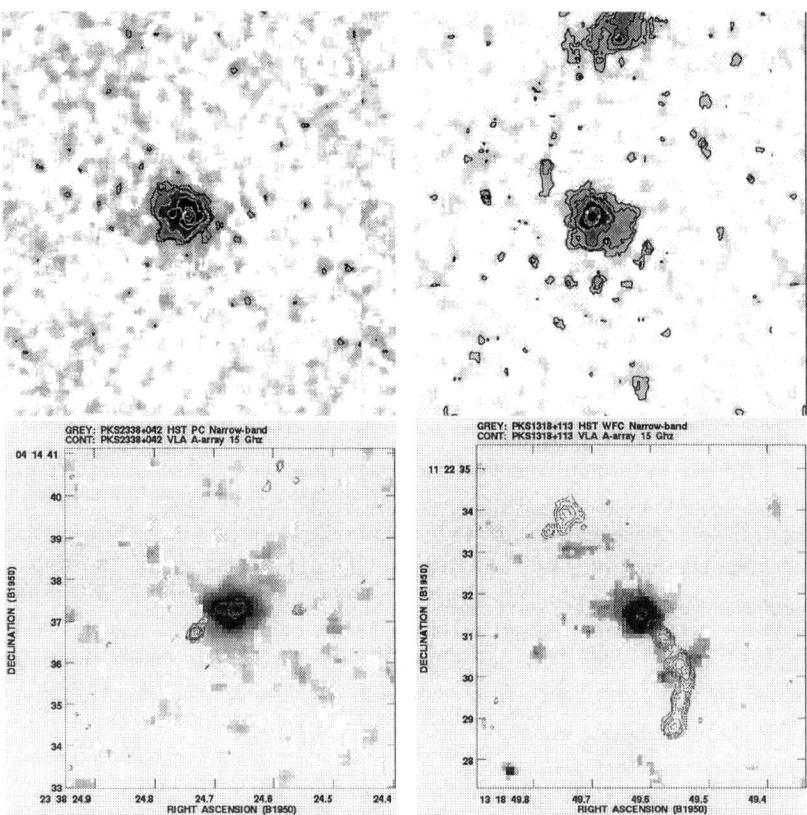

Figure 1 — F555W image of PKS2338+042 (z=2.594; top left) and F675W image of MRC1114-220 (z=2.282; top right). Both images have north at the top and east to the left, are 6.8"×6.8" in projected size, and have had the central quasar subtracted. We also show the narrow-band Lyα images of PKS 2338+042 (bottom left) and PKS 1318+113 (bottom right) with radio maps overlayed (from Lonsdale et al 1993). The continuum image of PKS2338+042 shows a "hole" ≈1" to the east of the nucleus and the Lyα image shows a "hook shape" to the south and east of the nucleus in PKS 2338+042. These two morphological features occur at the bright inner radio emission and where it bends to the southeast. We see a very similar effect in PKS1318+113 but in this case, the radio bends in a region of high surface brightness Lyα emission ≈2" to the southwest of the nucleus.

or evolutionarily 'young' galaxies: gas-rich systems in an unrelaxed state (cf. Rees [3]).

6 Summary and Future Prospects

1. The 'fuzz' around quasars during the 'Quasar Epoch' *can* be spatially-resolved with HST and in at least some cases has properties that are *consistent with* those of young or forming galaxies. In view of the lessons learned from the study of the high-redshift radio galaxies, we need to be cautious about primarily attributing the UV continuum emission to a young stellar population. Starlight from the quasar host galaxy, scattered light from the quasar, nebular emission, and even intervening galaxies can also contribute.

2. High-z radio-loud quasars have significantly brighter and larger fuzz than radio-quiet quasars at similar redshifts. This may mean that radio-loud quasars are hosted by more massive host galaxies (as is also hinted from ground-based near-IR imaging - cf. Lowenthal et al [12]). It also seems to imply that the gaseous halos of radio-quiet quasars are highly optically-thin in the Lyman limit. Could this difference be the result of radio-loud quasars residing in regions of higher gas pressures (and hence higher gas densities) possibly provided by the radio source's 'cocoon' (cf. Begelman & Cioffi [14]) or a 'cooling flow' (cf. Crawford & Fabian [15])? The differences between the radio-loud and radio-quiet quasars do underscore the potential dangers in generalizing the inferences made concerning high-z radio galaxies to the AGN phenomenon as-a-whole.

Lest we close on an overly pessimistic note, it is important to emphasize that while an enormous amount of HST time has been devoted to observational cosmology (and rightly so), to date only a trivial amount of time (about 20 orbits through Cycle 5 by our reckoning) has been spent studying the environments of the cosmically-dominant AGN population: radio-quiet quasars at high- redshift.

The new capabilities of HST following the servicing mission - the broadening of the wavelength coverage to the near-IR (rest-frame visible) with NICMOS and the capability for true 2-D spectroscopy with STIS - mean that the time is now ripe for a dedicated attack on this problem. As we have argued in section 1, progress on unraveling the mystery of the rise and fall of the quasar population in the early universe would have important implications for understanding the reheating of the IGM, the formation and early evolution of galaxies, and of course the quasar phenomenon itself.

Acknowledgments

We thank Kavan Ratnatunga and Chris Burrows for useful discussions concerning the WFPC2 PSF. We thank Alan Stockton for sharing some of the results of his HST 3CR mini-survey with us prior to publication. We also acknowledge the support of the HST Guest Observer program.

References

1. B. Boyle in *The Environment and Evolution of Galaxies*, ed. J. M. Shull & H. Thronson (Kluwer, Dordrecht, 1993).
2. E. Turner, AJ, 101, 5, 1991.
3. M. Rees, MNRAS, 231, 91P, 1988.
4. J. Kormendy and D. Richstone, ARA&A, 33, 581, 1995.
5. T. Heckman, M. Lehnert, W. van Bruegel, and G. Miley, ApJ, 370, 78, 1991.
6. J. Biretta *et al*, *Wide Field and Planetary Camera Instrument Handbook, Version 4*, (STScI, 1996).
7. M. Lehnert, T. Heckman, K. Chambers, and G. Miley, ApJ, 393, 68, 1992.
8. C. Steidel, M. Giavalisco, M. Dickinson, and K. Adelberger, AJ, 112, 352, 1996.
9. M. Giavalisco, C. Steidel, and F. Macchetto, ApJ, 470, 189, 1996.
10. M. Lehnert and R. Becker, in preparation.
11. T. Heckman, M. Lehnert, G. Miley, and W. van Breugel, ApJ, 381, 373, 1991.
12. J. Lowenthal, T. Heckman, M. Lehnert, and J. Elias, ApJ, 439, 588, 1995.
13. B. Wang and T. Heckman, ApJ, 457, 645, 1996.
14. M. Begelman and D. Cioffi, ApJ, 345, L21, 1989.
15. C. Crawford and A. Fabian, MNRAS, 239, 219, 1989.

Acknowledgements

We thank Kevin Rauch and Clint Burrows for useful discussions for sending the WFPC2 PSF. W. thank Alan Stockton for sharing some of the results of his PC1 narrow survey with us prior to publication. We also acknowledge the support of the HST Guest Observer program.

References

HUBBLE SPACE TELESCOPE IMAGES OF A SAMPLE OF TWENTY NEARBY LUMINOUS QUASARS [a]

JOHN N. BAHCALL, SOFIA KIRHAKOS, AND DAVID H. SAXE,
Institute for Advanced Study, School of Natural Sciences, Princeton, NJ 08540

DONALD P. SCHNEIDER
Department of Astronomy and Astrophysics, The Pennsylvania State University, University Park, PA 16802

Observations with the Wide-Field Camera of the *Hubble Space Telescope (HST)* are presented for a representative sample of 20 intrinsically luminous quasars with redshifts smaller than 0.30. These observations show that luminous quasars occur in diverse environments that include ellipticals as bright as the brightest cluster galaxies (2), apparently normal ellipticals (10), apparently normal spirals with H II regions (3), complex systems of gravitationally interacting components (3), and faint surrounding nebulosity (2). The quasar host galaxies are centered on the quasar to the accuracy of our measurements, 0.3 kpc. There are more radio quiet quasars in galaxies that appear to be ellipticals (7) than in spiral hosts (3), contrary to expectations. However, three, and possibly five, of the six radio loud quasars have detectable elliptical hosts, in agreement with expectations. The 20 objects studied in this paper show that luminous quasars occur preferentially in luminous galaxies. The median observed luminosity of the hosts is 0.4 magnitudes brighter than L*, a result that is inconsistent with the field galaxy luminosity function. The superb optical characteristics of the repaired *HST* make possible the detection of close galactic companions; we detect eight companion galaxies within projected distances of 10 kpc from quasar nuclei. The presence of very close companions, the images of current gravitational interactions, and the higher density of galaxies around the quasars suggest that gravitational interactions play an important role in triggering the quasar phenomenon.

1 Introduction to the Pictures

You may well wonder what a talk on luminous, small redshift quasars is doing in a conference devoted to HST and the high-z universe. The ostensible reasons for including this talk are: 1) the observations at small redshift are the basis for determining what evolution has occurred at the larger redshifts; and 2) the greater detail that is available in the images of small redshift quasars may be useful in interpreting the observations of larger redshift objects.

In fact, the real reason for including this talk was that the presenter(JNB) wanted to express in person his admiration and affection for Alec Boksenberg as

[a] Based on observations with the NASA/ESA Hubble Space Telescope, obtained at the Space Telescope Science Institute, which is operated by the Association of Universities for Research in Astronomy, Inc., under NASA contract NAS5-26555.

a creative scientist, an original instrument builder, an inspirational organizer, and a wonderful friend.

The talk summarized the results of our analysis of HST-WFPC2 observations of a representative sample of 20 of the most luminous ($M_V < -22.9$) nearby ($z < 0.30$) quasars. The goal of these observations was to help understand the quasar phenomenon by determining the environment in which quasars occur. The main result of this paper is that there is not one type of environment, but instead a wide range of environments in which the most luminous quasars appear to be embedded. The HST images also contain a number of extraordinary phenomena and some surprises, including: very close companions, host ellipticals for radio quiet quasars, spiral hosts with well developed arms and prominent H II regions, galaxies caught in the act of merging, apparently faint galactic hosts, and very extended emission.

The results of this work have been published partially in a series of papers in ApJ and ApJ Letters and, more recently, in a summary paper, with the same title as this talk, that has been submitted to ApJ and is available (with references to earlier work by ourselves and by many other researchers) at http://www.sns.ias.edu/~jnb (see Some Recent Preprints and Reprints).

Figure I shows the unprocessed images (without subtraction of the stellar quasar) for all 20 of our program objects. It is apparent from these images that luminous quasars have a variety of environments. More details are available in the complete paper on the subject described above.

Acknowledgments

This work was supported by NASA contract NAG5-1618, NAG5-3259, NASA grant number NAGW-4452 and grant number GO-5343 from the Space Telescope Science Institute, which is operated by the Association of Universities for Research in Astronomy, Incorporated, under NASA contract NAS5-26555.

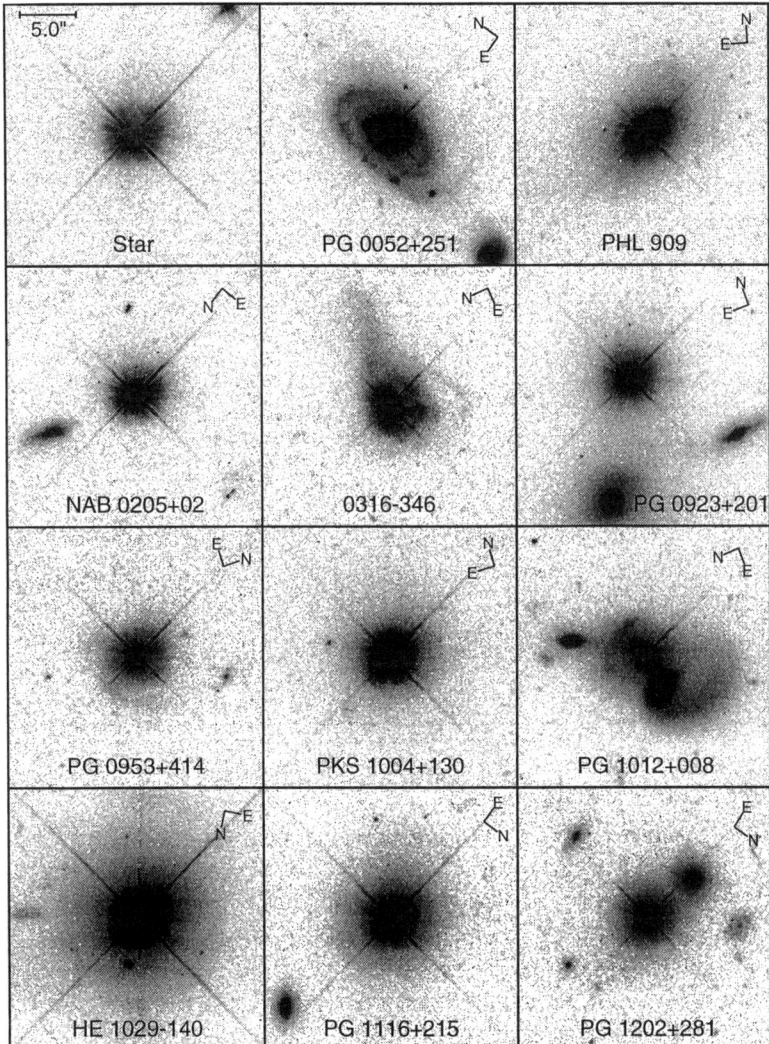

Figure 1: A $23'' \times 23''$ WF image of each one of the twenty luminous nearby quasars in our sample. A blue field star, MMJ 6490, is also shown for comparison (first panel). These images were obtained using the *HST* WF3 and the F606W filter. The exposure times are 1400 s or 1100 sec. Cosmic ray subtraction and pipeline STScI flatfielding are the only processing performed on the *HST* images shown here.

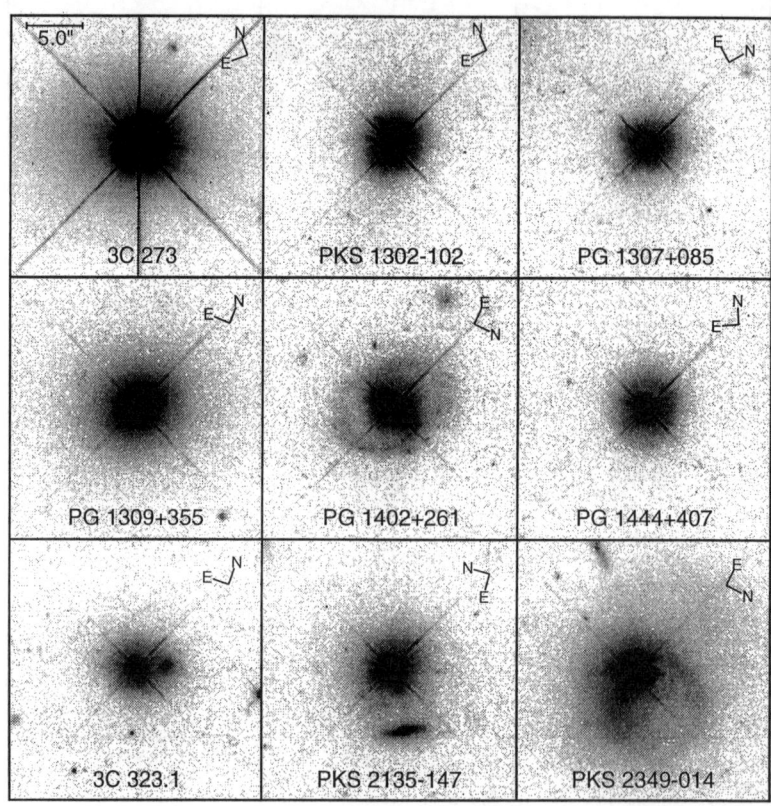

Figure 1: *Continued*

METAL ABUNDANCES IN HIGH REDSHIFT QSOS

F. HAMANN

Center for Astrophysics & Space Sciences, UC-San Diego, USA

J.C. SHIELDS

Steward Observatory, University of Arizona, Tucson, USA

C.B. FOLTZ & F.H. CHAFFEE

MMT Observatory, University of Arizona, Tucson, USA

We discuss preliminary results from a multi-faceted study of metal abundances in high redshift QSOs. One aspect of our study involves new measurements of the N V $\lambda1240$/He II $\lambda1640$ and N V/C IV $\lambda1549$ emission line ratios in $z > 4$ sources. These flux ratios are sensitive to the relative N abundance and the overall metallicity. Our measurements of large N V line ratios at $z > 4$ indicate super-solar abundances within $\sim$1 Gyr of the Big Bang (for $q_o \sim 0.5$). Independent study of the various intrinsic absorption lines in QSOs supports the emission-line results; typical broad absorption line (BAL) regions and at least some associated ($z_a \approx z_e$) systems have super-solar metallicities. A new theoretical analysis of the absorption-line data provides quantitative uncertainties and firm lower limits on the metal abundances, even when the ionization state is unknown.

1 Introduction

Broad emission lines and intrinsic absorption lines of heavy elements in QSO spectra require some degree of chemical enrichment out to redshifts of nearly 5. An important goal of QSO research is to understand the extent and origin of this enrichment. If QSOs at all redshifts reside in galactic nuclei, the abundance studies could yield vital information on galactic nuclear evolution that is not available by other means. Recently, several independent studies [1,2,3,4,5] of the emission and absorption lines have concluded that QSO metallicities are typically quite high – at least sometimes exceeding solar. Gas-phase abundances up to $\sim$10 $Z_\odot$ in QSOs can be attributed to the normal, rapid stellar enrichment expected in the cores of massive galaxies.[2,3] This interpretation implies, however, that massive galactic nuclei, or at least their seed progenitors, are already highly evolved at QSO redshifts. In fact, if the initial mass functions do not severely favor massive stars, most of the gas must be converted into stars and stellar remnants before the QSOs become observable. We are working now to test the QSO abundance results and extend both the emission and absorption line analyses to larger samples and higher redshifts.

2 Broad Emission Lines

Photoionization and chemical evolution calculations [1,2,4] have shown that the N v $\lambda1240$/He II $\lambda1640$ and N v/C IV $\lambda1549$ emission line ratios are useful indicators of the relative N abundance and the overall metallicity. Figure 1 shows preliminary measurements of these flux ratios for 15 $z \gtrsim 4$ QSOs measured at the MMT,[6] plus additional ratios at lower redshifts from the literature.[2] The data from the literature are incomplete for redshifts $2 \lesssim z < 4$ because lower limits were not available when only N v was detected. We expect that more large N v/He II ratios would appear at these redshifts in a well-measured sample. The dotted lines in the figure show the line ratios predicted by photoionization models using solar abundances and physical conditions that nearly maximize the ratios. Extensive calculations [2,4] show that much smaller line ratios can occur readily, but it is not possible to explain the upper envelope of measured ratios (most notably N v/He II) unless nitrogen is relatively enhanced and the overall metallicity is above solar. In some cases, the derived lower limits are $Z \geq 5 \, Z_\odot$.

3 Intrinsic Absorption Lines

Intrinsic absorption lines include the broad absorption lines (BALs) and at least some of the narrower associated ($z_a \approx z_e$) systems. These lines yield column densities in various ions that can be converted to relative abundances using model-dependent ionization corrections. For example, the abundance of any metal relative to hydrogen follows from,

$$\left[\frac{M}{H}\right] = \log\left(\frac{N(M_i)}{N(HI)}\right) + \log\left(\frac{f(HI)}{f(M_i)}\right) + \log\left(\frac{H}{M}\right)_\odot \tag{1}$$

where $(H/M)_\odot$ is the solar abundance ratio, and N and f are respectively the column densities and ionization fractions of H I and the metal M in ionization stage i. The ionization corrections, $f(HI)/f(M_i)$, can be large and uncertain for intrinsic absorbers because the ionization state is often high and poorly constrained. If the gas is in photoionization equilibrium with a QSO spectrum and optically thin at all (important) continuum wavelengths – both reasonable assumptions for intrinsic absorbers – the correction factors depend on just the shape of the ionizing spectrum and its intensity relative to the local gas density (i.e. the ionization parameter). For any given spectral shape, there is a minimum correction factor for each metal ion that yields a minimum metallicity [M/H] from Eqn. 1. Figure 2 shows the minimum correction factors normalized by solar abundance ratios (i.e. the last two terms in Eqn. 1) for C IV and Si IV

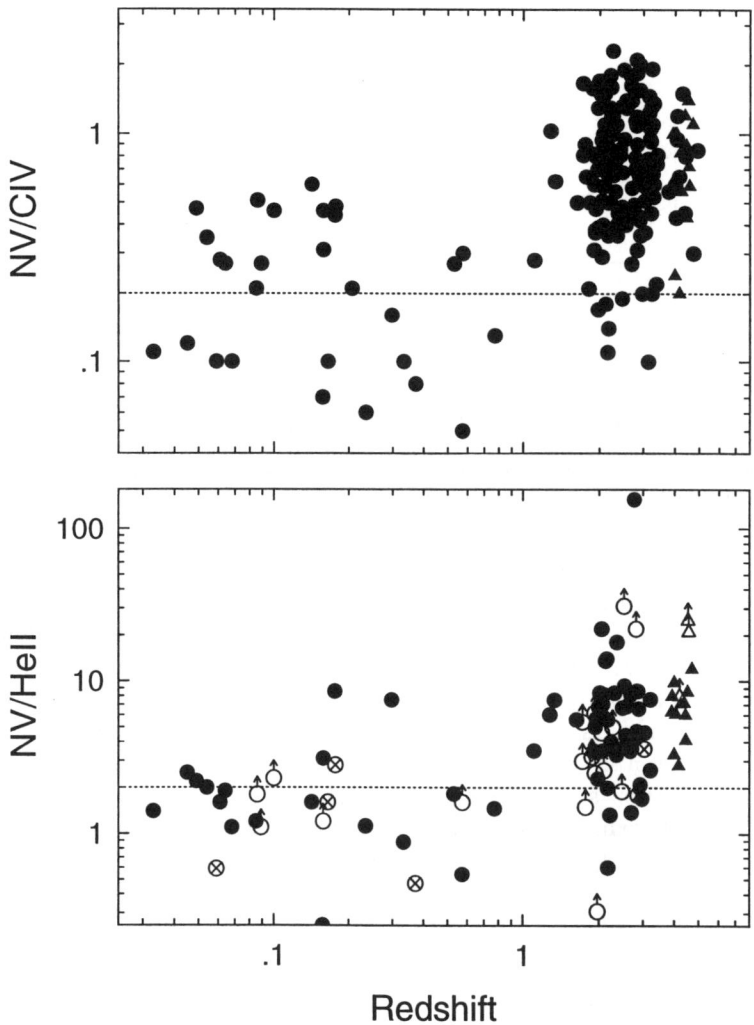

Figure 1: Measured emission line ratios in QSOs. The triangles are the new MMT data for $z > 4$. Open symbols indicate lower limits. The dotted lines define the threshold above which super-solar abundances are required by standard photoionization models.

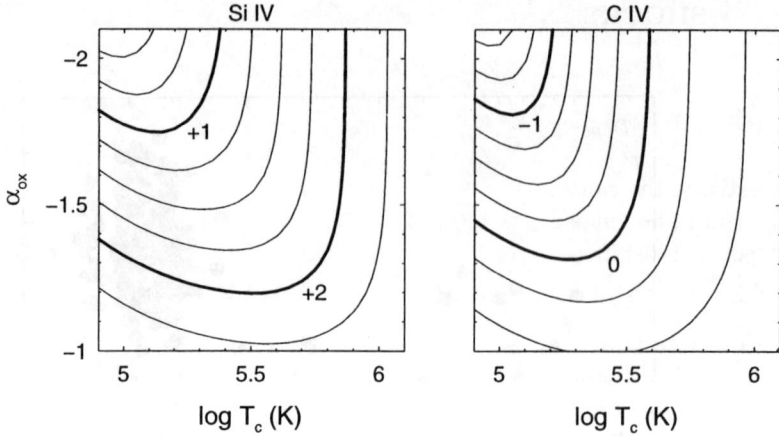

Figure 2: Contours of constant minimum ionization correction, normalized by solar abundances, $\log(f(\text{HI})/f(\text{M}_i)) + \log(\text{H/M})_\odot$, are plotted for Si IV (left) and C IV (right) in optically thin absorbing clouds photoionized by different QSO continua. The parameters α_{ox} and T_c define the continuum shapes. Bold contours appear every 1.0 dex and thin contours every 0.25 dex.

and a range of ionizing spectral shapes. The spectra are parameterized here by the power law index α_{ox}, which relates the flux densities (F_ν) at 2500 Å and 2 keV, and a cutoff energy kT_c that characterizes an exponential decline from the UV toward the X-rays. The optical-UV and X-ray spectra on either side of the cutoff have fixed power law slopes of -0.5 and -1.0, respectively. Applying the results in Figure 2 to measured column densities via Eqn. 1 shows that, even in cases where the ionizing spectrum is poorly constrained and the ionization state of the gas is unknown (for example, when only H I and C IV lines are measured), firm lower limits on the metal abundances (C/H) imply $Z \gtrsim Z_\odot$ for typical BALs and some $z_a \approx z_e$ absorbers.[3]

This work was supported by NASA grants NAG 5-1630 and NAG 5-3234.

1. F. Hamann & G.J. Ferland, ApJ, 391, L53, 1992.
2. F. Hamann & G.J. Ferland, ApJ, 418, 11, 1993.
3. F. Hamann, ApJS, in press, 1996.
4. G.J. Ferland et al., ApJ, 461, 683, 1996.
5. P. Petitjean, M. Rauch & R.F. Carswell, A&A, 291, 29, 1994.
6. J.C. Shields & F. Hamann, Rev. Mex. Astr. Astrof., (Conf. Series), in press, 1996.

SHOCKS IN POWERFUL RADIO GALAXIES

N.E. CLARK
Department of Physics, University of Sheffield

To investigate the cause of the alignment effect and the ionization mechanism in powerful radio galaxies, we have made a detailed study of a sample of low–intermediate redshift radio galaxies [1] [2] [3] [4]. The results obtained provide clear evidence that, for the sources studied, jet-induced shocks determine:

1. The **distribution** *of the extended line-emitting gas.* The extended emission line regions (EELR) which are associated with the radio hot spots at the head of the advancing radio jets have a bow-shock morphology; in all cases the spatial extent of the optical line emission is defined by the boundaries of the expanding radio source.

2. The **kinematics** *of the extended line-emitting gas.* Velocity splitting of ± 600 km s^{-1} is seen in the EELR associated with the radio knots; the line profiles are broad along the radio axes (FWHM ~ 1000 km s^{-1}); and there is a clear anti-correlation between linewidth and ionization state which is consistent with a shock structure.

3. The **physical conditions** *of the extended line-emitting gas.* The electron temperatures are higher in the EELR than in the inner nuclear regions, and are consistent with shock-heating; the pressures measured in the EELR are too high for confinement in the host galaxy haloes, but are similar to the minimum pressures of the associated radio hot spots, consistent with ram-pressure confinement of the line-emitting gas.

There is also evidence that shocks have an ionizing effect in these sources: in particular, although most of the line ratios are consistent with either shock-ionization [6] or photoionization by an $\alpha \sim -1.5$ power-law continuum [5], the low HeII(4686)/Hβ ratio measured in the extended gas is more consistent with shock-ionization than AGN-photoionization (Figure 1). Although a model has recently been proposed which can explain a spread in HeII(4686)/Hβ ratios in terms of AGN-photoionization [7], shock-ionization is the most likely explanation for these sources in which shocks are clearly present.

The sources discussed above lie at redshifts such that the full range of optical diagnostic emission lines can be accessed, but they share many of the properties of—and appear to be the nearby prototypes of—high redshift radio galaxies (HZRG). The radio-optical associations and extreme emission-line kinematics observed in HZRG indicate that shocks are also important at

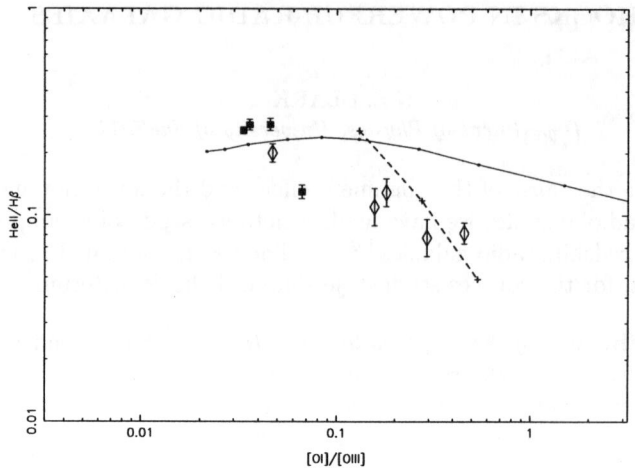

Figure 1: Line-ratio diagnostic diagram showing the HeII(4686)/Hβ ratios and the corresponding [OI](6300)/[OIII](5007) ratios for the inner nuclear regions (squares) and the EELR associated with the radio lobes (diamonds) for the four sources studied. The $\alpha = -1.5$ power-law photoionization model (solid line) is in agreement with the observations in the nuclear regions, and also successfully reproduces the line ratios measured in low-redshift radio galaxies. *The HeII/Hβ ratios measured in the extended gas are too low to be consistent with AGN-photoionization models, but are consistent with fast-shock models (dashed line).*

high redshifts. Unfortunately, these distant sources are difficult to study in detail—particularly as the important optical diagnostic lines are shifted into the infrared—but recent evidence suggests that the UV line ratios measured in HZRG are inconsistent with the photoionization models which reproduce the optical line ratios at low redshifts.

References

1. C.N. Tadhunter *et al*, A&A, 288, L21, 1994.
2. N.E. Clark & C.N. Tadhunter, in *Cygnus A – Study of a Radio Galaxy.*
3. N.E.Clark, PhD thesis, 1996.
4. N.E. Clark *et al*, submitted to MNRAS.
5. A. Robinson *et al*, MNRAS, 227, 97, 1987.
6. L. Binette *et al*, ApJ, 297, 476, 1985.
7. L. Binette *et al*, in press.
8. M. Villar-Martin *et al*, in prep.

THE KINEMATICS AND MORPHOLOGY OF THE PECULIAR ACTIVE GALAXY IRAS 04210+0400

A.J. HOLLOWAY, W. STEFFEN, J. MEABURN
Department of Physics and Astronomy, University of Manchester, Manchester M13 9PL, England

A. PEDLAR
Nuffield Radio Astronomy Laboratories, University of Manchester, Jodrell Bank, Cheshire SK11 9DL, England

D.J. AXON
Space Telescope Science Institute, Baltimore MD 21218, USA

J.A. LOPEZ
Instituto de Astronomia, UNAM, Apartado Postal 877, Ensenda, Mexico

1 Introduction

IRAS 04210+0400 has a redshift of z=0.0462 (Beichman et al. 1985) implying a distance of 185 Mpc, at which 1″, corresponds to 900 pc (assuming H_0=75 km s^{-1} Mpc). Hill et al. (1988) showed radio structure consisting of a central double source and large scale ($\sim$ 25 kpc) radio lobes. Hill et al. (1988) suggested that the 'spiral arms' may be associated with the radio ejecta. The 20 cm radio luminosity of 2.4×10^{23} Watts Hz^{-1}, the large radio lobes and the narrow emission line spectrum (Beichman et al 1985) fits the definition of a Narrow Line Radio Galaxy (NLRG) which are associated with ellipticals. Models by Hill et al. (1988) and Steffen et al. (1996a) explain the spiral features as photoionized remnants from the passage of a radio jet through the ISM.

2 Observations

Narrow band [O III] 5007-Å and V-band images have been made using the AAT with the Manchester Echelle Spectrograph used in its imaging mode, San Pedro Martir telescope and JKT. In the [O III] 5007-Å image the previously continuous spiral arms are now resolved into more discrete features, a clear knot in the northern arm and a splitting double structure to the south (Holloway et al, in prep.). Long-slit spectra were obtained using the IDS on the 2.5m INT. The two exposures were taken at PA 0°, along the radio axis. At the position of the radio lobe hotspots we see 'V'-shaped structures, which is indicative

397

of an expansion at these points. These observations are presented in more detail in Holloway et al(1996). To complement the new optical observations we obtained radio maps with the VLA A-array at 1425 Mhz and 4860 Mhz. Whilst the 1425 Mhz data detects both of the lobes, the 4860 Mhz data only sees the nucleus and southern lobe (Holloway et al, in prep.). HST WFPC-2 observations were made on 1995 January 31 using the F547M, F675W and F814W filters for proposal no. 5746, P.I. F.D. Macchetto (paper in prep.). The central elongated emission line structure is seen to align with the central radio double. The paths of the emission line spiral structures do not appear to have any associated radio emission until the location of the radio lobe hot spots. In the detailed map of the southern lobe we see the radio emission starting just before the location of the fork in the optical structure, where the [O III] 5007-Å emission halts and in the radio lobe Hα emission is dominant. Extended filaments are seen from the nucleus to the radio hotspots, and also in individual discrete clouds surrounding the nucleus, consistant with ionization within a conical UV radiation field from the core.

3 Outflow Modelling

Norman et al. (1988) and Loken et al. (1995) have modelled the structure of wide angle tail radio galaxies in terms of a jet (Mach number $2 - 5$) passing through a shock in the ambient medium. We suggest that a similar scenario applies to IRAS 04210+0400 at the position of the radio hot spots. We model the longslit emission line spectra using a simple parameterised description of the emission and velocity field of the ionized gas flow. A full account of the theoretical model is given in Steffen et al. (1996b)

References

1. Beichman C. *et al*, ApJ, 293, 148, 1985.
2. Hill G.J., Wynn-Williams C.G., Becklin E.E., MacKenty J.W., ApJ, 335, 93, 1988.
3. Holloway A.J., Steffen W., Pedlar A., Axon D.J., Dyson J.E., Meaburn J., Tadhunter C.N., MNRAS, 279, 171, 1996.
4. Loken C., Roettiger K., Burns J.O., Norman, M., ApJ, 445, 80, 1995.
5. Norman M.L., Burns J.O., Sulkanen M.E., Nature, 335, 146, 1988.
6. Steffen W., Holloway A.J., Pedlar A., MNRAS, in press, , 1996a.
7. Steffen W., Holloway A.J., Pedlar A., MNRAS, in press, , 1996b.

SIMULATED SPECTRA OF QUASAR HOST GALAXIES OBSERVED BY SUBARU CORONAGRAPHIC SPECTROMETER

M. IYE

National Astronomical Observatory, Mitaka, Tokyo 181, Japan

The scientific objectives, the design concept, and simulated performance of a coronagraphic spectrometer for 8m Subaru telescope are described.

1 Spectrographic Mode of a Coronagraphic Imager with Adaptive Optics for Subaru

Observations of faint diffuse objects around a bright object are of importance in the studies of protoplanetary disks, emission line regions of AGNs, gravitational lensed objects, and quasar host galaxies. Recent HST imaging study looking for host galaxies of nearby luminous quasars recalled the importance of removing the scattered light from the central point source [1]. One obvious way to probe the properties of diffuse component is to perform a spectroscopic study. The scattered light from a bright quasar would show the spectrum identical to the quasar itself. Whereas a faint halo consisting of stars and/or ionized gas would show distinctly different spectrum.

What we propose here is a coronagraphic spectrometer, designed as one of the four modes of operation of the Coronagraphic Imager with Adaptive Optics (CIAO) [3], an F/12 Cassegrain instrument of Subaru telescope. The occulting mask removes most of the light from the central bright point source. The light diffracted by structural components will be suppressed by an apodizing mask. By using a grism and a focal reducing optics, one can obtain low resolution spectra of spatially contracted images of these diffuse extended objects.

2 Simulated Performances

Figure 1 shows simulated spectra of a quasar host galaxy as will be obtained by CIAO/SPM. In this simulation, the template spectra for the quasar [4], the underlying galaxy [5], and the atmospheric OH airglow emission [6] were taken from literatures. A $V = 19$ quasar at $z = 1.5$ is assumed as a target to search for a host galaxy. At this z, the $H\alpha$ line falls in the H band and the $H\beta$ and the $[OIII]$ lines fall in the J band. With a full correction attained by the AO system, the quasar image in the infrared is expected to be smaller than 1 pixel. As for the host galaxy, assumed here is a galaxy with the surface brightness

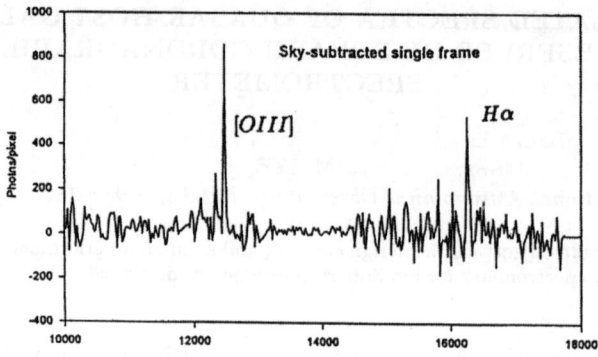

Figure 1: Simulated spectra of a quasar host galaxy as will be obtained with the CIAO/SPM.

of $V = 24$ mag/pixel, covering about 25 pixels. Only the Poisson photon shot noise was taken into account in this simulation. Possibly significant pixel nonuniformity of an IR array was not included.

It is shown, however, that the host galaxy with emission line regions will be revealed by appropriately post-processing the shot-noise limited spectra taken with the CIAO/SPM, if the quasar light is blocked effectively and the host galaxy surface brightness is not much fainter than about 3 % of the averaged atmospheric OH airglow brightnes. Such a spectroscopic observation will give an unambiguous confirmation on the existence of quasar host galaxies. Even negative results in detecting host galaxy spectrum will place a stringent restriction on the formation scenario of quasars and galaxies.

References

1. J.N.Bahcall, *et al.*, ApJ, 457, 557, 1996.
2. H.Takami, *et al.*, *Scientific and Engineering Frontiers for 8-10m Telescopes* eds. M.Iye and T.Nishimura, (Universal Academy Press, Tokyo), 327,1995
3. M.Tamura, *et al.*, *Scientific and Engineering Frontiers for 8-10m Telescopes* eds. M.Iye and T.Nishimura, (Universal Academy Press, Tokyo),339,1995
4. J.E.Baldwin, in *Active Galactic Nuclei,* eds C.Hazard and S. Mitton (Cambridge Univ.Press, Cambridge), 51,1977
5. F.Durret and J.Bergeron, A&AS, 75, 273, 1988.
6. T.Maihara, *et al.*, PASP, 105, 940, 1993.

THE SPACE DENSITY OF QUASARS AT $z > 4$ USING THEIR K-BAND MAGNITUDES

JULIA D. KENNEFICK and PATRICK S. OSMER

Astronomy Dept., The Ohio State University,
174 W. 18th Ave., Columbus, OH 43210, USA

MICHAEL A. PAHRE and S. GEORGE DJORGOVSKI

Palomar Observatory, Caltech, Pasadena, CA, 91125, USA

Kennefick *et al.* have recently reported quasar space densities at $z > 4$ and $M_B < -27^m$ using quasars discovered in an ongoing multicolor survey. Here we compute the spectral indices of the Kennefick *et al.* quasars using their optical spectra and K-band magnitudes. The median computed spectral index is -0.87, resulting in values of M_B that are on average $\sim 0.5^m$ brighter than previously computed (with an adopted spectral index of -0.5). Quasar space densities from the Kennefick *et al.* survey are reevaluated in light of the new values of M_B.

Initial results from an ongoing multicolor survey for $z > 4$ quasars using POSS II plates have been reported by Kennefick *et al.* [1]. The initial investigation led to the identification of 10 quasars at $z > 4$ ($\bar{z} = 4.35$). Computed quasar space densities at $M_B < -27^m$ were found to be down by a factor of seven from $z = 2.0$ to $z = 4.35$. In order to compare quasar space densities at $z > 4$ to those at lower redshift, it is necessary to compute M_B for the quasars from their r band magnitudes, adopting a value for the spectral index. Most groups assume $\alpha = -0.5$ where $f_\nu \propto \nu^\alpha$. If the actual spectral index of a quasar is different from this value, M_B will change by an amount $\Delta M_B = -2.5 \Delta \alpha \log(\lambda_{obs}/(1+z)/4400)$. A better way to measure M_B for a quasar is to measure the flux at $(1+z)4400$Å directly. For quasars at $z = 4.35$, this corresponds to wavelengths of $\sim 2.4 \mu m$, very close to the K_s band, which has $\lambda_{eff} = 2.15 \mu m$ (Figure 1). A spectral index can also be computed for the quasar by using the K band magnitude and the flux at $(1+z)1450$Å measured from the optical spectrum by assuming $f_\nu \propto \nu^\alpha$.

We have observed the 10 $z > 4$ quasars from the Kennefick *et al.* sample in the K_s band using the near-infrared camera on the Palomar 60 in. telescope [2] during 1994 December and 1995 July. The median computed value of the spectral index is $\bar{\alpha} = -0.87$. The values of M_B computed using the K_s magnitudes are on average 0.5 magnitudes brighter than those computed assuming $\alpha = -0.5$. Cumulative space densities are shown in Figure 1.

To compare quasar space densities at $z \sim 4$ to those at $z = 2$, one must decide what value of α to adopt at $z = 2$. In studies computing spectral indices

402

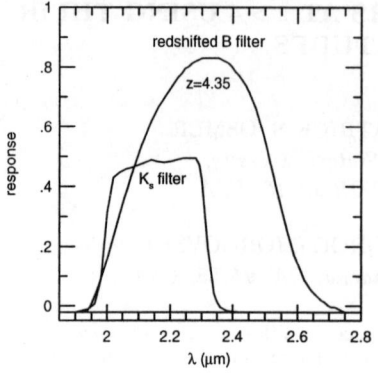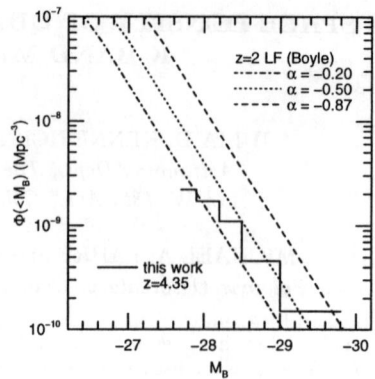

Figure 1: (*left*) Transmission curves for the K_s filter and the Johnson B filter redshifted to $z = 4.35$. The K_s band magnitudes are sampling the quasar spectra at restframe B. (*right*) Cumulative quasar space densities (solid line). See the text for details.

from optical spectra having a small range of frequencies, the median value of α was typically found to be -0.5. However, Neugebauer *et al.* [3] observed the complete set of quasars found in the Palomar-Green survey from 0.3 to 2.2μm, and found $\bar{\alpha} = -0.2$ for quasars at $1.0 < z < 2.1$. Shown in Figure 1 is the Boyle *et al.* [4] $z = 2$ QLF assuming several different values of α. If one takes α at $z = 2$ to be $\alpha = -0.87$, the same as that at $z = 4.35$, then again, one finds a drop in quasar space densities from $z = 2$ to $z = 4.35$. If $\alpha = -0.5$ (the Boyle *et al.* adopted value), the space density of quasars at $z = 4.35$ does not show a significant drop from $z = 2$. For an $\alpha = -0.2$, space densities at $z = 2$ and $z = 4.35$ are comparable. At $z = 2$, restframe B is in the J band. The measurement of a sub-sample of quasars at this redshift in J would give a direct measure of M_B and would help determine what evolution in the QLF is real and what is introduced by the adoption of a spectral index.

This work was supported in part by the NSF PYI award AST-9157412 and the Bressler Foundation. JDK also acknowledges an AAS travel grant.

References

1. J. Kennefick, S.G. Djorgovski and R. de Carvalho, AJ, 110, 2553, 1995.
2. D.M. Murphy *et al.*, PASP, 107, 1234, 1995.
3. G. Neugebauer *et al.*, ApJS, 63, 615, 1987.
4. B.J. Boyle, T. Shanks and B.A. Peterson, MNRAS, 235, 935, 1988.

THE SPECTRUM OF MONOPOLAR HYDROGEN

D. LYNDEN-BELL

Institute of Astronomy,
Madingley Road,
Cambridge.

Hydrogen with a Dirac Magnetic Monopole attached to the proton has a radically different spectrum from normal hydrogen. The ground state has only half the binding energy and the excited states are split into n lines where n is the principal quantum number. Furthermore the selection rule for dipole transitions becomes $\Delta J = \pm 1$ *or* 0 which replaces the $\Delta \ell = \pm 1$ of the usual hydrogen spectrum so straight down transitions are allowed. Figure 1 shows the 2 Lyman Series and the 5 Balmer series while Tables 1 and 2 give the wavelengths in Å. The final entries are the series limits.

While it is unlikely that Monopoles exist it would be most exciting if they were discovered spectroscopically. Since current particle physics predictions are that monopoles should be heavy but stable, the spectra have been calculated for a fixed nucleus. Although it has not yet been excluded that monopoles form most of the mass of the universe the expectation is that monopoles are very rare. The best place to look is probably in supernovae or in high energy environments such as AGN or quasars. In very high resolution stacks of many quasars it is worth looking for the Lyman lines 2774.62 & 2733.78 in absorption but be warned the second almost coincides with an HeII line.

Table 1: N=1 Lyman Series: $n \to 1$

	$J = 0 \to 0$	$J = 1 \to 0$
2	2774.62	2733.78
3	2199.98	2190.74
4	2030.99	2027.32
5	1955.97	1954.13
6	1915.65	1914.58
7	1891.33	1890.66
8	1875.20	1875.05
∞	1822.52	1822.52
n	$J = 0 \to 0$	$J = 1 \to 0$

Table 2: N=1 Balmer Series: $n \to 2$

3	10622.40	10410.50	11266.90	11028.80	10935.20
4	7577.90	7527.15	7900.30	7845.16	7822.92
5	6629.30	6608.14	6874.73	6851.97	6842.71
6	6187.80	6176.68	6401.10	6389.20	6384.34
7	5941.13	5934.48	6137.50	6130.40	6127.49
8	5787.67	5783.34	5973.86	5969.25	5967.30
∞	5311.20	5311.20	5467.58	5467.58	5467.58
n	$J = 0 \to 0$	$J = 1 \to 0$	$J = 0 \to 1$	$J = 1 \to 1$	$J = 2 \to 2$

(Fine structure transition $n = 2 \to 2$ 185695.00)

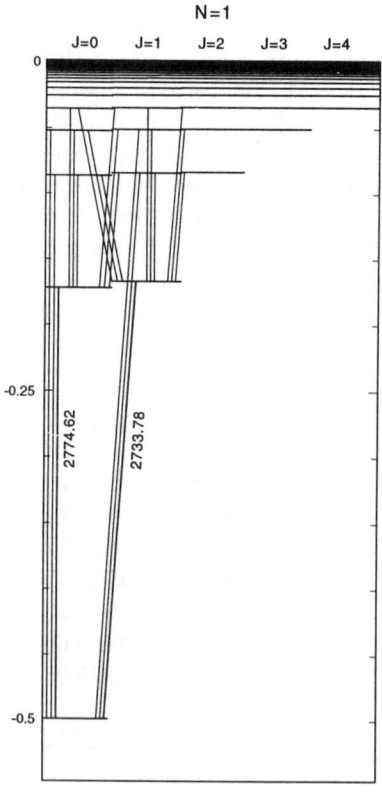

Figure 1:

SPECTROSCOPIC OBSERVATIONS OF QUASAR CANDIDATES FROM A DEEP MULTICOLOR SURVEY

PATRICK S. OSMER and JULIA KENNEFICK

Astronomy Dept., The Ohio State University, 174 W. 18th Ave., Columbus, OH 43210, USA

PATRICK B. HALL

Steward Observatory, University of Arizona, Tucson, AZ 85721, USA

RICHARD F. GREEN

National Optical Astronomy Observatories, P. O. Box 26732, Tucson, AZ 85726, USA

We have made spectroscopic observations of 57 additional quasar candidates from the survey of Hall et al (Papers I and II). To date we have identified 13 new quasars with $0.3 < z < 2.8$ and $16.8 < B < 21.9$, all from the group of 43 candidates with ultraviolet excess (UVX). No new quasars with $z > 3$ were found among the 9 BRX and 5 VRX candidates observed. As a result, there are now 58 confirmed quasars in the survey: 45 with $0.3 < z < 2$, 9 with $2 < z < 3$, 3 with $3 < z < 4$, and 1 at $z = 4.3$. The new data indicate that the excess of quasars with $z > 3$ in the survey will be less than suggested in Paper II.

1   Background and Introduction

We have carried out a deep multicolor survey with the KPNO 4-m telescope using six filters (U,B,V,R,I75,I86) covering the range from 0.34 to 0.86μm with magnitude limits of 22.1 to 23.8. We covered 0.83 deg^2 in six fields at high galactic latitude. We have produced a catalog of 21,375 stellar objects that contains the positions, magnitudes, and error estimates for each of the six filters (Hall et al [1], Paper I). The motivation of the survey is to search for lower luminosity quasars at $z > 3$ than have previously been studied and to search for lower luminosity quasars with $z \approx 2$ to constrain the nature of the evolution of the luminosity function. In addition, the survey is valuable for the study of faint field galaxies and stars in the galactic halo.

Initial spectroscopic observations and analysis of the data (Hall et al [2], Paper II) led to the identification of more than 40 quasars, including one with $z > 4$ and a comparable number of compact, emission-line galaxies plus a number of unusual and interesting stars.

¿From these results together with the extensive modeling and simulations presented in Paper II, we estimated that the survey contained more quasars with $B < 21$ and $z < 2.3$ than expected from the Koo & Kron [3] survey and more quasars with $z > 3$ than expected from the results of Warren, Hewett & Osmer [4], which predict 7 quasars deg^{-2} to R = 21. Consequently we have continued spectroscopic observations of quasar candidates from the survey.

2 Observations and Results

We observed 57 additional candidates from the survey with the KPNO 4-m telescope and its Cryocam and R-C spectrographs: 43 candidates were selected as ultraviolet excess objects (UVX), 9 as being very red in B - R (BRX), and 5 as outliers according to their UBVR magnitudes (VRX). The UVX objects were expected to have $z < 3$ while the BRX and VRX objects are good candidates for having $z > 3$.

We have identified 13 new quasars, all from the UVX candidates. They have $0.3 < z < 2.8$ and $16.8 < B < 21.9$. In this group, the quasar with z = 0.29 and B = 16.8 is the most notable for being the brightest and lowest redshift one in the sample to date. Its spectrum shows prominent Fe II emission. None of the BRX and VRX objects are identified as quasars.

The absence of new quasars with $z > 3$ among the BRX candidates reduces the expected total number of such quasars in the survey from 8.4 to the range 5.6 to 6.9. This is still a significant excess over the predicted number of 1.9 (allowing for efficiency of detection), but it is clearly important to obtain definitive spectra of the remaining candidates. The results for the UVX candidates indicate that the number of quasars in the survey with $z < 2.3$ and $B < 22$ will be also be less than estimated in Paper II and in better agreement with the predictions of the Koo & Kron [3] values.

Next steps are to complete the spectroscopy of the BRX candidates so that a definitive value of the surface density of faint $z > 3$ quasars in the survey is established; complete the spectroscopy of the UVX candidates with $B > 22$ to constrain the evolution of the faint end of the luminosity function at $z < 2.3$; make the entire catalog available to the community in electronic form; and identify and analyze the field galaxies and faint stars.

References

1. P.B. Hall, P.S. Osmer, R.F. Green, A.C. Porter and S.J. Warren, ApJS, 104, 185, 1996. (Paper I)
2. ———, ApJ, 462, 614, 1996. (Paper II)
3. D.C. Koo and R.G. Kron, ApJ, 325, 92, 1988.
4. S.J. Warren, P.C. Hewett and P.S. Osmer, ApJ, 421, 412, 1994.

HIGH RESOLUTION HST IMAGING OF BL LACERTAE OBJECTS [a]

J.E. PESCE, C.M. URRY, R. SCARPA

Space Telescope Science Institute, 3700 San Martin Drive,
Baltimore, MD 21218, USA

R. FALOMO

Osservatorio Astronomico di Padova, vicolo dell'Osservatorio 5,
35122 Padova, Italy

M. GIAVALISCO [b]

Carnegie Observatories, 813 Santa Barbara St.,
Pasadena, CA 91101-1292, USA

A. TREVES

SISSA/ISAS, Strada Costiera 11, 34014 Trieste, Italy

BL Lacertae objects (BL Lacs) are members of the blazar class of active galaxies, objects which emit tremendous amounts of energy from radio through gamma-ray wavelengths. We now know that, on average, BL Lacs are found in the centers of poor clusters of galaxies. However, for all but a few of the nearest objects, the study of their host galaxies and near environments from the ground is hindered by the bright nuclear emission. Using the Hubble Space Telescope (HST) WFPC2, we are studying a small sample of radio-selected BL Lac objects ($0.2 \lesssim z \lesssim 1.0$) at spatial resolution considerably higher than is possible from the ground. The goals of this project include the analysis of BL Lac host galaxies, a search for close companions and signs of interaction, and investigation of the more extended regions around the BL Lacs. Here we present results for the first objects of our sample.

1 Introduction

A way to understand the BL Lac phenomenon that is independent of model assumptions is through an analysis of the host galaxies and near environments of these objects. However, studies of BL Lac host galaxies from the ground have been hindered by the bright nuclear point source, with most targets relatively nearby ($z \lesssim 0.3$), and even with sub-arcsecond seeing they do not yield useable data within a few arcseconds of the bright nucleus. The unprecedented

---

[a] Based on observations made with the NASA/ESA Hubble Space Telescope, obtained at the Space Telescope Science Institute, which is operated by the Association of Universities for Research in Astronomy, Inc., under NASA contract NAS 5-26555.

[b] Hubble Fellow.

resolution of the refurbished *Hubble Space Telescope (HST)* allows detection and study of resolved flux an order of magnitude closer, and thus *HST* is a powerful instrument for studying the host galaxies of BL Lac objects, as well as for finding close companion galaxies.

Here we describe our *HST* observations of our first three radio-selected BL Lacs, 0814+425 ($z = 0.258$), 1823+568 ($z = 0.664$), and 2254+074 ($z = 0.19$). We used the F814W (basically I band) filter and centered each object on the high resolution Planetary Camera (PC) chip (0.046 arcsec pixel^{-1}). The objects were observed with a series of exposure times (to avoid saturation) and at two different roll angles (to facilitate the search for asymmetric features and nearby companions). The combined, total exposures are 1500 - 2100 seconds.

2 Analysis Methods

We modelled the nuclear point source with the Tiny Tim PSF software and subtracted the model from the PC image to determine whether the objects are resolved. There are various methods of PSF subtraction but all have intrinsic errors (e.g., the undersampled PSF causes normalization and centering problems). To obtain host galaxy properties, the unsubtracted images are fitted with a point source and a host galaxy convolved with the PSF. We derive isophotes and fit them with ellipses giving us morphological and photometric properties of the galaxy (e.g., surface brightness profiles, ellipticity and position angle). We also computed azimuthally averaged radial profiles and find the results to be similar. Various galaxy models (i.e., exponential disk, de Vaucouleurs) are then fitted to the profiles.

3 Results

The host galaxy of 0814+425 is unresolved and is either underluminous (by approximately 0.5 mag) or the quoted redshift for the object is wrong and it is really found at $z > 0.6$. For 1823+568 we find a host galaxy at the high luminosity range for BL Lac objects. This is apparently the highest redshift BL Lac host galaxy ever resolved. There is a possible companion galaxy which is highly disturbed and probably interacting with the BL Lac.

While the host galaxy of 2254+074 was previously studied from the ground, we show for the first time that it is unequivocally an elliptical. Our *HST* observations agree well with ground-based observations for all three objects, matching precisely in the regions beyond approximately 2 arcseconds.

EMISSION-LINE PROPERTIES OF $z > 4$ QSOs

J.C. SHIELDS[1], F. HAMANN[2], C. B. FOLTZ[3], & F. H. CHAFFEE[3]

[1] *Steward Observatory, University of Arizona, USA*
[2] *Center for Astrophysics & Space Sciences, UC-San Diego, USA*
[3] *MMT Observatory, USA*

1 Introduction

The number of QSOs with redshift $z > 4$ now exceeds 70. While substantial effort has been invested to find these objects, only modest attempts have been made so far to study their emission spectra. We have consequently undertaken a program of spectroscopy of QSOs at $z > 4$, using the Multiple Mirror Telescope, in order to quantify their emission properties in the rest-frame UV bandpass (1100 – 1700 Å). Preliminary results of this survey are reported here. Sample members are taken from the literature, largely from color-selection surveys.

2 Results

A preliminary average spectrum for 13 QSOs at $z \geq 4$ is shown in Figure 1. In most ways this spectrum is similar to mean spectra for AGN samples[1] at lower z. The N v $\lambda1240$ feature in this composite is relatively strong compared with that in spectra of AGNs at low z and luminosity, but is similar in strength when compared with QSOs at $z \approx 2 - 3$. The strong N v emission in luminous QSOs can be understood as a result of vigorous star formation and resulting chemical enrichment in the QSO host galaxy.[2]

The individual QSO spectra as well as the average in Figure 1 show evidence of unusually strong emission in O I $\lambda1304$, in terms of equivalent width as well as strength relative to other lines. This feature is believed to be dominated by fluorescence resulting from pumping via line coincidence with hydrogen Lyβ. The anomalous strength of this line may stem from elevated O/H abundance ratios, or other causes.

The $z > 4$ QSOs also exhibit relatively narrow profiles with large peak-to-continuum ratios and equivalent widths. Previous studies[3,4,5] of QSO ensembles have suggested that the broad emission-line region can be roughly decomposed into a high-ionization component with broad profiles and a low-ionization component with narrower profiles (the "intermediate-line region").

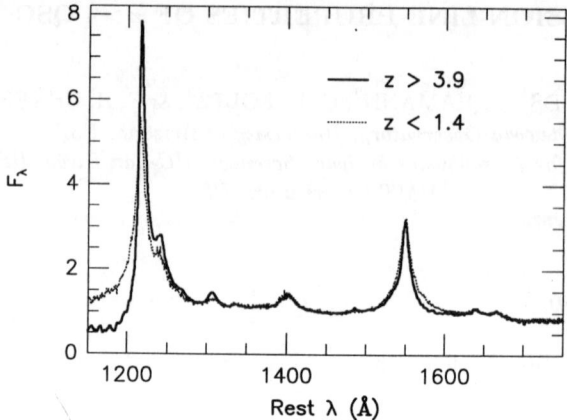

Figure 1: Average spectrum of 13 QSOs at $z \geq 4$, derived from MMT observations (solid curve). An average spectrum for 15 QSOs at $z < 1.4$ observed by *HST* is shown for comparison (dotted curve). Individual spectra are normalized in F_λ at 1440 – 1470 Å.

The narrow, peaky lines and strong O I may provide evidence that this low-ionization component is unusually dominant in these high-z, high-luminosity QSOs.

Selection effects may also account for part of this trend, however. Identification of many of these objects is based on measurements from UK-Schmidt plates of a very red $B - R$ color, which results as the Lyα line moves into the R bandpass. The R bandpass in this system is only ~600 Å wide, however, and observed EW(Lyα + N v) can exceed 500 Å, due to the $(1 + z)$ scaling. A selection bias results in magnitude-limited samples in favor of QSOs with large Lyα equivalent widths. Emission-line profiles for AGNs are correlated with equivalent width in the sense that narrower profiles are associated with larger equivalent widths. [3,4,6] A selection bias in favor of objects with large emission equivalent widths thus may preferentially result in selection of QSOs with narrow emission profiles.

References

1. A. Laor, et al., ApJS, 99, 1, 1995.
2. F. Hamann & G. Ferland, ApJ, 418, 11, 1993.
3. P.J. Francis, et al., ApJ, 398, 476, 1992.
4. B.J. Wills, et al., ApJ, 415, 563, 1993.
5. J.A. Baldwin, et al., ApJ, 461, 664, 1996.
6. P.S. Osmer, A.C. Porter, & R.F. Green, ApJ, 436, 678, 1994.

FAINT AGN/STARBURST NUCLEI IN MDS GALAXIES

V.L. SARAJEDINI

Steward Observatory, University of Arizona, Tucson, AZ 85721, USA

R.F. GREEN

National Optical Astronomy Observatories, Tucson, AZ 85719, USA

R.E. GRIFFITHS & K. RATNATUNGA

Johns Hopkins University, Phys. & Astron. Dept., Baltimore, MD 21218, USA

An important scientific aim of the Medium Deep Survey is to identify low luminosity AGNs from HST WFPC2 images to measure the faint end of the AGN luminosity function as well as to study the host galaxies of AGNs and nuclear starburst systems. We identify candidate objects based on morphology in 59 MDS fields and discuss the V-I colors of the nuclei. Spectra have been obtained for ~20 of our AGN/starburst candidates and a preliminary luminosity function is presented for these objects out to z=0.55.

1 Introduction

In this study, WFPC2 images from the Medium Deep Survey are used to search for galaxies containing an unresolved point source in their nucleus, possibly indicating the presence of an AGN or starburst. The detection of these galaxy types at moderate redshifts can help in our understanding of how the number density of such objects has evolved. Candidates are selected by fitting imaged galaxies with multi-component models using maximum likelihood estimatation techniques to determine the best model fit. Our candidates are selected as those galaxies best fit with an unresolved point source component in the galaxy nucleus in addition to a disk and/or bulge component.

2 Results

We find 84 galaxies requiring a nuclear, unresolved point source component to adequately model the galaxy light representing 10.2% ±1.1% of the 825 galaxies in our magnitude-limited survey to $I \lesssim 21.5$. Most comprise less than 20% of the total galaxy light. The majority of the host galaxies of these nuclei appear to be spirals with a small number (5%) residing in ellipticals.

The V-I color distribution of our nuclei (Figure 1, left panel) covers a large range and might be attributed to the presence of different object types in our sample. A broad, shallow distribution may result from starbursts seen at

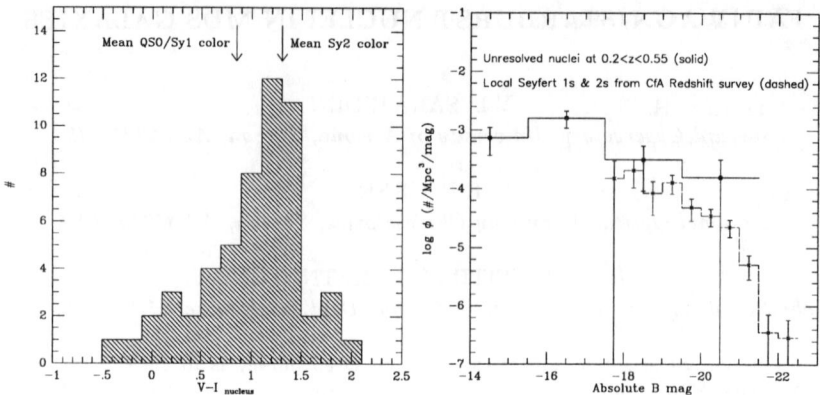

Figure 1: **Left panel:** (V-I) color histogram for our sample of unresolved nuclei. The Seyfert 1/QSO mean color from Elvis et al. (1994, ApJs 95, 1) for z≤0.5 is marked as well as that for Seyfert 2s based on data from Kalinkov et al. (1993, A&AS 98, 165). **Right panel:** The V magnitude luminosity function for point source nuclei in this study and that for all Seyfert galaxies (B magnitudes) from the CfA redshift survey. V magnitudes for our sample at the median redshift of z≃0.35 are comparable to rest-frame B magnitudes.

various stages of evolution or broadened by reddening due to dust. The blue tail seems to support this idea since the colors are bluer than those expected for QSOs/Sy1s and are probably due to the presence of young stars. Our peak falls close to the predicted Sy2 nuclear color with a blue side asymmetry biased toward Sy1 colors suggesting that the dominant population may be AGN-related

We have obtained redshifts for 21 of our 84 candidates and have calculated the luminosity function using the $1/V_a$ technique with q_o=0.5 and H_o=100 km/s/Mpc. For purposes of estimation, we multiply the LF derived from red-shift determination by a factor of 4 to make up for redshift incompleteness, while recognizing that the fainter objects may favor weaker absolute mag-nitudes. In the right panel of Figure 1 we compare our LF with that for Seyfert 1s and 2s in the CfA redshift survey from Huchra and Burg (1992, ApJ 393, 90). They find the percentage of AGNs in this survey of 2400 galaxies to be ~2% based on spectroscopic identification. If our nuclei are predomin-antly AGNs, we find no evidence for an increase in their number density out to z≃0.55 at -18 ≳ M_B ≳ -21. The faint extension of our LF appears to flatten fainter than M_B ≃ -17.

A more detailed discussion of this work can be found in Sarajedini et al. 1996, ApJL, in press.

RADIO GALAXIES FROM THE 5 GHZ MG CATALOG

D. STERN$^\diamond$, H. SPINRAD$^\diamond$, A. DEY$^\clubsuit$, M. DICKINSON$^\heartsuit$, D. SCHLEGEL$^\spadesuit$

$^\diamond$ *Deparment of Astronomy, Univ. of California, Berkeley, CA 94720*

$^\clubsuit$ *NOAO/KPNO, 950 S. Cherry Ave., P.O. Box 26732, Tucson, AZ 85726*

$^\heartsuit$ *Department of Physics, Johns Hopkins Univ., Baltimore, MD 21218*

$^\spadesuit$ *Department of Physics, Univ. of Durham, England*

We present the current status of our program to optically identify a complete sample of compact, moderately steep spectrum sources selected from the 5 GHz MIT–Greenbank survey. To date we have redshifts for 45% of the sample, with 11 sources identified at redshifts greater than two.

1 The Sample

We have selected a sample of 218 sources from the MIT–Greenbank (MG) survey of Bennett *et al.* (1986). The typical source strength is down by a factor of roughly four from the 3CR survey. From the MG sources with VLA "snapshot" maps, we have selected a subsample chosen according to these criteria: (1) $S_{5\,GHz} \geq 100$ mJy; a fainter flux limit thin in the 3CR survey implies more distant objects. Indeed, $\bar{z}_{3CR} = 0.87$, while $\bar{z}_{MG} = 1.17$, where we consider the subset of extragalactic 3CR sources obeying the MG selection criteria; (2) small angular size ($\theta \leq 10''$) biases the sample to the early Universe when the higher density of the IGM was more effective at confining radio lobes; (3) moderately steep radio spectral index ($\alpha \geq 0.75, S_\nu \propto \nu^{-\alpha}$ between 1.4 and 5.0 GHz) biases the sample against flat spectrum quasars and in favor of radio galaxies. This criterion is relatively non–constraining in comparison to several ultrasteep surveys currently in progress and will permit us to properly investigate correlations between α and other properties of the sample; and (4) simple (i.e. unresolved, double, or triple) morphology to select against low redshift targets.

Due to a dedicated effort at the Lick and Keck Observatories, we now have redshifts for over 45% of the sample. In Figure 1, we present the redshift distribution of our sample, with the spectroscopic distinction between quasars and radio galaxies indicated. We find a mild trend correlating more distant radio galaxies with steeper radio spectral indices.

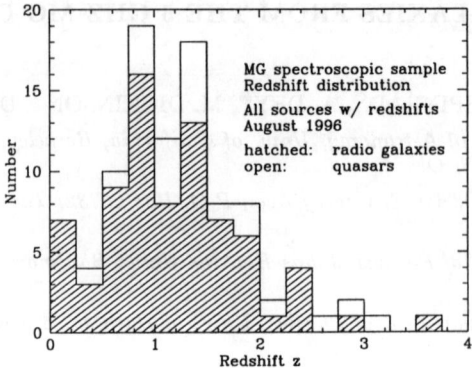

Figure 1: Redshift distribution of our sample of MG sources. The median redshift is $\bar{z} = 1.17$.

2 Keck Spectra

We have observed 14 radio galaxies from our sample with the higher signal-to-noise ratio afforded by LRIS on the Keck telescope. The data reveal a difference between the average 3CR and average MG radio galaxy spectrum: as radio flux density decreases, the ionization state of the gas appears to diminish as well.

3 Conclusions

In summary, intensive investigations of high redshift radio galaxies has established some patterns in their properties and suggested evolutionary scenarios for them, but considerable uncertainty remains (cf. McCarthy 1993). Our goals are twofold: (1) to sort out these mysteries and in particular to understand the detailed physical mechanisms which relate the radio source to the properties of its optical host galaxy, and (2) to make detailed investigations of individual interesting high redshift radio sources to probe galaxy evolution at early cosmic epochs. We are approaching both of these ends through a systematic optical survey of this radio sample selected at lower flux density levels than the 3CR with only a weak radio spectral index selection criterion.

References

1. C.L. Bennett, *et al*, ApJS, 61, 1, 1986.
2. P. McCarthy, ARA&A, 31, 639, 1993.

MgII & FeII EMISSION OF A HIGH-z QUASAR B1422+231

Y. TANIGUCHI[1,2], K. KAWARA[3,4], T. MURAYAMA[1], & N. ARIMOTO[4]

[1] *Astronomical Institute, Tohoku University, Aoba, Sendai 980-77, Japan*

[2] *Royal Greenwich Observatory, Madingley Road, Cambridge CB3 0EZ, UK*

[3] *ISO Science Operations Centre, ESA, Villafranca, 28080 Madrid, Spain*

[4] *ISAS, Yoshinodai Sagamihara, anagawa 229, Japan*

[5] *Institute of Astronomy, University of Tokyo, Mitaka, Tokyo, 181 Japan*

We present the rest frame UV/optical (2000−5300 Å) spectrum of the gravitationally lensed flat-spectrum radio-loud quasar B1422+231 system at $z = 3.62$ and discuss the formation epoch of its host galaxy.

1 Results and Discussion

In Figure 1, we show the rest frame UV/optical (2000−5300 Å) spectrum of the gravitaionally lensed flat-spectrum radio-loud quasar B1422+231 system at $z = 3.62$ which was obtained at the f/15 focus of the Kitt Peak National Observatory (KPNO) 4 meter telescope using the the long-slit CRSP (CRyogenic SPectrometer). The spectrum is similar to the LBQS (Large Bright Quasar Survey) composite spectrum, which indicates no significant UV/optical spectral evolution in some quasars at $0 < z < 3.6$. The Fe II(UV+optical)/MgII flux ratio is 12.2 ± 3.9, which is comparable to 8.9 for the LBQS composite spectrum of quasars at $z = 1-2$ (Francis et al. 1991), and 9.0 ± 3.5 for the quasars at $z = 0.15-0.63$ analyzed by Wills, Netzer, & Wills (1985).

Although it is not straightforward to derive the Fe/Mg abundance from the present data, the similarity in Fe II/Mg II between B1422+231 and low-redshift quasars (and the LBQS composite spectrum) suggests that the host galaxy of B1422+231 had already been in the late evolutionary phase of the Fe enrichment at $z = 3.6$. If the Fe enrichment started at 1.5 Gyr after the onset of the first star formation, the host galaxy of B1422+231 would have formed at $z \sim 9$ or earlier for $q_0 = 0.0$ and $H_0 = 75$ km s^{-1} Mpc^{-1}. The model with $q_0 = 0.5$ should have $H_0 \leq 30$ km s^{-1} Mpc^{-1} to be consistent with our results, but this is incompatible with the recent measurements of the Hubble constant.

References

1. Francis, J. P., Hewett, P. C., Foltz, C. B., Chaffee, F. H., Weymann, R. J., & Morris, S. L. ApJ, 373, 465, 1991.
2. Wills, B. J., Netzer, H., & Wills, D. ApJ, 288, 94, 1985.

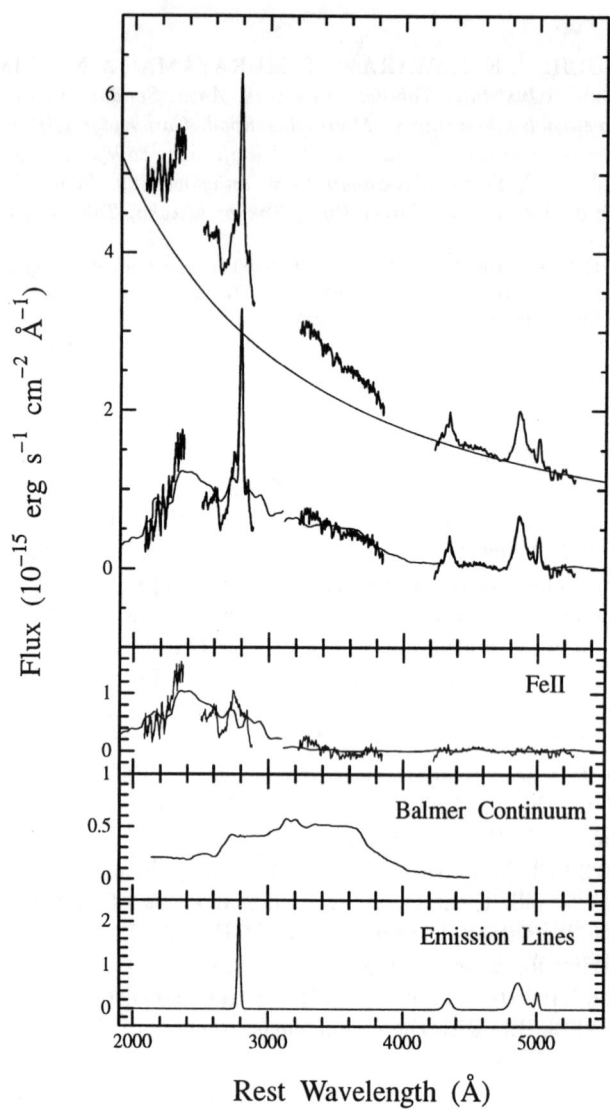

Figure 1: The rest frame spectrum of B1422+231.

WAVELENGTH DEPENDENT POLARIMETRY OF HIGH REDSHIFT QUASARS

WAYNE WEBB and MATTHEW MALKAN
Department of Astronomy, University of California, Los Angeles, CA 90024, U.S.A.

AIMO SILLANPAA and LEO TAKALO
Tuorla Observatory, SF-21500 Finland

Polarimetry has provided valuable information on the physical nature and geometry of the nonstellar continua in both Seyfert 2's and Blazars. However, the origin of the relatively small polarizations (P = 0.5 - 1.2%) in the more common Low Polarization Quasars (LPQs) is not so well understood.

We have obtained UBVRI polarimetry of a ten bright (mostly from the Palomar Greeen Survey) LPQs at high redshifts (z = 0.6 - 2.). The amplitude of polarization is flat in some objects and increasing with frequency in others. The position angle is wavelength-independent, suggesting the polarization in a given object originates in a single physical process.

These data have been fit with 8 different models for the polarization in normal quasars, including variations on electron scattering, dust, and scattering from an accretion flow. These polarization model fits are realized after separation of the spectral components in the multiwavelength continuum in the method of Malkan and Sargent (1982 ApJ 254, 22) and Malkan (1983 ApJ 268, 582). The long wavelength continuum is fitted with a power law. The estimated flux from the pseudocontinuum of Fe II and line and Balmer continuum emission is included, as in Edelson and Malkan (1986 ApJ 308, 59). The remaining ultraviolet excess was fitted with a two-parameter model of a relativistic, optically thick, geometrically thin, accretion disk (Sun and Malkan 1989 ApJ 346, 88).

Many of these high redshift quasars were found to have wavelength independent polarization (even to short UV rest wavelengths) – like the electron scattering model objects in Webb, Malkan, Schmidt, and Impey ("WMSI", 1993, ApJ 419, 494). These quasars had polarized flux proportional to the power law, or *all* non-stellar light, components. However, half of the objects had polarized fluxes which were proportional to the thermal (disk) or thermal and recombination light components. Two quasars had polarizations which were most consistent with dust scattering (i.e., 1151+117 and 2302+029).

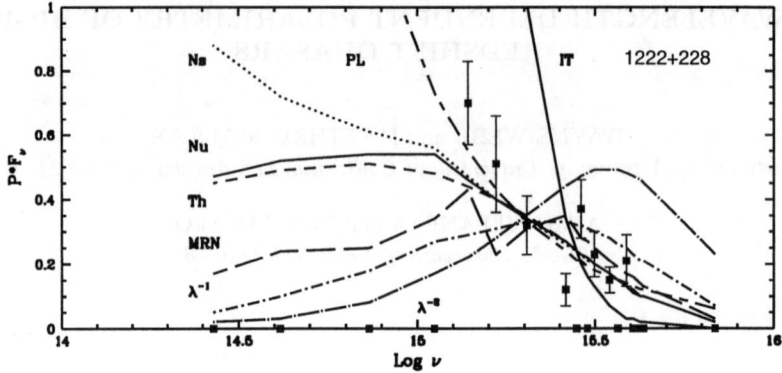

Figure 1: Eight Polarization Model Fits to 1222+228

We have previously obtained spectropolarimetry of three of these same quasars with HST (Impey, Malkan, Webb and Petry 1995, ApJ 440, 80). These data, down to rest wavelengths as short as 900A, are included in these polarization model fits. Those HST observations of 1222+228 were the first to detect the dramatic upturn in polarization (from 1% to 4.5%) at the shortest wavelengths. This unpredicted effect is now thought to arise from resonance scattering in outflowing winds (Blandford and Lee 1996 in press) or Faraday depolarization at longer wavelengths (Blaes and Agol 1996, ApJL 469, 41). It is notable that this abrupt rise in polarization was *not* seen in the U-band in any of this program's high redshift quasars, down to rest wavelengths shortward of Lyman-α or even the Lyman limit, which has implications for any model where the cut-on wavelength for the rise in polarization varies significantly.

We have previously obtained broad band polarimetry of one of these quasars (i.e., 1222+228). The March 1984 and April 1985 WMSI UBRI and the April 1992 LaPalma UBVR measurements reveal only one- to two-sigma variations in polarization, consistent with a constant percent polarization. The polarization position angles are similarly consistent within two- to three sigma. Both sets of polarimetry also independently produce the identical rank ordering of the 8 polarization models based on their goodness of fit to the data.

SECTION 6

THE FUTURE OF HST &

THE NEXT GENERATION SPACE TELESCOPES

THE FUTURE OF HST AND THE PLANNING FOR A NEXT GENERATION SPACE TELESCOPE

R. A. E. FOSBURY[a]

Space Telescope - European Coordinating Facility,
Karl-Schwarzschild-Str. 2, D-85748 Garching bei München, Germany

After the success of the refurbishment mission to HST in December 1993, the telescope has taken its place at the very leading edge of astrophysical research. The breadth and depth of the topics covered at the HST conference in Paris in December 1995 gave us a foretaste of the value of the data archive being accumulated and the supreme importance of future observations for guiding and enriching the programmes carried out with Keck and the very large groundbased telescopes nearing completion. Here I outline, as far as we know them, the plans for HST beyond the 1997 and 1999 servicing missions. The current planning for a Next Generation Space Telescope as part of NASA's *Origins* initiative is described with particular emphasis on possibilities for European participation.

1 Introduction

The question of the future of ultraviolet through infrared astronomy from space has recently been reviewed by the 'HST and Beyond' committee chaired by Alan Dressler of the Carnegie Observatories. Their report entitled 'Exploration and the search for origins: A vision for ultraviolet-optical-infrared space astronomy', available from:

http://saturn1.gsfc.nasa.gov/ngst/Background/HST_Beyond.PDF

follows and complements the report of the 'Toward other planetary systems' committee chaired by Bernie Burke of NASA's Solar System Exploration Division.

The principal recommendations are:

1. The operation of HST beyond its currently scheduled termination date of 2005 using a much lower-cost operation.

2. The development of a space observatory of 4m or larger aperture, optimised for imaging and spectroscopy over the 1–5 μm wavelength range.

3. The development of the capability for space interferometry.

---

[a] Affiliated to the Astrophysics Division, Space Science Department, European Space Agency

In addition, the committee noted how important it is for scientists to explain their motivations, goals and results to the public at large. It is in this spirit that these activities fall under the umbrella of NASA's *Origins* initiative.

This article sketches briefly the schedule for future HST servicing missions and then outlines the plans which are currently being made in the US for a Next Generation Space Telescope (NGST) with an aperture approaching 8m.

2   The future of HST

The future plans for HST servicing missions, currently planned for 1997, 1999 and 2002, are described by Chris Blades in these proceedings and details are given on the STScI Web pages at:

http://www.stsci.edu/servmiss/servmiss.html

The two new instruments to be installed during the second mission (STS-82, February 1997) are described here by David Axon and Rodger Thompson (Near Infrared Camera and MultiObject Spectrograph) and by Bruce Woodgate (Space Telescope Imaging Spectrograph).

In 1999, there will be a major reboost of the spacecraft to counteract the orbital decay induced by the Solar maximum. New solar arrays will be fitted and the ESA Faint Object Camera will be replaced by the Hopkins/Ball Advanced Camera for Surveys which is described here by Garth Illingworth.

In 2002, there is an opportunity for one or two new instruments to be installed. NASA are about to issue an announcement of opportunity with a due date for proposals several months after the second servicing mission. There is clear potential for international collaboration in the building of at least one of these instruments.

The plans beyond 2005 are currently uncertain and the possibilities for an extended, low-cost operation are being actively investigated. It is clear that the transition between the scientific operations of HST and NGST will have to be carefully managed within the budgetary constraints. It must be remembered, however, that the NGST will be optimised for longer wavelenghts than HST — which will remain the only major observatory for ultraviolet astronomy.

3   The planning for NGST

Following the first meeting about NGST, at STScI in December 1995, the study has gained considerable momentum and so this section is necessarily a snapshot of the developments which will lead to an interim report in November 1996 and a final report in mid-1997. Up-to-date information on the status of the various study components is available at:

`http://saturn1.gsfc.nasa.gov/ngst/`

In particular, the study outline at:

`http://saturn1.gsfc.nasa.gov/ngst/top/Studyp/STUDYP~1.html`
is a compact and interesting reference.

Three independent studies are being carried out and their results will be merged during August and September 1996. These are being carried out by:

- The Lockheed Martin Corporation

- TRW Civil & International Systems Division

- The Goddard-led Study Team

Of these, only the deliberations of the latter were accessible at the time of this meeting and so only these are summarised here.

3.1 Scientific drivers

Four categories of science drivers were considered in shaping the mission of which only priorities 1 and 2 are listed here:

Priority 1 (central to mission)

- See galaxy formation: implies large near-IR telescope optimised for 1–5 μm, radiatively cooled to less than 70K, zodiacal light sensitivity limited, 60–100 mas resolution at 1–2 μm

- Find high-z supernovae: implies wide field cameras with FOV greater than 3×3 arcmin with more than 4096^2 pixels

- 1.5 yr survey, 3.5 yr GO programme: implies lifetime of expendables

- Find first globular clusters: implies collecting area greater than 12m^2

- Measure many z's of 'red' galaxies: implies low spectral resolution '3D' multi-object spectrograph

Priority 2 (very important)

- See star formation (post-ISO/SIRTF): implies thermal-IR 5–30 μm camera/spectrograph

- Study foreground galaxies: implies wide field visible camera

- All-sky pointing once/yr for GOs: defines sunshield shape and need to look perpendicular to sun

- Monitor supernovae for 2.5 months: constrains sunshield shape and angle with respect to sun

- Follow-up for 3 GO cycles: implies 10 year lifetime goal

- $R = 1000$ spectroscopy at $z = 5$ for 5σ sources: implies large aperture, 8m goal

3.2 The working groups

The Goddard-led study has set up several working groups which meet independently but regularly coordinate their activities. These include: an Engineering/Management Study Group; an Industry Advisory Board and groups looking at Operations Systems; the Optical Telescope Assembly; the Science Instrument Module; the Science Performance; the Spacecraft Support System and the NGST Systems. In addition, there is a Science Advisory Committee which provides input to all three studies.

The budgetary guidelines set by NASA for the study are for a spacecraft cost of $M 500 (phase C/D) and a total mission cost (to include launch and operations) of some $M 900. It is clear from this that the mission has to be very different in concept from HST and in particular has to be built quickly using proven technology. The absence of the need for astronaut access, while greatly reducing costs, means a somewhat greater mission risk. The operational costs will also have to be kept strictly controlled.

3.3 A spacecraft concept

At the time of this meeting, the Goddard-led study was concentrating on a passively cooled 8m (almost filled) aperture, diffraction limited at 2 μm. The orbit would be chosen to minimise environmental disturbances, to allow stable communication links and stability for the nominal mission lifetime with minimum observation interruptions and maximum observational opportunities. The trajectory to orbit would be chosen to keep the launch requirements within the limits of available launchers and to keep the transfer time as short as possible. The most promising solution appears to be an 'L2 halo' orbit (in the Sun/Earth system) at a distance of some 1.5 million km from Earth. Such a location has enormous advantages over low-earth orbit by removing many observing constraints. Communications are easier than with heliocentric orbits and there is full-sky coverage with opportunities for long continuous observations. The stable thermal environment facilitates passive cooling.

The required sun-shield limits the instantaneous pointing of the telescope to a ring with an axis (V3) along the spacecraft-sun vector and with a width which is set by the detailed shield shape but would typically be $+5°/-57°$ from the plane perpendicular (V1-V2) to this vector around the V2 axis (nb. V1 is the optical axis of the telescope).

As a general tool for studying the effect of variations of NGST design on the achievement of the scientific goals (as presented in the 'HST and Beyond' report), the STScI has developed a computer program which computes the fraction of the science programme achieved in a given mission lifetime as a function of telescope aperture. This work is described at:

`http://augusta.stsci.edu/`

The current design of the Optical Telescope Assembly (OTA) is for a segmented 8m aperture with an f/1.25 primary, an f/24 OTA and a 5×5 arcminute FOV. The 4-mirror, centred design has small M3 and M4 mirrors located behind the primary. The primary is re-imaged onto a flat, but deformable quatenary mirror for fine figure control. With the fast primary and off-axis field, there is very little obscuration. A fine steering mirror — fed with an error signal from the main camera — would provide astrometric pointing, avoiding the necessity of providing milliarcsecond *spacecraft* stability. Various concepts for ultra-thin, lightweight mirror panels are being investigated.

The baseline launch vehicle is an Atlas IIAs but it is clear that a larger launcher would allow sufficient aperture for the science programme with a much simpler deployment scheme.

3.4 Instruments

The science instrument suite currently being studied consists of:

- a NIR camera with a 4×4 arcmin FOV using 64 1024^2 InSb (0.6–5 μm) arrays in four optical assemblies — rather like the HST WFPC

- a NIR multi-object spectrograph with a 3×3 arcmin FOV: this and the NIR camera would be passively cooled to less than 40K

- a thermal-IR camera with a 2×2 arcmin FOV using a 1024^2 Si:As array

- finally, a thermal-IR spectrograph would feed the same detector as the TIR camera: these last two instruments would work from 5–26 μm and would need to be actively cooled to about 8K

These four instruments would be packaged together and would use adjacent parts of the focal plane.

3.5 Communications and operations

The data transmission requirements are important because of the relatively large distance from Earth (10s round-trip light travel time). At an estimated 40,000 frames per year (5 Tbyte/yr), the average sustained data rate would be approximately 160 kbyte/s. A single dedicated 11m aperture ground antenna working with dual s-band and x-band frequencies would provide 8h/day coverage. The NASA Deep Space Network would be used only in emergencies.

There are many considerations which determine the operational modes and the nature of the orbit means that they would be different, and generally simpler, than for HST. Using common microprocessors throughout and employing a workstation/network environment would avoid much parallel development of different systems. The number of available observing modes per instrument/detector would be limited to about four and there would be easy mode changes for flexible scheduling. One of the major simplifications compared with the current HST operation would be the use of an autonomous guide star acquisition using a sub-array of the NIR camera. This avoids the need for costly pre-selection of guide stars but may result in the inability to make repeated pointings to exactly the same part of the sky.

This opportunistic pointing control would allow an adaptive scheduling scheme which means that, after observing programme design, there need be relatively little intervention from the ground except for error/problem resolution.

4 European involvement

The NGST feasibility study in the US is proceeding very fast and, judging by the excellent quality of the reports and presentations, it is a serious effort with participants clearly having confidence that the project will proceed. An obvious limitation of the current concept is the small size of the baseline launcher (the Atlas IIAs). A larger launcher would make NGST simpler and cheaper — although not necessarily larger. Since the low cost of the mission will demand a very rapid construction schedule, it is clear that all the required technologies will need to be in place before the building phase of the project. There will be a number of technological studies which need to be carried out and these may be in common with those needed for different components of the *Origins* and other programmes. It is clear also that there will be opportunities for participation in the operational phase.

Given the very large investment in groundbased facilities by Europe, a substantial involvement both in the future of HST (beyond 2001) and in the NGST is vital for the health and vigour of European astronomy. Such facilities are so mutually complementary that they enormously amplify each others value.

Acknowledgments

I should like to acknowledge the warmth and openness of colleagues at NASA, the STScI and other US organisations who have welcomed — without any formal guarantees of contribution — the participation of ST-ECF staff in these studies.

The viewgraphs prepared for this talk are available on the WWW at:
`http://ecf.hq.eso.org/~rfosbury/Beyond_HST/`

USE OF THE SPACE TELESCOPE IMAGING SPECTROGRAPH (STIS) FOR STUDYING THE HIGH REDSHIFT UNIVERSE

B, WOODGATE and THE STIS TEAM

Goddard Space Flight Centre, Greenbelt, MD 20771, USA

The STIS instrument is planned to be installed on the HST during the second servicing mission, currently planned for February 1997. The wavelength range is 115-1000nm over 4 bands, with spectral resolving powers between 50 and 100,000. Camera modes are used for target acquisition. Correction for spherical aberration and astigmatism is included. The 115-170nm range is covered by a CsI MAMA (Multi-Anode Microcahannel Array) detector, and the 165-310nm range by a Cs2Te MAMA, each with a format of 2048x2048 pixels; while the 305-555nm and 550-1000nm ranges are covered by a single CCD with a format of 1024x1024 pixels. The multiplexing advantage of using these two dimensional detectors compared to the 1 x 512 pixel detectors of the first generation spectrographs is one or two orders of magnitude, depending on the mode used.

1 Observing modes

STIS capabilities include:

1. low resolution spectral imaging (spectral resolution $R \sim 1000$; long slits and wide apertures; 115–1000 nm covered in four bands)

2. medium resolution spectral imaging ($R \sim 10000$; long slits; 115–1000 nm covered in smaller simultaneous wavelength intervals)

3. medium resolution echelle spectroscopy ($R \sim 23000$; short slits; only in the ultraviolet)

4. high resolution echelle spectroscopy ($R \sim 105000$; short slits; only in the ultraviolet)

5. objective spectroscopy ($R \sim 50$; wide field; one ultraviolet band)

6. broadband imaging (ultraviolet or visible; a few filters available for identification and acquisition of line or continuum sources or for extending dynamic range of acquisition capability).

In these imaging spectroscopy and camera modes, the spatial resolution achieved by STIS is 0.06 arcseconds FWHM in the ultraviolet and 0.12 arcseconds FWHM in the visible.

A description of the observing with the supported modes of STIS is found in the STScI STIS Handbook, and the available but unsupported modes are found on the same Web site (Baum et al, 1996).

2 Scientific Program

The scientific program of the STIS IDT is representative of the many important problems that can be investigated with STIS by the larger HST user community. It includes studies of the dynamics and mass distributions of normal and active galactic nuclei, the evolution of galactic halos, protogalaxies, and the intergalactic medium, the composition, kinematics, and physical processes in various phases of the interstellar medium, the magnetospheres, aeronomy, and stratospheric chemistry of the outer planets, bow shocks and other dynamic phenomena in protostars, the temperature structure within magnetic flux tubes in the outer atmospheres of late-type stars, and a deeper search for planetary systems forming around nearby stars.

3 Methods of using STIS for the study of the high redshift universe

3.1 Echelle absorption spectroscopy of QSOs

Evolution of Galactic Halo Regions and Proto-Galaxies. The long light travel paths from distant quasars to the Earth pass through many intervening clouds of gas, observed in absorption against the quasars' strong ultraviolet continuous energy distributions. Those systems with high column densities of neutral hydrogen contain heavier elements at fractions around 0.01 of solar abundance. Mean free path arguments suggest that the comoving space density of such absorbers is constant. If they are referred to a common class of objects, such as the L* galaxies of the Schechter luminosity function, then the cross section is very large compared to the extent of observed halos in low redshift objects. The stronger absorption systems show velocity and ionization complexity most easily accounted for by regions of active star formation and galactic stellar radiation fields. The strongest systems at $z = 2$ that show Lyman alpha absorption profiles with radiation damping wings have been demonstrated by Wolfe and collaborators to contain most of the baryon density measured in galaxies at $z = 0$.

The goals of studying these systems are to obtain t he values and evolution of the physical quantities of temperatures (through curve of growth analysis and doublet ratios), densities (from sensitive line ratios such as Mg I/Mg II,C II*/C II and C III 97.7/117.6 nm), origin of ionization (local photoionization or shocks, diffuse ultraviolet background), abundances and depletion patterns (Mn II, Cr II and Zn II strengths), velocity structure, absorption cross-sections, and implied star formation rates. STIS will investigate the redshift range $z = 0$ to 1.6 in the UV, for evolutionary comparison with higher redshift objects observed from the ground.

Evolution of the Intergalactic Medium. Along with the metal-line clouds seen in quasar spectra, there are systems of lower column density for which only hydrogen Lyman lines are detectable, known as the Lyman alpha forest. These systems are observed to have a power law distribution of column densities and to evolve as a population with larger cross sections and/or greater co-moving density toward earlier cosmic time. The discovery of Lyman alpha clouds along the line of sight to the nearby quasar 3C273 and several other quasars with both FOS and GHRS promises fruitful searches for nearby systems along many sightlines that can be searched for optical and radio counterparts. One physical scenario suggests a population of clouds in pressure equilibrium with a hot intergalactic medium, with temperatures about 10^4 K and neutral fractions of 10^{-4}. Some systems may have cooler temperatures and may depart significantly from spherical geometry.

The goal of studying the Lyman alpha forest with STIS is to determine the evolution of these clouds: in stability against gravitational collapse and evaporation by deriving the z and column density distributions for $z < 1.6$; in temperature by observing the higher Lyman series and He I; in neutral fraction; in spatial clustering and correlation with observed large-scale structure features at low redshift; and in implied environment in terms of hot confining medium or ionizing radiation background. The study of metal-line and Lyman alpha forest absorbers was given key project status for first-generation instruments and is likely to remain a driving problem for HST science.

STIS will provide a significant improvement over first generation instruments in studying the properties of the intergalactic medium at low z, since it can offer a wavelength resolving power of 3×10^4 along with a very broad wavelength coverage in a single exposure, a redshift range of 0.5, or a cosmic time span of over 3 billion years, using modes E140M and E230M.

3.2 Low background spectroscopy

For sources too faint to observe using the echelle modes, the STIS UV low resolution modes have an advantage compared to the first generation spectrographs for long exposures by providing very low backgrounds. The MAMA detector background rates are 1–2×10^{-5} counts s^{-1} per resolution element, so that observations remain signal limited for 15–30 orbits. These values are measurements on the ground. The increase seen on orbit using the Digicon detectors due to Cerenkov radiation is not expected with the MAMAs. This is due to their event position encoding method which rejects simultaneous events separated by more than 6 pixels, while many Cerenkov photons from a cosmic ray in the detector window would strike the cathode simultaneously. Significant spectral binning to lower resolution for continuum shape measurements, broad line or edge detection is possible for observations of several orbits. For example spectral edges in QSO spectra from the Gunn-Peterson effect would be more easily detectable.

3.3 Wide field spectroscopy

Faint compact emission line galaxies or sub-galactic clumps may be studied using wide field spectroscopy. With the CCD a 50×50 arcsec field may be used in the modes G750L, G430L and G230LB, covering the wavelength ranges 550–1000nm, 305–555nm and 165–310nm respectively. These CCD modes may be used for prime or parallel observing. With the MAMAs a 25×25 arcsec field can be covered in the modes G140L and G230L, covering the wavelength ranges 115–170nm and 165–310nm, for prime and coordinated parallel observing. A small fraction of the observing time could be spent with direct imaging, to provide a positional reference for wavelength determination and for the morphology of the galaxies. Broad band imaging without filters, 200–1000nm for the CCD, can provide detection of faint galaxies in a relatively short time. See also the companion paper at this conference by Gardner *et al.*

In the visible, the low sky background per resolution spatial element, due to the intrinsically low surface brightness in space and to the high spatial resolution, makes wide field spectroscopy a competitive technique for compact emission line sources. However, the sky brightness does form the dominant background, since wavelengths from the full waveband can contribute to the background for each resolution element, unlike for narrow slit spectroscopy. In the UV, the dominant sky background source of the visible from space, scattered sunlight, is much weaker because of the shape of the solar spectrum, so that this method remains sensitive, and of course space is required for the UV.

In emphasizing compact emission line sources, would we be observing the dominant population of objects at early times in the universe, or just a few peculiar objects? Observations increasingly show that the proportion of objects that contain emission lines increase as a function of redshift beyond $z = 0.3$ (eg Lilly *et al.*,1995). At higher redshifts, Lyman alpha emitting galaxies are being found in increasing numbers, with redshifts between 2 and 4.8 (Lowenthal *et al.*,

1991,1995; Moller and Warren, 1993; Machetto *et al.*, 1993; Francis *et al.*, 1996; Hu *et al.*, 1996; Pascarelle *et al.*, 1996). Furthermore, Pascarelle et al showed that the objects they observed at $z = 2.4$ with WFPC2 have typical sizes of 0.1 to 0.2 arcsec, sub-galactic clumps of less than 2 kpc size. Many of these objects also have CIV in emission, suggesting AGN emission may have been more common early on.

The increasing frequency of compact emission line objects at high redshifts suggests that the techniques of emission line spectroscopy and broader band imaging will be able to extend studies such as the Hubble Deep Field. The combination of spectral bands in the UV and visible gives access to redshifts in dominant emission lines of $0 < z < 7$ for Lyman alpha, $0 < z < 1.7$ for [OII]3727, and $0 < z < 0.5$ for H alpha.

3.4   UV imaging

The Hubble Deep Field contains many objects of sufficiently high redshift that they are seen by WFPC2 in the rest frame UV. Their peculiar morphology compared to low redshift galaxies in the visible is due to some combination of selection o f hot stars and gas emitting in the UV and evolution of the galaxies over the majority of the Hubble time. UV imaging is required to trace the UV emission of galaxies as a function of redshift to separate these effects.

Very sensitive UV imaging for the study of star forming galaxies may be obtained by using the MAMA detectors without filters. This maximizes the throughput and spectral coverage. The CsI MAMA (mode MIRFUV) is sensitive over 115–170nm, while the CsTe MAMA (mode MIRNUV) is sensitive over 115–310nm, each with a 25×25 arcsec field of view. Also UV filters are available for further discrimination. The low zodiacal light in the UV, combined with the low dark rate of the detectors and their zero readout noise, will enable UV imaging observations to remain signal limited for integration times of many orbits.

References

Baum,S. et al 1986, "STIS Instrument Handbook", Version 1.0.
URL:http://www.stsci.edu/ftp/instrument_news/STIS/topstis.html
Francis,P.J., Woodgate,B.E., Warren,S.J., Moller,P., Mazzolini.M., Bunker,A.J., Lowenthal,J.D., Williams,T.B., Minezaki,T., Kobayashi,Y. and Yoshii,Y., 1996, ApJ., 457,490.
Gardner,J. *et al.*, this conference.
Hu,E. *et al.*, this conference.

Lilly,S.J., Tresse,L., Hammer,F., Crampton,D., and LeFevre, 1995, ApJ.,455,108.

Lowenthal,J.D., Hogan,C.J., Green,R.F., Caulet,A., Woodgate,B.E., Brown,L. and Foltz,C.B., 1991, ApJ, 377,L73.

Lowenthal,J.D., Hogan,C.J., Green,R.F., Woodgate,B.E., Caulet,A., Brown,L. and Bechtold,J, 1995, ApJ, 451,484.

Machetto,F., Lipari,S., Giavalisco,M., Turnshek,D.A. and Sparks, W.B., 1993, ApJ., 404, 511.

Moller,P. and Warren,S.J., 1993, A&A, 270, 43.

Pascarelle,S.M., Windhorst,R.A., Odewahn,S.C. and Keel,W.C., 1996, Nature, 383

THE SCIENTIFIC CAPABILITIES OF NICMOS

D.J. AXON

Affiliated to the Astrophysics Division of ESA, Space Telescope Science Institute, 3700 San Martin Drive, Baltimore, MD21218, USA[a].

We describe the scientific capabilities of NICMOS and the technical state of the instrument following its ground testing, concentrating on those aspects of the instrument which are relevant to extragalactic astronomy and the cosmological study of galaxies in particular.

1 Introduction

The properties of galaxies at high redshift hold the key to understanding galaxy evolution and the large scale structure of the universe. Much of what we have learnt in the last decade has been based on a combination of deep optical photometry [6][7], ,as faint as B = 28, and large spectroscopic surveys [2][5], at somewhat brighter limits. The near-IR holds vital information, [4][3], as it is here that the bulk of the stellar population which make up the masses of typical galaxies radiate. Indeed, beyond redshifts of 2–3 the light from the the 4000$\mathring{A}$ break is shifted into the near IR, and at redshifts of above 7 even the Lyman limit is in the IR. It is precisely in this range that current theories would have us believe that L_* galaxies are forming most of their stars. Furthermore, even though the advent of large ground based telescopes (e.g. Keck) is now making it possible to study the Spectral Energy Distributions (SED's), and hence the star formation histories of faint galaxies for the first time, at the faintest magnitude levels, we must rely on colour information to identify distant galaxies using *photometric redshifts* [8] and examine their SED's (c.f. Dunlop's contribution to these proceedings). However, discriminating genuinely high Z galaxies from high extinction systems at intermediate Z is not straightforward and here also deep IR photometry is essential.

Between 1 and 2.5 μm on the ground the infrared background is dominated by telluric absorption and emission which limits the depth of astronomical imaging. The conventional near-IR bands of JHK were designed to sit in the gaps between these opaque molecular absorption bands. Outside the absorption features there is also considerable background emission in both lines and continuum, mainly from OH and O_2 emission produced in the atmosphere at

---

[a]On leave from, Nuffield Radio Astronomy Laboratory, University of Manchester, Jodrell Bank, Macclesfield, Cheshire SK11 9DL

an altitude $\sim$ 87 km.

The Near Infrared Camera and Multi-Object Spectrometer, *NICMOS*, is an axial replacement instrument which will be available on the Hubble Space Telescope (HST) in 1997. Above atmosphere, on board HST, the terrestrial background is removed, leaving only the *Zodiacal light* at short wavelengths, and *thermal emission* from the telescope at long wavelengths. As shown in Figure 1 the sum of these two is a minimum at 1.6 μm($\sim$H band) and all the *NICMOS* cameras carry broad-band filters centered on this spectral range. Because of this much reduced background *NICMOS* provides higher sensitivity between 1.1 to 1.6 μm, than an 8m class telescope, allowing deep photometry of distant galaxies.

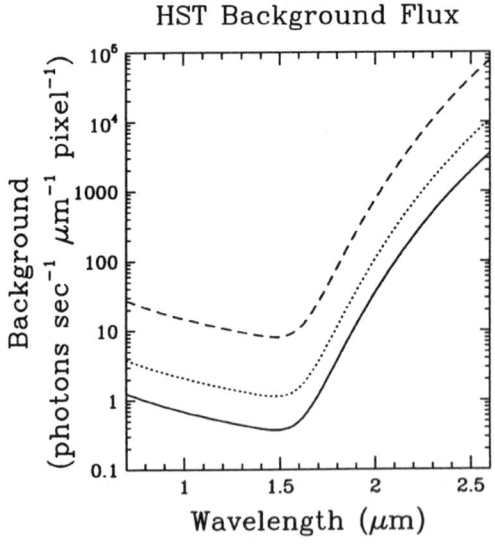

Figure 1: The expected background as a function of wavelength for each of the *NICMOS* cameras on HST

The photometric performance predicted for *NICMOS* is summarized in Table 1. To put this in context, with comparable integration times to those used for the previous WFPC2 study of the Hubble Deep Field [9] it should be possible to obtain photometry down as faint as $H = 30$, some 5–6 magnitudes fainter than the deepest existing ground based IR imaging (c.f. the contribution by Esther Hu in these proceedings).

As well as the aspects we have already touched on the uncloaking of the

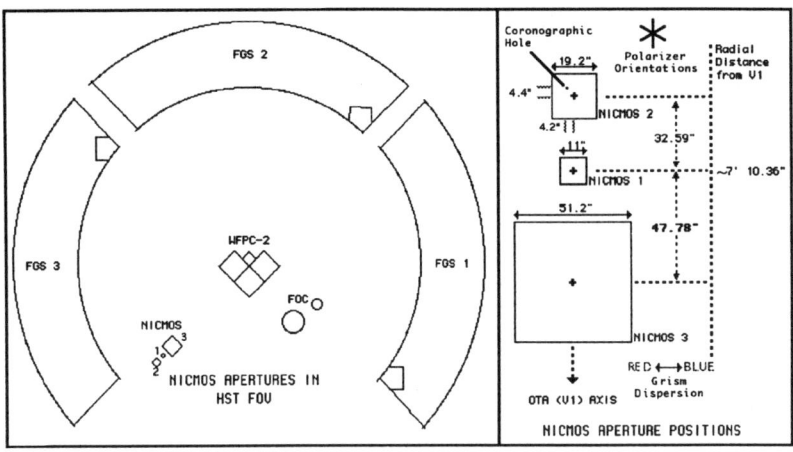

Figure 2: The NICMOS field arrangement. The left hand panel shows the placement of NICMOS in the HST Field of View relative to WFPC-2 and FOC. The right hand panel shows both the relative locations of each of the cameras, and their fields of view.

intervening galaxies responsible for QSO absorption line systems, and gravitational lenses are just two of the many other related areas in which deep IR imaging is crucial. Here we will therefore give a short account of the scientific capabilities of *NICMOS*, concentrating on those which are of prime interest to the study of distant normal and active galaxies.

Table 1: Limiting Sensitivities in jansky per pixel for 5 sigma in 60 minutes.

Camera	Filter			
	F110W	F140W	F160W	F240M
Bandwidth (μ m)	0.8 to 1.35	0.8 to 1.8	1.4 to 1.8	2.3 to 2.5
NIC1	1.5×10^{-7}	1.7×10^{-7}	4×10^{-7}	-
	(J~ 25)	(J~ 24)	(H~ 23.5)	-
NIC2	7×10^{-8}	-	1.8×10^{-7}	1.2×10^{-5}
	(J~ 25.7)	-	(H~ 24.3)	(K~ 19.2)
NIC3	6×10^{-8}	-	1.1×10^{-7}	7×10^{-6}
	(J~ 25.8)	-	(H~ 24.7)	(K~ 19.9)

2   Instrument Capabilities

NICMOS has three cameras designed for simultaneous operation, each containing a dedicated 256 × 256 pixel NICMOS–3 HgCdTe detector whose basic characteristics are given in Table 2. The instrument covers the wavelength range from $\lambda = 0.8$ to 2.5 μm, and contains a variety of filters, grisms, and polarizers. The *NICMOS* optics correct the spherically aberrated HST input beam and present each detector with three adjacent but not spatially contiguous fields-of-view of different image scales. The cameras are in a straight line pointing radially outward from the center of the telescope focal plane. The basic configuration is illustrated in Figure 1. As can be seen the gaps between the cameras are large, and therefore getting good positioning for all cameras may be rather difficult. Each camera carries a complement of 19 optical elements (Tables 3, 4, & 5), selected through independent filter wheel mechanisms, one per camera. The *NICMOS* dewar uses solid nitrogen as a cryogen for a projected on-orbit lifetime of approximately 4.5 +/- 0.5 years. *NICMOS* will achieve diffraction limited performance in Camera 1, *NIC1* for $\lambda > 1.0$ μm, and in Camera 2, *NIC2* , for $\lambda > 1.75\mu$m.

NICMOS does not have a physical shutter mechanism. The detector reset in which all pixels are set to bias level acts as an electronic shutter. The individual pixels of the arrays are strictly independent and can be read-out non-destructively. Because the array elements are independently addressed, the arrays do not suffer from the artifacts such as charge transfer smearing and *bleeding* which afflict CCDs, when overexposed. If, however, they are illuminated to saturation for sustained periods they retain a *memory* of the object in the saturated pixels. To counter this problem *NICMOS* automatically goes into an idle mode, in which the detector is continuously reset when not observing. As with all existing IR arrays of this type, careful removal of the dark current, and detector reset signatures, using matched science and dark frame exposures is essential.

Table 2: Flight Array Characteristics

Camera	Dark Current $(e^-/second)$	Read Noise (e^-)	Bad Pixels	Gain (e^-/ADU)	Saturation (e^-)
1	< 0.2	30	86(0.1%)	6 − 8	184000
2	< 0.1	28	74(0.1%)	6 − 8	173000
3	< 0.2	33	37(0.1%)	6 − 8	205000

Table 3: The Camera 1 Filter Set.

Name	Central λ (μm)	Bandwidth (μm)	Comment
Blank	N/A	N/A	Blank
F110W	1.025	0.8 − 1.35	
F140W	1.3	0.8 − 1.8	Broad Band
F160W	1.55	1.35 − 1.75	Minimum background
F090M	0.9	0.8 − 1.0	
F110M	1.1	1.0 − 1.24	
F145M	1.45	1.35 − 1.55	Water
F165M	1.6	1.55 − 1.75	
F170M	1.7	1.6 − 1.8	
F095N	0.953	1 %	[S III]
F097N	0.976	1 %	[S III] continuum
F108N	1.083	1 %	He I
F113N	1.13	1 %	He I continuum
F164N	1.644	1 %	[Fe II]
F166N	1.66	1 %	[Fe II] continuum
F187N	1.87	1 %	Paschen α
F190N	1.90	1 %	Paschen α continuum
POL0S	1.1	0.8 − 1.3	Short λ Polarizer
POL120S	1.1	0.8 − 1.3	Short λ Polarizer
POL240S	1.1	0.8 − 1.3	Short λ Polarizer

2.1 Grism Spectroscopy

Camera 3 has 3 grisms which can be used for low resolution slitless area spectroscopy between 0.8 - 2.5 μm and whose optical parameters are summarized in Table 6. Since the grisms are slitless every pixel will see the background emission summed over the entire passband of the grism, whereas the source spectrum will be dispersed over many pixels. To minimize the overlap of orders and reduce the thermal background from the telescope the grisms are coated with interference filters. However, the large bandpass of the Grism C filter means that 2nd order is also transmitted. Grisms A and B are in spectral ranges with very little background while Grism C is in the range of strong thermal emission from the HST. The resulting increase in the background flux for grism C is dramatic, around a factor of 10^3, and exposures have to be limited to $\sim$ one minute to avoid saturation by the background. Thus, spectra of sources with Grism C requiring longer integrations must be built by com-

Table 4: The Camera 2 Filter Set.

Name	Central λ (μm)	Bandwidth (μm)	Comment
Blank	N/A	N/A	Blank
F110W	1.1	0.8 − 1.4	
F160W	1.6	1.4 − 1.8	Minimum background
F187W	1.8756	1.75 − 2.35	Broad Band
F205W	1.9	1.75 − 2.0	Broad Band
F165M	1.7	1.55 − 1.75	Planetary continuum
F171M	1.715	1.68 − 1.75	HCO_2 and C_2 continuum
F180M	1.80	1.76 − 1.83	HCO_2 and C_2 bands
F204M	2.04	1.9 − 2.09	Methane imaging
F207M	2.1	2.0 − 2.15	
F222M	2.3	2.15 − 2.30	CO continuum
F237M	2.375	2.3 − 2.45	CO
F187N	1.87	1 %	Paschen α
F190N	1.9	1 %	Paschen α continuum
F212N	2.121	1 %	H_2
F215N	2.15	1 %	H_2 and Brγ continuum
F216N	2.165	1 %	Brγ
POL120L	2.05	1.9 − 2.1	Long λ polarizer
POL0L	2.05	1.9 − 2.1	Long λ polarizer
POL240L	2.05	1.9 − 2.1	Long λ polarizer

bining several such short exposures. Therefore, whenever possible, one should use grisms A and B in preference.

An image through an appropriate Camera 3 filter at the same pointing as the grism observation is neccesary to locate the spectrum of each object in the field. To prevent the spectra of individual objects from overlapping in complex or crowded fields the dispersion direction projected onto the sky can be controlled by *"rolling"* or moving the spacecraft, or both. This technique, of using two role angles, is illustrated in 3 which shows the pointings at which the Guaranteed Time Observers propose to obtain grism spectroscopy and photometry in the Hubble Deep Field. For more difficult fields or extended sources several such roll/pointing combinations may be needed to de-contaminate individual spectra. Unfortunately, specifying an absolute orientation for a grism observation creates constraints on the number of visibility windows available for scheduling. If many different orientations are needed to unscramble the source spectra, this can become a serious problem, and may make the observations

Table 5: The Camera 3 Filter Set.

Name	Central λ (μm)	Bandwidth (μm)	Comment
Blank	N/A	N/A	Blank
F110W	1.1	0.8 − 1.4	
F160W	1.6	1.4 − 1.8	Minimum background
F175W	1.75	1.2 − 2.3	
F222M	2.3	2.15 − 2.3	CO continuum
F240M	2.4	2.3-2.5	CO band
F108N	1.0830	1 %	He I
F113N	1.13	1 %	He I continuum
F164N	1.644	1 %	[Fe II]
F166N	1.66	1 %	[Fe II] continuum
F187N	1.875	1 %	Paschen α
F190N	1.9	1 %	Paschen α continuum
F196N	1.962	1 %	[Si VI]
F200N	2.0	1 %	[Si VI] continuum
F212N	2.121	1 %	H_2
F215N	2.15	1 %	H_2 continuum
F150W	1.5	1.11.9	Grism B continuum
G096	0.9673	0.8-1.2	GRISM A
G141	1.4I4	1.1-1.9	GRISM B
G206	2.067	1.4-2.5	GRISM C

un-schedulable.

2.2 Coronography

Camera 2 provides a coronographic imaging mode which will be particularly important for studies of the fuzz around QSO's. The coronographic spot imaged on to the focal plane results in a circular occulted region 0.3 arcseconds in radius which corresponds to the $\sim$ 93% encircled energy point of the 1.6μm PSF. The local contrast gain obtainable depends on the accuracy of the target centering and repositioning. While a simple automatic on–board acquisition is provided in complex fields separate acquisition images will often be needed to ensure that the target is correctly positioned behind the occulting spot.

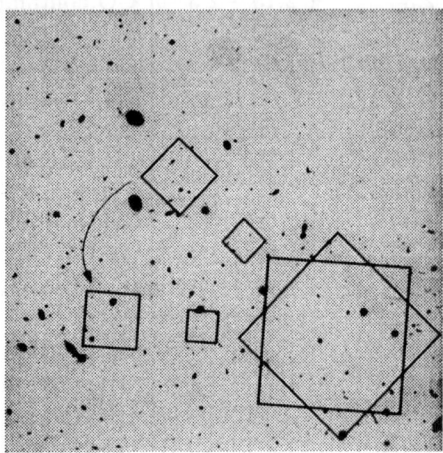

Figure 3: A grey scale WFPC2 image of the Hubble Deep Field (HDF) showing the outlines of the three NICMOS cameras at the locations to be studied with the grisms in the NICMOS GTO programme.

Table 6: Grism Parameters

Grism	Resolution Per pixel	Central Wavelength	Bandpass (μm)	Lines Per mm
A	200	0.964	0.8-1.2	45.0
B	200	1.401	1.1-1.9	30.769
C	200	2.058	1.4-2.5	21.05

2.3 Polarimetry

Polarimetry is possible only in Cameras 1 and 2, whose filter wheels contain 3 polarizers at relative orientations of 120°. Observations in all three polarizers provide the Stokes parameters of linearly polarized light. The spectral coverage is fixed for each camera, Camera 1 - 0.8-1.4μm, Camera 2 - 1.9-2.2μm and the polarizers cannot be crossed with any other optical elements.

3 Detector Readout Modes

The non-destructive nature of the *NICMOS* array readouts has enabled a variety of readout–modes to be implemented to ensure that optimal performance is obtained for a wide range of scientific projects. The simplest of these, *ACCUM two sample readout*, emulates the basic imaging capabilities of CCD's generating a single image. This mode is useful for short integrations on relatively bright objects. The minimum exposure time is $\sim$ 0.6 sec, and the minimum time between successive exposures is $\sim$ 8-12 seconds. The allowed *ACCUM* mode integration times are restricted to 173 tabular values, only 32 of which will have matched dark calibration frames. In order to reduce the read noise associated with the initial and final reads for faint sources the observer also has the option of requesting up to 25 *Multiple Initial and Final* reads. In theory the read noise should be reduced by $1/(n)_{1/2}$ where n is the number of reads. Due primarily to amplifier glow no further reduction in read noise is seen when the number of readouts exceeds about 9 (i.e. a S/N gain of around 3). For all programmes which need deep integrations or integrations on fields with objects of quite different brightness the best choice is *MULTIACCUM*. In this mode a sequence of non-destructive read-outs (each consisting of a single readout) is carried out at intermediate stages of an exposure to create images of various integration times. Up to 25 readouts (linearly or logarithmically spaced), spanning a total integration time from 0.215 seconds to 8590.0 seconds can be obtained at observer specified times. In practice the actual values which can be used are somewhat more restricted as dark frames at matched integration times are needed, and these will only be provided at certain predetermined times described in the Phase II proposal instructions. Subsequently an optimized image can be reconstructed by processing the stack of readouts removing the effects of cosmic rays and saturation.

4   Mapping & Background Removal

The thermal background from the telescope, which may be variable, needs to be removed from all NICMOS observations longward of 1.6 μm by obtaining off-source images. While *NICMOS* has a built in Field Offset Mirror (FOM) which can carry out motions of up to 26 arcseconds this will be too small a chop for many programmes. In addition the FOM performance is not yet know and it will therefore **not be available in cycle 7**. These background images must therefore be obtained by offsetting the telescope from the target to an "*empty*" region of the sky. It is advisable to use several different background positions to filter out contaminating sources. Due to the required telescope motions significant time overheads may be incurred by observations which need background subtraction or propose to map extended regions of the sky. A set of pre-defined observing *patterns* have been built to make the task of constructing such sequences of target and offset exposures easier. Two distinct type of telescope motions are supported and can be used in combination:

Dither: Individual motions, limited < 40 arcsecs, intended to be used to accomplish sequences of overlapping exposures for the construction of mosaics.

Chop: Motions, < 1800 arcsecs, intended for the measurement of the background at one or more locations significantly removed from the target pointing.

In addition to background subtraction at long wavelengths, the patterns are useful for creating maps of extended targets at any wavelength.

5   Concluding Remarks

A more comprehensive account of the many technical aspects of *NICMOS* which we have not been able to cover in this article can be obtained either from the *NICMOS* World Wide Web page, at STScI,

$http://www.stsci.edu/ftp/instrument_news/NICMOS/topnicmos.html$

which is continuously updated with the latest information, or by reading the NICMOS Instrument Handbook [1], which can also be downloaded from that address.

Acknowledgements

I would like to thank Chris Skinner, John Mackenty and Roger Thompson for their contributions to the Nicmos Instrument Handbook, which formed the foundation of much of the technical material presented here, and Christine Ritchie for preparing Figure 3.

References

1. Axon, D., *et al*, NICMOS Instrument Handbook, Version 1, Baltimore:STScI, 1996.
2. Cowie, L., Songaila, A, & Hu, E. Nature, 354, 460, 1991.
3. Cowie, L., *et al*, ApJ, 434, 114, 1994.
4. Glazebrook, K.,*et al*, MNRAS, 266, 65, 1994.
5. Lilly, S., *et al*, ApJ, 460, L1, 1996.
6. Maddox, S., *et al* , MNRAS, 247, 1P, 1990.
7. Metcalfe,*et al* MNRAS, 273, 257, 1995.
8. Steidel, C,,*et al* ApJ, 462, L17, 1996.
9. Williams., R., *et al*, AJ, 112, 1135, 1996.

PARALLEL OBSERVATIONS WITH STIS AND NICMOS

J.P. GARDNER, (Durham/NASA-GSFC), B.E. WOODGATE, (NASA-GSFC),
R.I. THOMPSON, (Arizona), S.R. HEAP, (NASA-GSFC), D.W. McCARTHY Jr.,
(Arizona), E.M. MALUMUTH, (NASA-GSFC), and R.J. WEYMANN, (Carnegie
Obs.)

The STIS and NICMOS teams have combined the parallel observations to be made
during their guaranteed time observations into a coordinated scientific program.
We will obtain parallel observations with STIS, NICMOS, and WFPC2 for the
dual purposes of (1) investigating star-formation in galaxies at moderate to high
redshift and (2) discovering and investigating low-mass stars and brown dwarfs in
the field.

Imaging of random fields at high galactic latitude is a powerful tool for investigating galaxy formation and evolution and the cosmological geometry. While much has been learned, two important questions still remain. There appears an excess over simple model predictions of intrinsically faint blue galaxies with irregular morphologies at intermediate to faint apparent magnitudes. The nature of this population remains unexplained, but it has become clear that the key to this problem lies in the ultraviolet. We plan to obtain parallel images with STIS and WFPC2, which will disentangle the effects of the redshift on the bandpass and galaxy evolution by constructing a statistically complete description of the properties of galaxies in the near-UV: number counts, colours and morphology.

Many observations point to the redshift range $1 < z < 3$ as a period of significant activity in galaxy formation. However, galaxies in this redshift range are difficult to identify with optical spectroscopy. We plan to obtain parallel observations using slitless spectroscopy with STIS and with NICMOS, which will specifically address this redshift range, while simultaneously providing a sensitive search for brown dwarf stars.

Redshifts available with slitless spectroscopy

camera:	STIS 3.1	STIS 4.1	NICMOS 096	NICMOS 141
wavelength:	3050-5550Å	5500Å-1μm	0.8-1.2μm	1.1-1.8μm
Lyα 1216Å	1.51 - 3.56	3.52 - 7.22	5.58 - 8.87	8.05 - 13.80
OII 3727Å	0.00 - 0.49	0.48 - 1.68	1.15 - 2.22	1.95 - 3.83
Hβ 4861Å	0.00 - 0.14	0.13 - 1.05	0.64 - 1.47	1.26 - 2.70
OIII 5007Å	0.00 - 0.11	0.10 - 1.00	0.60 - 1.40	1.20 - 2.59
Hα 6563Å	.... -	0.00 - 0.52	0.22 - 0.82	0.68 - 1.74
Hα & Hβ	.... -	0.13 - 0.52	0.64 - 0.82	1.26 - 1.74

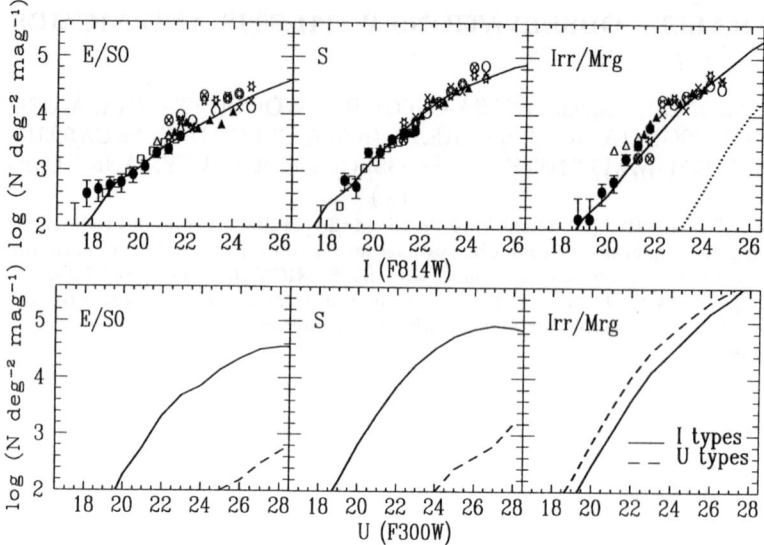

Figure 1: Number counts of galaxies as a function of morphological class. The top panels are the counts in I_{814}, with the types assigned in the same filter. The bottom panels are the counts in U_{300}, with the types assigned in I_{814} (solid lines) and with the types assigned in U_{300} (dashed lines). It is clear both that morphological type changes greatly when determined in different filter bandpasses, and that the ultraviolet is the best place to study the irregular/merger type of faint blue galaxies. The dotted line in the upper right panel is the number of galaxies currently undergoing merging. This figure has been kindly provided by Carlton Baugh.

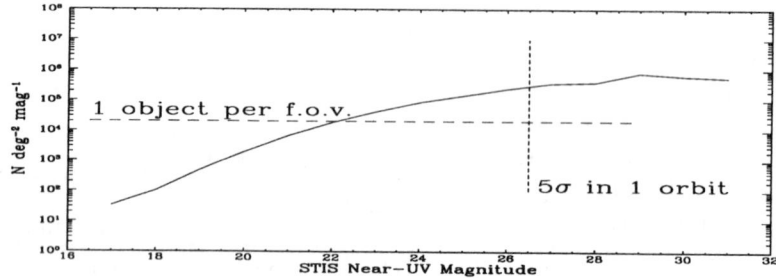

Figure 2: Number counts of galaxies expected in the STIS Near-UV camera. Also plotted are lines indicating the $25'' \times 25''$ field of view of the STIS NUV camera, and the 5σ detection depth for a point source in a 1 orbit exposure. Traditional passive-evolution models are subject to much uncertainty in the near-UV, and our data will provide strong constraints on the effects on those models of absorption by dust, the Lyα forest, and the amount of star-formation in the universe.

THE ADVANCED CAMERA
FOR THE HUBBLE SPACE TELESCOPE

G.D. ILLINGWORTH

UCO/Lick Observatory, University of California,
Santa Cruz, California 95064 USA

H. FORD

Department of Physics and Astronomy, Johns Hopkins University,
Baltimore, Maryland 21218 USA

The Advanced Camera for Surveys (ACS) will increase HSTs capability for surveys by a factor of ~10. The ACS will have three cameras. The first, the Wide Field Camera (WFC), will be a high throughput (45% at 700 nm, including the HST optical telescope assembly (OTA)), wide field ($200'' \times 204''$), optical and I-band camera that is half critically sampled at 500 nm. The second, the High Resolution Camera (HRC), is critically sampled at 500 nm, and has a $26'' \times 29''$ field of view and 25% throughput at 600 nm. The HRC optical path will include a coronagraph which will improve the HST contrast near bright objects by a factor of ~10. The third camera is a far ultraviolet, Solar-Blind Camera (SBC) that has a relatively high throughput (6% at 121.6 nm) over a $26'' \times 29''$ field of view.

1 Design Philosophy for the Advanced Camera

The ACS team is building an Advanced Camera for the Hubble Space Telescope which will have a high throughput, wide field, optical and I-band camera (WFC), a critically sampled high resolution camera (HRC), and a high throughput, moderate field of view far ultraviolet, solar-blind camera (SBC). See Ford et al. (1996) for a more complete discussion of the ACS. The Advanced Camera for Surveys (ACS) will increase HST's capability for surveys and discovery by at least a factor of ten. The key characteristics of the ACS are listed in Table 1. The WFPC2 (Wide Field Planetary Camera 2) characteristics are given for comparison; compared to the SBC, the FOC and COSTAR have full sampling, but a field of only $7'' \times 7''$ and throughput that is typically only ~1.5%.

Our approach to the ACS design was driven by two facts. First, images obtained with the FOC/COSTAR and WFPC2 cameras demonstrate that HST can deliver superb spatial resolution. Second, HST imaging with the present cameras is seriously limited by low throughputs (FOC and WFPC2), small fields of view (FOC), and undersampling (WFPC2). These three factors work together to severely limit HST's utility for deep, high-resolution imaging and for deep surveys. The time required to survey an area of the sky to a given

Table 1: Key Characteristics of the Advanced Camera for Surveys

Prioritized Features	WFC		HRC		
	ACS WFC	WFPC2 f/12.9	ACS HRC	WFPC2 f/28.3	ACS SBC
Throughput incl. OTA (%)	45@700nm	14@600nm	25@600nm	14@600nm	6.1@121.6nm
at 800nm	36	6.6	14	6.6	6.1@121.6nm
at 400nm	34	5.3	20	5.3	5.3@130nm
at 200nm	—	0.012	12	1.2	4.1@140nm
					2.9@150nm
					1.7@160nm
Equiv. FOV	$200'' \times 204''$	$143'' \times 143''$	$26'' \times 29''$	$35'' \times 35''$	$26'' \times 29''$
Sampling (500nm)	Half	Quarter	Full	Half	Half*
Polarization	$< 1\%$	$\sim 5\%$	$< 5\%$	$\sim 5\%$	$< 1\%$

*@295nm

limiting magnitude and a specified signal-to-noise ratio (SNR) is inversely proportional to product of the detector area and the net DQE when the SNR is limited by shot noise in the signal. When the SNR is limited by the shot noise in the background, the time is inversely proportional to Area $\times$ DQE^2. The most conservative measure of HST's survey or discovery efficiency is thus Area $\times$ DQE. We aim to improve HST survey efficiency in the optical, the far UV, and especially the I-band, by at least a factor of 10. To achieve this goal we are paying careful attention to: i) an optical design that minimizes the number of reflections, ii) high-reflectivity coatings on the mirrors, filters, and windows, and iii) obtaining CCDs with high quantum efficiency in the optical and the near IR for the WFC, and in the near UV and optical for the HRC.

The ACS must be built in approximately one-third the time and for one-quarter of the cost of the first and second generation HST instruments. Our primary strategy for building the ACS within the cost and schedule constraints set out in the NASA Announcement of Opportunity is reliance on the Space Telescope Imaging Spectrograph (STIS) design and technology. The SBC and HRC detectors and electronics are STIS design, and, in fact, for the SBC likely will be a STIS flight spare. The two 2K×4K WFC I-band detectors and electronics derive directly from STIS CCDs. Whenever possible, we use STIS electronics and STIS mechanisms. Approximately 70% of our electronics modules and mechanisms are "build to print" from STIS drawings. The ACS flight software derives from STIS and NICMOS software whenever possible.

The second aspect of cost control is a strongly integrated team consisting of the Johns Hopkins University and the ACS Science Team, the Goddard

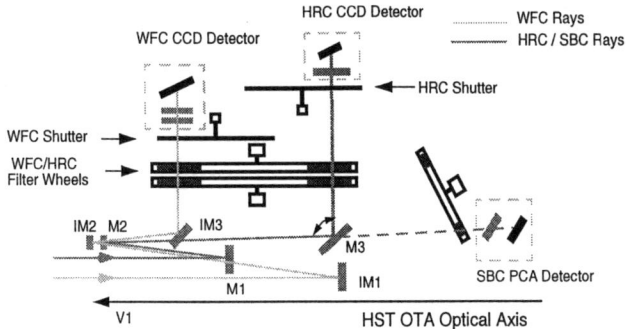

Figure 1: An optical schematic of the Advanced Camera

Space Flight Center, and Ball Aerospace. This ACS team minimizes management costs and maximizes scientific, managerial, and engineering resources which can be applied to building the ACS. The team members are Holland Ford (PI), Frank Bartko, Pierre Bely, Tom Broadhurst, Robert Brown, Chris Burrows, Edward Cheng, Mark Clampin, Jim Crocker, Paul Feldman, David Golimowski, George Hartig, Garth Illingworth, Randy Kimble, Carolyn Krebs, Tom La Jeunesse, Michael Lesser, George Miley, Susan Neff, Marc Postman, Marc Rafal, Bill Sparks, Zlatan Tsvetanov, Rick White, and Bob Woodruff.

The final aspect of ACS cost control is the science team's philosophy of "keep it simple." We prioritize our science goals and balance technical decisions on cost and performance against our most important science goals. We are not trying to achieve the unrealistic goal of building an Advanced Camera that will do everything for everyone.

2   Instrument Design

2.1   Overview of the Advanced Camera

The optical layout of the ACS is shown in Figure 1. The WFC and HRC have separate optical paths and mirrors. The WFC mirrors are designated IM1, IM2, and IM3. The HRC mirrors are designated M1, M2, and M3. The light in each optical path first encounters a spherical mirror (IM1 or M1) which images the HST pupil onto the IM2 (M2) mirror. These mirrors are anamorphic aspheres which are figured with the inverse spherical aberration on the HST primary mirror, and thus correct the spherical aberration in the HST primary and the field dependent astigmatism of the HST at the center of the ACS field of view. The light from the WFC IM2 mirror is reflected by a Schmidt-like plate (IM3) through the two filter wheels to the WFC CCDs. The Schmidt plate corrects astigmatism over the WFC field of view (FOV).

The HRC and the SBC use the same M1 and M2 mirrors. The MgF_2 overcoatings on these two aluminum-coated mirrors are optimized for maximum reflectivity at 121.6 nm. Because of reflection losses in the far UV, we chose to have only two reflections in the optical path to the SBC. A flat mirror M3 is inserted into the light path to direct the light through the two filter wheels to the HRC 1024^2 CCD. The MgF_2 coating on M3 is optimized for maximum reflectivity at wavelengths >200 nm.

2.2 The Wide Field Camera

The WFC features a $200'' \times 204''$ field of view optimized for sky-limited V and I-band imaging. The WFC employs two butted SITe 2048×4096 CCDs with 15 micron pixels, with a plate scale of $0.051''$/pixel and near critical sampling in the I-band. The high red throughput of the WFC is achieved by a unique optical design employing only three mirrors with protected silver coatings, and the use of thinned, backside-illuminated CCDs with anti-reflection coatings optimized for red wavelengths. The reflectivity of three silver coated mirrors is 50% higher than three aluminum coated mirrors would be at 800 nm. The mirror coating, as well as the geometry, yield less than one percent polarization sensitivity. Overcoated silver mirrors maintain high reflectivity for several years with exposure to typical laboratory air quality and humidity.

Figure 2 shows expected net WFC and HRC net efficiencies (including HST OTA) compared to the WFPC2. We have specified that the quantum efficiency (QE) at 800 nm must be $\geq 59\%$, with a goal of 68%. The maximum peak throughput must be $\geq 74\%$ with a goal of 77%.

2.3 The High Resolution Camera

The HRC provides spectral coverage from 200–1000 nm with a $26'' \times 29''$ field of view. The detector is a 1024×1024 STIS-like CCD manufactured by SITe. This CCD will be optimized for high quantum efficiency in the near UV by Mike Lesser at the University of Arizona. Lesser's process uses a platinum flashgate (PPtF) and an AR coating optimized for the 200–300 nm bandpass. If Lesser's process is not perfected in time to meet the ACS schedule, we will use a SITe CCD with a new SITe ultraviolet anti-reflection coating. Figure 3 shows the QE expected from Lesser's process, the QE from the new SITe UV coating, and the QE of the STIS CCD. The net HRC plus HST OTA efficiency is shown in Figure 2.

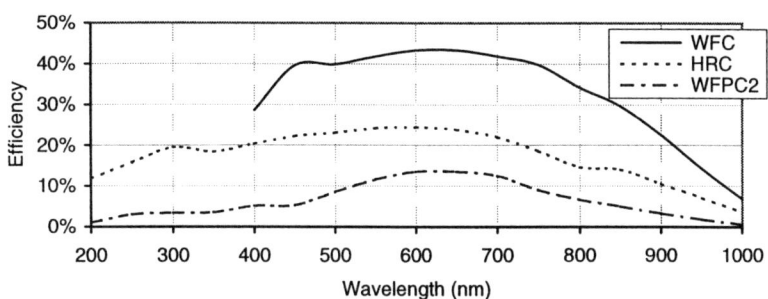

Figure 2: The expected WFC and HRC net efficiencies (including the HST OTA) compared to WFPC2.

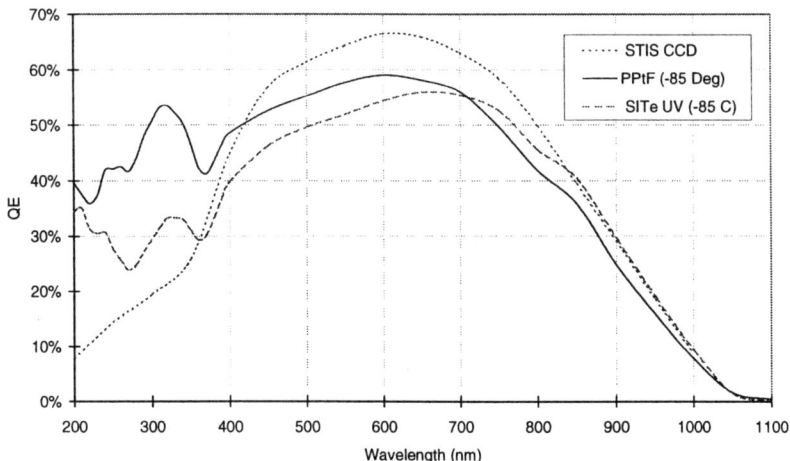

Figure 3: Expected HRC QE with SITe's and Lesser's processes

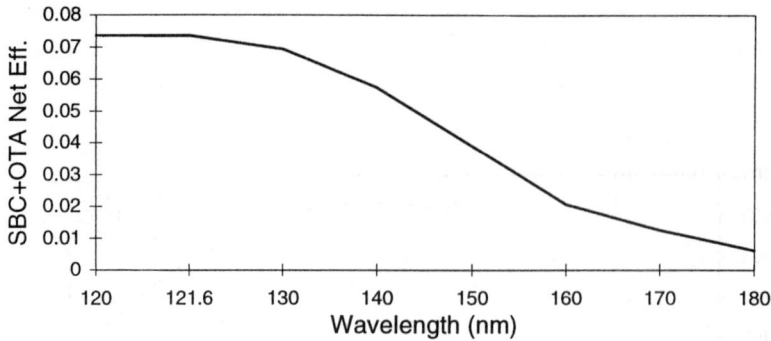

Figure 4: Expected SBC net efficiency (including the HST OTA) versus wavelength

The HRC image is critically sampled for wavelengths greater than 500 nm, with an average plate scale of 0.025″/pixel. The HRC uses only two powered mirrors and a fold mirror to relay the HST field image to the CCD for highest throughput and lowest scatter. The HRC mirrors are coated with aluminum plus MgF$_2$ to achieve the highest possible reflectivity between 122 to 1000 nm.

2.4 The Solar Blind Camera

The SBC is optimized for highest throughput from 115 nm to 170 nm. The image is sampled at 0.030″/pixel. The SBC uses a STIS-based PCA photon-counting detector with an opaque CsI photocathode and a C-plate micro-channel plate (MCP). The relay optics and corrector mirror mechanism are shared with the HRC camera. A full 26″ × 29″ FOV is available. The resultant image quality at the PCA MCP surface is better than λ/8 everywhere in the FUV, and nearly uniform with field. The design is optimized from 121.6 to 160.8 nm. The expected QE of the SBC detector, including the HST OTA, is shown in Figure 4.

3 Filters and Dispersers

We have simplified the optical design and reduced cost by sharing the filter wheels for the WFC and HRC, as well as sharing the corrector optical path for the HRC and SBC. The WFC and HRC filter wheel mechanism is comprised of two separate filter wheels, each driven by an independent motor and sharing a common housing. These filter wheels are populated with 17 bandpass filters (13 of which are spectrally compatible with both the WFC and HRC), a set of five linear ramp filters, a grism for the WFC, and a prism and three visible and three ultraviolet polarizers for the HRC. The SBC has five longpass filters, two prisms, a narrow band Lyman-α filter, and four opaque positions. Altogether the ACS incorporates a total of 38 spectral elements. The large WFC filters can be used with the HRC. Figure 5 shows the bandpasses of the filters. Five of the broadband filters are based on the Sloan Digital Sky Survey filter set (g, r, i, z, and a modified u called wfcu).

Narrow-band filters for the WFC and HRC take three forms: standard narrow-band filters which cover the whole field of view of both cameras, linear ramp filters similar in concept to those on WFPC2, and an HRC methane band filter for observations of the giant planets. The standard narrow-band filter complement is Hα, [O III] and [Ne V]. The four narrow-band ($\sim$2%) and one broadband ($\sim$9%) linear ramp filters provide an imaging capability for redshifted emission line sources. The desired bandpass is selected by a combination of target positioning and filter wheel rotation. The central strip of each of the five linear ramp filters can be used with the HRC.

Two sets of polarizers, one for the UV and the other for the visible, also are included for imaging with the HRC. Each set of polarizers consists of three separate elements to give relative polarizing angles of 0, 60, and 120 degrees. The SBC filter wheel contains a set of short-cut filters which step across the SBC bandpass. The broadest filter, MgF$_2$, transmits geocoronal Lyman-α, while the CaF$_2$ filter blocks geocoronal Lyman-α and transmits light longward of 125 nm. The ACS includes three prisms and a grism for deep, low spectral resolution imaging from 120 nm to 1000 nm (see Figure 5); the resolution of these is typically R$\sim$100.

4 Coronagraph

The ACS coronagraph will have two high contrast options. Both options provide for suppression of the light from a bright object which is positioned behind a mask using autonomous on-board target acquisition. The first option is a "Fastie spot", which is a 0.8″ diameter reflecting spot deposited near the corner of the HRC CCD entrance window (where the beam size is 0.5″). It will allow unvignetted high contrast imaging (including with the ramp filters) from 0.7″ to 4″ around a bright source (e.g., imaging of a quasar host galaxy.) Vignetted, but still useful data will be available all the way in to 0.4″ from the source.

The second option is a commandable coronagraphic mask that can be positioned in the aberrated HST focal plane, together with a Lyot stop that is simultaneously positioned at the pupil image close to the M2 mirror. This

456

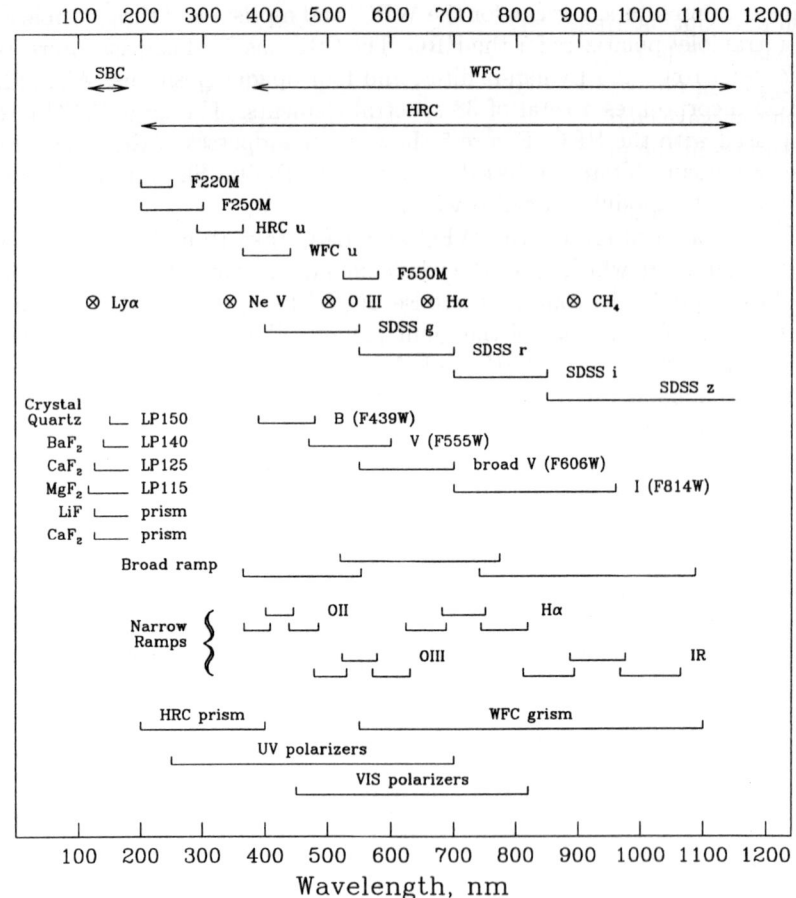

Figure 5: ACS filter summary. The wavelength range covered by each channel is indicated at the top of the chart. The horizontal bars represent the bandpass of each filter; the filters name is given at the right hand side of the bandpass. The narrow band (~1%) filters are shown by a ⊗ symbol with the name of the emission line next to it. A set of broad band filters similar to the ones currently used in WFPC2 (names in parentheses) provide continuity between the two instruments and a clean transformation to the classical Johnson and Cousins systems. The ACS is equipped with a set of narrow (~2%) ramp filters for emission-line imaging of redshifted objects. A broad (~9%) ramp filter is also included for off-band continuum imaging. Dispersing elements include two prisms for the SBC, one for HRC, and a grism for WFC.

option does not interfere with normal camera operation in any way. The Lyot stop has the effect of removing the diffracted light halo that surrounds the target, giving almost an order of magnitude higher contrast improvement at the longest wavelengths. The residual halo that is present is caused by scatter from ripples in the primary and secondary mirrors.

5 Operational Modes

Two major operational features of the ACS improve its science productivity and provide efficient operation. The first feature allows the observer to simultaneously use two cameras for parallel imaging. This is made possible by two filter wheels (shared by the WFC and HRC for cost saving), containing a total of 30 filter positions, each of which can be used by the WFC and HRC, and a third 12-position filter wheel dedicated to the SBC.

The second feature allows simple WFPC2-like target acquisition because the fields of view of each camera are large compared to HST pointing uncertainties. Additionally, although each optical channel of the Advanced Camera is optimized for specific types of science, a number of operational features have been designed into the ACS to allow the observer to select permitted combinations of optical channels, to choose a particular readout mode, and to configure filter elements to enhance their specific science programs. For example, the large format and layout of the WFC CCD's make it particularly well suited for large-scale, deep mapping surveys, by making it possible to obtain contiguous exposures with single guide star acquisitions. The rows and columns of the WFC and HRC CCDs are parallel to the V2 and V3 axes, making it possible to efficiently "tile" a survey area while stepping the guide stars along the long directions of the FGS sensors.

Acknowledgments

The Advanced Camera is funded by NASA through the HST Project Office at the Goddard Space Flight Center.

References

Ford, H. et al. 1996, in Space Telescopes and Instruments II, SPIE Vol. 2807 (Bellingham, WA: SPIE), in press.

GRAVITATIONAL WAVE DETECTION: AN OPTICAL APPROACH

Ch. BRACCO

Observatoire de Haute-Provence, F-04870 Saint Michel l'Observatoire, France

Plane wavefronts from a distant background light source become rippled when they travel through Gravitational Waves (GW). We give an approximate expression for the generated distorsion. The ripples appear on top of the static lensing deformation and would result in a scintillation effect for a distant observer. However, this effect does not seem to be directly observable with a single monochromatic GW source, as it was expected [a]. Nevertheless, the ripples do exist.

---

[a] The conclusions in the poster displayed were rather optimistic regarding the direct detection of the scintillation effect. Refined calculations now lead to a much smaller effect, which may be undetectable in the situation considered.

Plane wavefronts of a distant background light source become rippled [1] when they cross a region where they interact with GW. The GW effects on electromagnetic waves have classically been described as a modulation of the refractive index of vacuum. The wavefront profile W is calculated [1] as the integral of the refractive index modulation along the optical path. We use classical optical approximations, neglecting diffraction effects, and consider small deflection angles. In addition, we deal exclusively with the isotropic part of the GW emission. We focus our investigations on monochromatic sources of GW. This eliminates burst phenomena, like supernovae explosions and final stages of binary coalescences. We use the GW source barycentric frame of reference, with cylindrical coordinates (ρ, α, z), where the z-axis is the line of sight. ρ is the impact parameter of the beam in the sky plane, λ_g is the spatial GW wavelength and L is the distance from the GW source to the observer.

The wavefronts exhibit a *bump* in the *near zone*, where $\rho < \lambda_G$, and show [1,2] ripples in the radiation zone, where $\rho \gg \lambda_G$.

In the *near zone*, the GW effects are strongly affected by the Newtonian dynamical terms due to the time dependance of the Newtonian potential. In addition, the near zone is too small on the sky plane for a reasonable probability of finding a usable background light source within it.

In the radiation zone, the wavefronts are rippled under pure GW effects and when $\rho \gg \sqrt{L\lambda_g}$, their profile is given by

$$W(\rho, \phi_0) = \gamma \times \frac{4\lambda_G}{\pi\rho} \times \frac{2y_\rho}{(1+y_\rho^2)^2} \left[\sin(\alpha_\rho y_\rho + \phi_0) - y_\rho^2 \sin(\alpha_\rho/y_\rho + \phi_0)\right]$$

$$\text{where} \begin{cases} \alpha_\rho &= 2\pi\rho/\lambda_g \\ \\ y_\rho &= \frac{L}{\rho} + \sqrt{1 + \left(\frac{L}{\rho}\right)^2} \end{cases}$$

with ϕ_0 the phase shift of consecutive wavefronts with respect to the reference and where γ is a multiplicative factor depending upon the GW source characteristics.

Since first derivatives are generally negligible, curvatures add as the second derivatives of the wavefront profile and the GW effects simply add as a perturbation on top of the static lensing deformation.

The amplitude of the ripples decreases linearly with L, and the smaller ρ is, the greater the wavelength. As these wavelength are greater than λ_g, the ripples are *superluminal*.

The time dependance of the wavefront curvature may cause a *scintillation effect* for a distant observer, if the background light source appears smaller than the GW wavelength angular extension on the sky plane. Intensity modulations ΔI may be directly related to variations in the wavefront curvature ΔC [1,2], calculated as the second derivative of W, as

$$\frac{\Delta I}{I} = 10^{-2} \times \frac{L\Delta C}{1 + k_E^2}$$

where $k_E = \rho_E/\rho$ and ρ_E is the GW source Einstein radius.

For the more likely GW sources expected, as millisecond pulsars and fast binaries, the scintillation effect is of the order of [2] 10^{-9} in the best cases. We cannot expect to observe so weak intensity modulations in the light of distant background objects.

However, the ripples do exist, though the scintillation effects from a single monochromatic GW source seem to be unobservable [a]. Their amplitude, though very small, is λ_g times greater than the expected metric perturbation at the distance L from the GW source. These ripples are the manifestation of a long range interaction. Their study is pursued.

1. A. Labeyrie, A&A, 268, 823, 1993.
2. Ch. Bracco, A&A, submitted

---

[a]The ripples were thought to have a damped sinusoidal profile, at the time I diplayed the poster. The intensity modulation was expected at percent level within 30 *arcsec* from a millisecond pulsar. A direct detection with Hubble STIS camera in the photon counting mode was expected if the background light source was a quasar or a white dwarf. Actually, this would have been the case if the GW effects before and after the GW source added. A careful analysis showed that there is only a partial cumulative effect. Increasing the integration domain even tends to blur the distorsion.

INDEX OF AUTHORS

Note: Page numbers in **bold** refer to first author.